T0255767

Railway Geotechnics

Railway Geotechnics

Dingqing Li
James Hyslip
Ted Sussmann
Steven Chrismer

CRC Press
Taylor & Francis Group
Boca Raton London New York

CRC Press is an imprint of the
Taylor & Francis Group, an **informa** business

A SPON PRESS BOOK

CRC Press
Taylor & Francis Group
6000 Broken Sound Parkway NW, Suite 300
Boca Raton, FL 33487-2742

First issued in paperback 2019

ISBN-13: 978-0-415-69501-5 (hbk)
ISBN-13: 978-0-367-86659-4 (pbk)

Library of Congress Cataloging-in-Publication Data

Li, Dingqing.
  Railway geotechnics / Dingqing Li, James Hyslip, Ted Sussmann, and Steven Chrismer.
    pages cm
  Includes bibliographical references and index.
  ISBN 978-0-415-69501-5 (alk. paper)
  1. Railroad tracks--Design and construction. 2. Geotechnical engineering. 3. Railroad tracks--Foundations. 4. Earthwork. I. Hyslip, James P., 1964- II. Sussmann, Theodore R., 1970- III. Chrismer, S. M. (Steven Mark), 1955- IV. Title.

TF220.L47 2015
625.1--dc23   2015011731

Visit the Taylor & Francis Web site at
http://www.taylorandfrancis.com

and the CRC Press Web site at
http://www.crcpress.com

Contents

2 Loading

3 Substructure

5 Design   197

Preface

Railway Geotechnics is written by four colleagues who studied at the University of Massachusetts, Amherst, in an academic program advised by Professor Ernest T. Selig. Our collective time at the university spanned over a decade, during which we were individually inspired by Professor Selig to work on and further advance the subject of railway geotechnology, which he pioneered and developed into a rigorous field of study. Since graduation, the aggregate of our professional experience includes railway operations, consulting, research, and education.

The field of railway geotechnology was in its infancy when we were in our early careers. Because the engineering behavior of track substructure was not well understood up to that point, perspectives on the causes and cures of substructure instability were often informed by anecdote rather than by verifiable fact. Mystique surrounded the subject in the absence of critical thinking, often resulting in costly applications of remedial methods that did not address the root causes of track substructure problems.

Exemplifying the inquisitive nature of Professor Selig, we strive to question the behavior of railway track simply by asking questions related to loading and material behavior and their effects on track performance. The resulting answers have allowed us to further develop railway geotechnology to understand the causes of track substructure instability and the cost–benefit trade-offs of potential remedial options. By asking informed questions and discerning answers that are supported by evidence, we apply the scientific method along with experience and creativity to advance the state of the art.

Advancing the field of railway geotechnology by the writing of this book is a natural step for each of us in our careers. The book continues the work *Track Geotechnology and Substructure Management* by Selig and Waters (1994) and provides an update to this field of study so that current railway engineers and managers have easier access to new and emerging best practices.

During years of writing and discussions, we each had moments that challenged some of our beliefs while we debated the merits of emerging technology and practices. The book that has taken shape is the result of this process, and it serves to document our collective understanding of railway geotechnics. While other books advance a more detailed theoretical analysis or computational technique, our experience has shown the great value of focusing on understanding root causes of problems and drawing conclusions about expected performance. As computers advance our ability to analyze the problem, often the answer lies not in more analysis but in better understanding of the problem and the materials involved.

The goal of this book is to provide a better understanding track substructure in order to enable more effective design, construction, maintenance, and management of railway track so as to ensure the vitality of rail transportation. We hope that this work will prove useful to current railway engineers and managers as well as college students pursuing careers in the field of railway engineering.

We believe that our book covers the best practices in railway geotechnics, but the opinions and statements given in this book are ours and not those of the respective organizations that we work for.

We have many people to acknowledge for their support. First, our wives (Ajin Hu, Kim Hyslip, Mary Beth Sussmann, and Kathryn Chrismer) and families have tolerated this book for several years, and have shown great encouragement and patience with us through that time. Next, we thank our representative Tony Moore from Taylor & Francis, who provided continuous encouragement through the writing process. And finally, we recognize a group of professionals who have supported us by reviewing the chapters and providing critiques and suggestions: Jay Campbell, Darrel Cantrell, Gary Carr, John Choros, Dwight Clark, Hannes Grabe, Francois Heyns, Carl Ho, Hai Huang, Aaron Judge, Mike LaValley, Nick O'riordan, Alan Phear, David Read, Jerry Rose, Mario Ruel, Phil Sharpe, Dave Staplin, John Tunna, Erol Tutumluer, Peter Veit, and Rainer Wenty.

Chapter 1

Track

Track is the most fundamental component of the railway infrastructure (Figure 1.1). Track supports the rolling stock by distributing wheel loads from the track superstructure to the track substructure. The substructure of the track, which includes ballast, subballast, subgrade, and drainage arrangements, has the dominant influence on track performance. *Railway Geotechnics* is written to address the challenges associated with designing, constructing, and maintaining a well-performing and long-lasting railway track. This book presents methodologies and technologies that can be used for the development of a state-of-art railway track substructure and complements the previously published book *Track Geotechnology and Substructure Management* by Selig and Waters (1994).

1.1 INTRODUCTION

The structure of railway track (Figure 1.2) has evolved greatly since its first development more than 150 years ago. The evolution to a strong and durable track structure is the result of overcoming ever-increasing challenges associated with moving more, faster, or heavier trains. Safety and cost have always been the primary concerns. The quest for safe and cost-effective railway transportation have driven improvements in materials and processes that have provided a track structure with increased reliability and durability.

Modern-day track superstructure and substructure components have changed greatly from the materials used historically. An in-depth discussion of changes in track structure elements through the years is provided by Kerr (2003). The superstructure of the track has benefitted from many years of research, development, and testing that have resulted in a good understanding of materials, inspection, and maintenance options. Rail steel has become very durable, and rail inspection technologies as well as preventive maintenance strategies have advanced significantly. Tie (sleeper) materials have improved in strength and durability, with improvements in

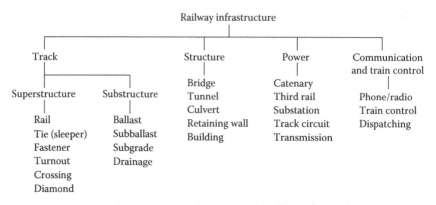

Figure 1.1 Elements of railway infrastructure.

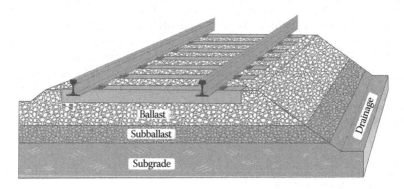

Figure 1.2 Railway track structure.

hardwood tie treatment and use of other materials such as concrete, composite, and steel. The fastening system has evolved from cut spike and tie plate to elastic systems that improve rail restraint from movement relative to the ties.

Railway ballast gradually evolved from the early practice of simply using locally available gravel to the use of coarse and open-graded hard crushed rock. Appreciation for ballast and other components of the track substructure has grown throughout the history of the railway industry. In North America, track support and substructure stresses were the focus of the Talbot Reports that were published near the turn of the twentieth century (Talbot 1918). *Railroad Engineering* by Hay (1982) is a mid-twentieth-century perspective. Later research by the industry with dedicated funding provided by the Association of American Railroads and the Federal Railroad Administration culminated in the book *Track Geotechnology and Substructure Management* by Selig and Waters (1994). Research in track substructure has continued not only in North America but in several other countries as well. In 2005, Indraratna and Salim from Australia published the book *Mechanics of Ballasted Rail Tracks: A Geotechnical Perspective* that covers their latest research results and findings in track substructure.

Railway Geotechnics is intended to cover the basic elements of railway geotechnics shedding light on some of the challenging issues in designing and maintaining a reliable and cost-efficient railway track. The authors' goal is to present the geotechnical complexities related to railway track in a manner that can be useful to researchers and practitioners alike.

1.2 HISTORICAL PERSPECTIVE

Railway track exists all over the world, and track from each region was built conforming to the constraints and practicalities of the time. The US expansion between 1850 and 1950 provides the historical perspective of railway track and how it evolved over time. Track built with these early track construction methods and materials, some of which are still in service today, often bring legacy problems due to shortcomings and limitations of these early track construction practices.

It is important to identify the root cause of track performance problems to develop a repair that corrects the problem to avoid continued and repeated maintenance. Sometimes, the diagnosis and remediation of a section of track requires an understanding of the historic construction procedures and material availability.

1.2.1 US railway expansion

In the early days, railway engineers who selected routes for railway lines were mainly concerned about identifying routes with the minimum grade, lowest need for bridges or tunnels, and the least amount of earthmoving. Workers cut trees to clear the right-of-way, and the resulting logs and lumber were used for railway construction often as cross-ties (Figure 1.3). Lacking mechanical equipment, the major geotechnical challenge was earthmoving for cut-and-fill operations (Ambrose 2000).

An example of early track construction demonstrating track substructure support consideration is shown in Figure 1.4, where the tie supports are embedded in individual gravel beds spaced 3 ft (1 m) apart. Even at this early stage of track construction, the critical aspect of tie support is clearly understood as shown by the gravel placement.

Trestles were commonly used to bridge ravines or bodies of water (Figure 1.5). Railroad trestles were prevalent due to the challenges of building a railroad embankment using hand labor to excavate and transport fill. The construction of a trestle with periodic bents (supports) was popular, because the major structural elements and connection details could be standardized to provide a reliable structure constructed mainly from wood (Ambrose 2000) that was often available along the right-of-way.

Figure 1.3 Early track structure with hand hewn railroad ties. (Courtesy of Library of Congress.)

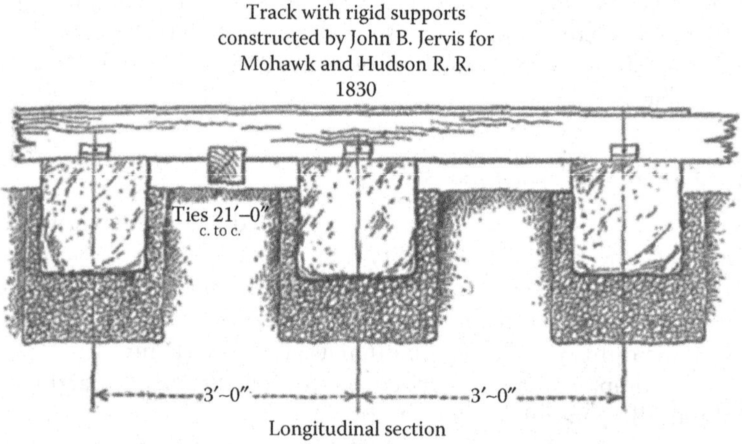

Track with rigid supports
constructed by John B. Jervis for
Mohawk and Hudson R. R.
1830

Ties 21'–0"
c. to c.

◄------3'–0"-------►◄-------3'–0"-------►

Longitudinal section

Figure 1.4 Early track construction showing individual tie support detail. (Data from Stevens, F.W., *Beginnings of the New York Central Railroad*, Knickerbocker Press, New York, 1926.)

Construction of track transitions is often the most critical part of building railway track. Transitions have at least two dissimilar structural support components, which affect track stiffness and the deformational characteristics (elastic and plastic) of the track's top of rail position. Transitions have always required special maintenance attention and, consequently, always cost more money to maintain compared with that of open track construction.

Figure 1.5 Early track construction at a trestle transition. (Courtesy of Library of Congress.)

Figure 1.6 Trestle construction and related fill operations. (Data from Hungerford, E., *The Modern Railroad*, A. C. McClurg, Chicago, IL, http://www.gutenberg.org/files/40242/40242-h/40242-h.htm, 1911.)

   Wooden trestles exposed to the elements of nature could have a relatively short life span requiring frequent repair or upgrade of deteriorating components. Eventually, the trestles need to be replaced with upgraded trestles using treated lumber, replaced with more traditional bridges, or filled in with soil to construct an embankment (Palynchuk 2007) (Figure 1.6).

The type of soil used for fill on embankments was historically dictated by the location of the borrow area and the ability to transport the fill material. Sometimes the lack of efficient transportation prohibited the import of suitable fill material. Local soil, which may have been problematic, was used, if readily available at low cost.

Mechanized construction equipment made a big impact on track construction and maintenance. The first large-scale mechanized earthmoving was made possible by railway track technology and by steam-powered excavation and fill operations. The "flexibility" of track construction and realignment allowed track to be laid into a borrow area, where rail cars were loaded by steam shovel and then moved by steam locomotive to the fill area where they were emptied. The Panama Canal construction project is an example of early mechanized excavation with railway technology to transport cut spoils. The track construction and relocation operations (Figure 1.7) provided the degree of flexibility needed to locate tracks so that empty train cars could be placed adjacent to the steam shovel for loading (McCullough 1977).

Many of the early railroad routes underwent major realignment to increase track speed by reducing the degree of curvature and percentage of grade by constructing large-scale cut/fills, tunnels, and bridges. Construction practices

Figure 1.7 Track realignment to facilitate mechanized earthmoving in Culebra Cut of the Panama Canal, January 1912. (Courtesy of Harris and Ewing Collection, Library of Congress.)

developed to allow track to be constructed over poor natural subgrade, including displacement fill operations, where very soft deposits were displaced by filling from one end of the soft soil zone to another. However, the results of early fill displacement were variable, and some embankments that were constructed over soft soil or peat deposits may still be compressing over time, producing track settlement from this unseen source.

Another method of constructing track over soft soil and peat or muskeg was the use of corduroy, which consisted of timbers laid perpendicular to the direction of the track, continuously on the poor ground with embankment soil placed above. This is an early application of tensile reinforcement to create a stable embankment over unstable soil (Figure 1.8).

Jointed track (Figure 1.9) with periodic rail joints typically every rail length, 11.9 m (39 ft), was standard railway construction practice until the introduction of continuous welded rail (CWR). Many railway lines still have jointed track, although most main lines have CWR. The rail joint is a discontinuity that generates dynamic wheel loads from both the size of the gap and the differential vertical track support conditions local to the joint. For these reasons, rail joints require high maintenance.

CWR reduced maintenance due to the elimination of the joint, which reduced dynamic wheel load and associated settlement, rail end batter, and tie and ballast deterioration. However, even after eliminating the joint through the use of CWR, the location of the joint can continue to be a maintenance issue due to *joint memory*: a term used to describe the already deteriorated track substructure layers due to past high dynamic wheel loads. Past high dynamic wheel loads deformed each layer such that the joint area would be a local depression in the subgrade and would not drain

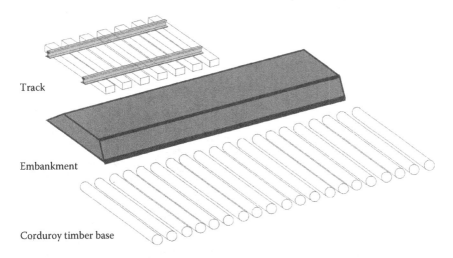

Track

Embankment

Corduroy timber base

Figure 1.8 Corduroy construction for track support.

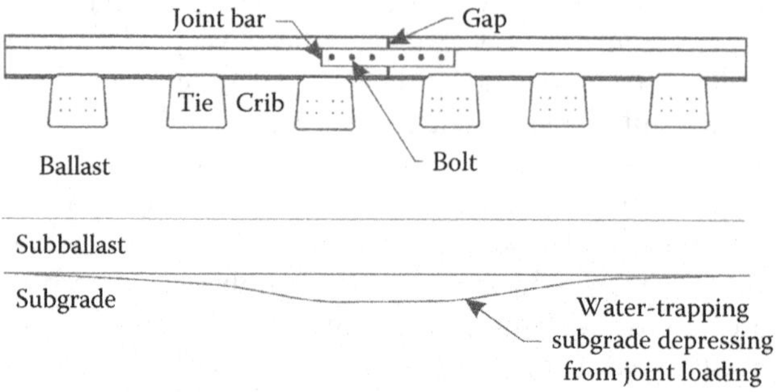

Joint bar — Gap

Tie | Crib

Ballast — Bolt

Subballast

Subgrade — Water-trapping
subgrade depressing
from joint loading

Figure 1.9 Rail joint.

well (Figure 1.9). The geotechnical challenges of this dynamic loading-related stability problem often continue to affect track performance decades after installation of CWR, if the underlying layer depression and poor drainage is not corrected.

The benefits of using CWR are many, but CWR can also present new challenges. The main challenge is the development of large thermally induced forces in the rail that could cause the track to buckle in the summer (Figure 1.10) and the rail to pull apart in the winter. In trying to reduce the risk of track buckling, the rail is often installed at an elevated rail installation temperature to minimize extreme rail compression in the summer. Ballast plays an important role in providing lateral and longitudinal resistance to track movement. Curves in CWR can tend to "breathe" (move in and out) due to changes in rail temperature, and the crib and shoulder ballast together provide lateral resistance to this movement.

Figure 1.10 Rail thermal compression-induced track buckle. (Data from Kish, 1997.)

An adequate ballast shoulder not only provides the needed lateral resistance to avoid track buckling but also provides confinement to retain ballast in place for vertical track support. An adequate shoulder width is typically in the order of 0.3 m (12 in.) or more. When a shoulder is too small, perhaps <0.2 m (8 in.), the confinement may be inadequate and ballast could displace laterally resulting in track settlement.

Concrete ties (sleepers) have become common on both passenger and freight railway lines. Concrete ties can withstand higher loads and speeds. However, there are examples of concrete ties that have not met the design life mainly due to poor track support conditions and concrete quality. With proper quality control and design, including track maintenance to provide good track support, concrete ties may provide a good track structure with the needed longevity.

1.2.2 Development of modern mainline track substructure

As wheel loads and train speeds increased along with the rationalization of lines, demands on the track structure increased significantly. The increase in speed for both passenger and freight traffic was accompanied by higher horsepower and heavier locomotives with higher adhesion leading to larger track forces. Heavier railway freight cars further contributed to an increase in car loading (Figure 1.11 for North America). Railway research and testing departments worked to develop track structure components

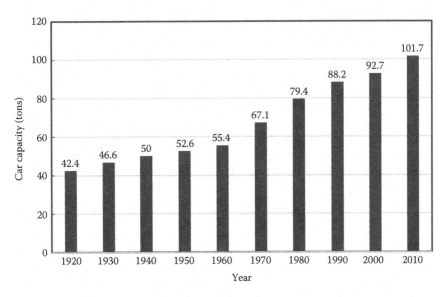

Figure 1.11 Historic changes in the average North American railroad car capacity. (Data from AAR, 2012.)

and design standards that could withstand the applied loads throughout much of the twentieth century (PRR 1917; Gilchrist 2000). The conclusion of the Pennsylvania Railroad (PRR) study indicated that 0.2 m (8 in.) of ballast was inadequate and 0.6 m (24 in.) of combined thickness of trap-rock ballast and subballast of either gravel or cinders was recommended to support 258 kN (58,000 lb) locomotive axle loads on 85 lb/yard rail. This recommended ballast and subballast thickness actually exceeds the currently minimum required cross section of 0.3 m (12 in.) of ballast and 0.15 m (6 in.) of subballast for North American heavy axle load (HAL) track design.

As railways streamlined operations to increase efficiency, unit trains (trains that carry the same commodity in all cars) with ever-increasing payload capacities emerged, continuing the increase in track loads. Some economic analyses have shown a cost advantage of using higher capacity and heavier railroad cars up to a certain axle load (Martland 2013), considering operation benefits as well as damages to the track and bridge structures.

The railway industry was not fully prepared for the onset of 100 ton freight cars in the 1960s and 1970s. A combination of deferred track maintenance coupled with a large increase in car loading from 70 to 100 ton capacity freight cars drove a large increase in the rate of track degradation. From this lesson, the Association of American Railroads HAL research was initiated in the 1980s to help prepare the industry for car capacity increases to 110 and 125 ton. Much of the track damage resulting from the loading increase from 70 to 100 ton cars was in the track substructure. Although increased loading has led to locations of poor track support, it is also true that some of the old roadbed under historic rail lines are still in operation, capable of bearing increased loading with little track deterioration. The reason for variable performance of old roadbeds is often due to the variable quality of track substructure materials used. Decades of compaction under loading often made roadbeds extremely hard and dense, sometimes referred to as the "hardpan" layer. This hardpan is often a combination of the old cinder and slag-type materials mixed with local materials that have stabilized to provide a strong and stable platform for the track. In many sections, investigation of track support problems can be a form of archaeology where knowledge of historic construction practices, realignment projects, company standards, and operations can provide insight.

Finally, a change in the railroad regulatory climate occurred in 1980 with the enactment of US federal law known as the Staggers Act. The Staggers Act provided for considerable deregulation of railroad industry in the United States, which eventually allowed the railroads to reinvest in their infrastructure. The law made it easier for railroads to consolidate and streamline their track network, for example, turning a four-track right-of-way into a two-track right-of-way. After 1980, the US railroads saw an increase in revenue traffic, which now was operated over fewer miles.

This resulted in increased manifestation of track substructure issues, but at the same time, the increase in revenue to the railroad allowed them to maintain and upgrade their lines.

1.2.3 Future for track substructure

As the industry evolves, it is increasingly important that track engineers determine the most effective corrective method for recurring track substructure-related problems by identifying the root cause and then prescribing an appropriate remedial action. The tendency to apply routine maintenance to problems often does not lead to successful repair, if the maintenance does not address the root cause. Track engineers must also meet the challenge of performing remedial substructure work in ever more restrictive time constraints. High-density traffic on mainlines leave little time for maintenance, and the lack of available work windows can result in substandard maintenance if time is not available to properly complete necessary repairs.

Conventional ballasted track has a long history of excellent performance under a variety of loading conditions and can be the most cost-effective track structure for most applications, provided that it is built and maintained with proper construction and maintenance techniques. In addition to being the most common choice for heavy haul freight operations around the world, recent observations have shown that ballasted track can perform well for high-speed passenger operations in excess of 250 km/h. The trade-offs between ballasted track and other choices of track structure such as slab track should be considered from the perspective of life-cycle cost (LCC) to ensure that the most cost-effective solution is implemented.

1.3 SUPERSTRUCTURE

The track superstructure includes the main load-supporting elements of the track that react and transfer train load to the track substructure. The superstructure includes the rail, tie (sleeper), and fastening system that work together to support train loading by reducing the large stresses at the wheel–rail interface to levels that are tolerable for the substructure layers. The superstructure is required to resist both vertical and lateral loads with only small elastic and permanent deformation.

The rail is the most visible element of the track structure and directly supports passing wheels. The rail has changed substantially over the years, from smaller rail cross sections that were installed as individual rail lengths (most recently in 39 ft lengths) to larger rail cross sections welded into continuous strings common in track today. The use of high-strength steels and head hardening have resulted in superior rail performance, while the head of the rail has been made larger by proportion to the rest of the rail section to provide a larger area for wear. Improvements in rail welds are needed and will benefit track substructure performance if the weld can provide increased yield strength in the heat-affected zone, which will result in reduced plastic deformation of the rail, less variations in rail profiles, and lower dynamic loading and track deterioration.

Improved rail can improve track performance and life in well-supported track sections, but heavier and stronger rail is no cure for track support problems. Some have assumed that heavier rail can accommodate, or bridge over, track support problem areas. Although stronger and stiffer rail will distribute the load over a larger area, the effect is minimal, and once the track support has deteriorated, the only choice is to improve the track support because heavier rail will not have a noticeable effect.

Cross-ties (sleepers) are generally wood or concrete. Although composite/plastic and steel ties have been developed for specific applications or to obtain desired properties, these types of ties have not yet been widely adopted. Wood ties tend to be most common in the United States not only because of lower cost but also because of the superior resilience of wood, making it more forgiving of dynamic track interaction. Both concrete and wood ties are affected by abrasion from the ballast and wear at the rail–tie contact area. Wood ties are more sensitive to drainage problems mainly due to their susceptibility to rot and decay.

Fastening systems retain the rail in place by restraining lateral, longitudinal, and vertical relative movement between the rail and tie. The rail fastening system, including the rail seat pad (if any), affects track behavior, and pad stiffness influences dynamic load attenuation and vibration transmission, both of which influence track substructure performance. The plate and cut spike fastener is still the most common fastening system on wood ties, which must be coupled with appropriate rail anchoring to restrain rail longitudinal movement.

Elastic fasteners are the rail–tie fastening system that provide resilient lateral, longitudinal, and vertical rail restraint. The resilience of this fastening system develops with deformation similar to spring stiffness, although not all fasteners provide a linear stiffness variation with deflection. Elastic fasteners are the only fastener type used with concrete ties and these are sometimes used on wood tie track. In concrete tie installations, the rail seat pad is an important consideration to control dynamic loads and rail seat abrasion, while ensuring adequate longevity of the pad. In addition, concrete tie track requires insulators to provide electrical insulation of track signals and a degree of resilience to the lateral rail–tie interaction. Elastic fasteners and fastening systems have many common elements whether applied to concrete ties or as direct fixation fasteners on slab track.

Under-tie pads have been used successfully on the bottom face of concrete ties (Singh et al. 2004). One problem with concrete ties has been the deterioration in the tie–ballast contact zone. The under-tie pad has improved tie–ballast contact, reduced abrasion in this zone, as well as added resilience to the very stiff concrete tie, all of which could help to improve ballast life in concrete tie track.

(a)

(b)

Figure 1.12 Fouled ballast in special trackwork: (a) crossing diamond and (b) turnout.

Special trackwork is also part of the track superstructure and includes turn-outs, crossing diamonds, insulated joints, and along with grade-crossings and other locations of unique track construction are locations that often experience more rapid degradation. Each of these track features can generate large dynamic loads that are transferred to the substructure. For instance, Figure 1.12 shows two examples of special trackwork with fouled ballast around the ties, presumably due to excessive ballast degradation from dynamic loading. The support of special trackwork becomes a critical issue in track design, as dynamic loads from rail discontinuities can be more than twice as high as open track.

1.4 SUBSTRUCTURE

Track substructure consists of the foundation layers that support the track superstructure and the drainage arrangements. The foundation layer consists of ballast, subballast, and subgrade. In the case of slab track, the track structure is ballast-less, and the track substructure includes subbase and subgrade.

1.4.1 Ballast

Ballast is the granular material placed as the top substructure layer and is the layer that is in direct contact with the ties (sleepers). Ballast is generally large, angular, and uniformly graded (similar size) granular particles (Figure 1.13) derived from crushed hard rock material such as granite and basalt. Figure 1.13 shows images of both AREMA 3 and 25, the widely used North American ballast gradations. The images show the limited contact area of clean ballast and the tendency of smaller particles to occupy the voids between larger particles in more broadly graded ballast (AREMA 25).

Ballast has many functions that are requirements for good, well-supported track. The most important are the following:

- Supports the rail–fastener–tie track panel by providing adequate vertical, lateral, and longitudinal resistance.
- Transmits and reduces wheel/rail forces.
- Facilitates surfacing and lining operations.
- Provides drainage.
- Provides resilience and damping of dynamic wheel/rail forces.

Ballasted track is not sealed like highway pavement but is designed as an open structure to provide rapid drainage and facilitate maintenance. Ballast-related track maintenance methods include surfacing and lining, stoneblowing, track renewal and superstructure replacement, ballast undercutting to restore drainage and resilience, fouled ballast removal including

(a)   (b)

Figure 1.13 Example ballast gradations: (a) AREMA 3 and (b) AREMA 25 gradation.

crib excavation with backhoe or vacuum, and ballast compaction often with a dynamic track stabilizer. Maintenance is required when one of the following problems occurs: excessive vertical or lateral track deformation, excessive ballast degradation (ballast fouling), or drainage is no longer effective.

To fulfill the ballast functions, there must be adequate ballast layer thickness and proper particle size and gradation. Much of the research on ballast material has focused on developing standards for a strong and stable ballast layer. The ballast layer must have a proper gradation to provide a large void space that facilitates both drainage and storage of fouling material. Small materials that accumulate in the void spaces within the ballast layer are called fouling material. Fouling material can originate from abraded or broken ballast, material that spills or blows into the track, or material that flows onto the track from water drainage or flooding.

The thickness of the ballast layer should be specified based on the structural capacity of the track to ensure that it can withstand and distribute the applied train loading at a stress level that will not deform the subgrade over the expected life of the track. Ballast layer thickness-related experiments were conducted by the PRR in the early 1900s. The experimental test setup is shown in Figure 1.14. The photo of the experiment shows a two-axle bogie loaded with iron ingots that runs back and forth over a track section that was built to allow the buggy to slow to a stop and reverse direction for each pass. Power is applied by an overhead power system in the middle of the track section under the shed to maintain the desired speed

Figure 1.14 Pennsylvania railroad experiment on ballast thickness. (Data from PRR, 1911.)

over the test section. This example captures the basic elements of all subsequent research, including applying a number of axle loads to the track structure, counting the number of load applications to develop the applied tonnage over the test section, and monitoring the deformation of the track (PRR 1911).

While the study by PRR (1911) may not have been the first test of ballast thickness specifications, the elements of this test track set very high standards for subsequent investigations over the years. A similar arrangement was used at the Association of American Railroads Chicago Technical Center test track, which remained in service until the mid-1990s. Research subsequent to this 1911 study has done a better job of measuring and predicting stress distribution through the track structure and has refined this initial work, but the results of this 1911 study experiment developed a strong framework for track and roadbed construction standards. Track built to standards developed in this 1911 study remains in service, including portions of the Northeast Corridor.

In addition to thickness requirements, ballast research has focused on material properties, gradation specifications, and the impact of maintenance operations. One area of significant research was ballast compaction and the associated increase in strength over time as the compaction of ballast increases under traffic.

Ballast compaction is essential to provide lateral track resistance. Although ballasted track is normally very stable with a considerable lateral strength reserve to resist imposed stresses, when ballast is newly placed or if it has been disturbed by maintenance (tamping or tie replacement), lateral resistance of the track may only be half of its value compared to when it is compacted (Li and Shust 1999; Sussmann et al. 2003b). Resumed traffic following the maintenance will lead to gradual ballast recompaction and eventual restoration of its lateral strength and track buckling resistance.

1.4.2 Subballast

The subballast, or the *blanket* (another term often used), is the granular layer between the ballast and the subgrade (Figure 1.15). This granular layer often consists of broadly graded gravel and sand. The subballast complements the ballast by further distributing the applied loads and reducing the stresses on the subgrade, and it provides protection from frost action. However, the subballast has some important functions that cannot be fulfilled by ballast:

- Prevents penetration and mixing of subgrade and ballast.
- Prevents subgrade attrition by ballast, when the top subgrade is composed of clay stone or shale that may be abraded by large ballast stones.

The subballast has a critical role in drainage and has competing requirements to both drain readily and direct water away from the subgrade.

Figure 1.15 Subballast below ballast and above subgrade.

When shallow groundwater is present, the subballast should be capable of providing drainage of water that might be flowing upward from the subgrade. Design of the subballast layer also requires the analyses of layer thickness and gradation. Drainage requirements often conflict with separation and filtration criteria that must also be satisfied (Chapter 6) to prevent soil migration and piping.

Subballast saturation is a common problem that is often ignored. This results from inadequate drainage of the subballast layer from conditions such as the formation of subgrade settlement depressions that trap water or subballast gradation that makes the subballast inadequately permeable. Under repeated wheel/rail forces, significant deformation and even rapid failure can occur in the subballast layer when drainage is impeded. Not only does the accumulated water soften fine-grained subgrade, it also greatly reduces the stiffness of the subballast along with the ability of the subballast layer to distribute applied loads (Read and Li 1995).

In recent years, several techniques have been used to improve on conventional granular subballast, including the use of geosynthetics to reinforce the subballast layer and hot-mix asphalt underlayment as the subballast replacement, which are discussed in Chapters 3 through 5.

1.4.3 Subgrade (trackbed, formation)

Subgrade may consist of either soil or rock and is the platform on which the track structure from the subballast up is built, as shown in Figures 1.2 and 1.15. It can be part of an embankment that is built with fill materials or can be natural ground in cut sections of track where subgrade may consist of the natural soil or placed soil layer(s).

Trackbed, *roadbed*, *track foundation*, and *formation* are other terms used to describe track substructure. Trackbed is used mainly to refer to the tie/sleeper support layers inclusive of the entire track substructure below the bottom of tie, and this will be the manner in which the term *trackbed* (roadbed, track foundation) will be used in this book. *Formation* is another common term used to refer to the subgrade but includes any placed fill such as the subballast (blanket) layer directly under the ballast.

The primary function of the subgrade is to act as the track foundation by providing uniform and adequate support. In other words, subgrade should be capable of providing a suitable working base for the construction of ballast and subballast (or the subbase in the case of slab track), supporting the track structure, and accommodating the stresses due to traffic loads without failure or excessive deformation.

Although wheel loads are reduced significantly in the subgrade through the load distribution provided by the superstructure, ballast, and subballast, the influence of wheel loads can extend several meters below the bottom of the ties. As such, the subgrade over this depth should be strong to resist failure and excessive deformation due to wheel loads as well as the weight of track and substructure. In addition, a stable platform must be maintained over time and should not be excessively affected by environmental conditions (e.g., wet/dry and freeze/thaw cycles). Subgrade should also be designed so that surface water and ground water drain away from the track. A poorly designed and constructed subgrade can cause many track performance problems, so it is imperative that design and construction of subgrade be done according to the best practices available.

Under repeated wheel loads, an overstressed subgrade can experience "cumulative plastic deformation" (ballast pocket) or "progressive shear failure" (subgrade squeezing, shown in Figure 1.16). These are the two most common failure modes for subgrade composed of fine-grained cohesive

Figure 1.16 Progressive subgrade deformation.

soils. These problems can lead to deterioration of track-geometry conditions, which in turn will increase dynamic wheel loads and accelerate deterioration.

Although soft subgrade failures are more common, problems on hard subgrade also exist. While track stability is not typically a problem on hard subgrade, abrasion of hard subgrade by coarse ballast particles in the presence of water can cause the abraded subgrade fines to infiltrate and foul the ballast, reducing its permeability, which affects track life, performance, and maintenance requirements.

1.4.4 Slopes

Fill and cut slopes are prevalent along most railway lines. The construction of embankments and the excavations for cuts of early railway lines were completed before geotechnical principles were widely understood as an engineering discipline. Problems can range from overall instability of the slope to embankment foundation-related problems to erosion. Overall embankment instability can be an issue where the embankment is built on soft soil and foundation problems develop. Figure 1.17 shows an embankment on soft soil where the track structure was settling into the soft embankment and the embankment has settled into the native foundation soil as indicated by the bulging of the soil at the bottom of the ballast section. Also, the severe tilting and rotation of the telephone poles indicate excessive subgrade displacement.

The evolution of geotechnical engineering to a recognized and well-founded discipline fostered the construction of more stable slopes. However, the majority of embankments and cut slopes have slopes that are steep according to current geotechnical engineering design principles. Steep slopes were used because of limited earthmoving capabilities in the early years of track

Figure 1.17 Embankment settlement problem.

Figure 1.18 Example of rail used for embankment slope reinforcement.

construction coupled with the proximity of neighboring tracks, buildings, and other infrastructure, which all acted together to prohibit widening the embankment to accommodate reduced slopes.

A common method that has been used in an attempt to stabilize fill slopes is driving rails vertically to provide a degree of confinement to steep embankment slopes. As shown in Figure 1.18, rails were driven vertically on each side of the track beyond the base of the embankment and connected across the embankment at the top to provide confinement. The remediation at this particular site was successful in improving track performance. However, ad hoc repair strategies are often applied without understanding the root cause. This uncertainty over applying the proper remediation is tolerated based on the hope that at least some stabilization will result and because of the relative ease of installation and low cost of performing the type of attempted remediation shown in Figure 1.18. The extra effort to fully diagnose the problem, develop a targeted design, and apply a repair that addresses the actual problem is often cost justified when considering the overall expense of repeated remediation attempts.

1.5 SUBSTRUCTURE EFFECTS ON TRACK PERFORMANCE

Proper support from track substructure is the single most critical element needed for good track performance. Track substructure significantly affects overall track performance, the performance and functions of other track components and that of passing vehicles. Good track support is characterized by strong resistance of the ballast, subballast, and subgrade layers to plastic deformation and by the elastic deflection under rolling load being small and reasonably uniform along the track. Support that is soft and variable along the track causes deformation and damage with each cycle of stress and strain for virtually every track component.

After substructure problems become manifest on the track surface (Figure 1.19), the track condition tends to decrease rapidly as each passing wheel load deforms the track more. The critical stage where track support becomes increasingly more problematic is when the ballast–tie interface is affected, in particular, when a tie cavity develops. Most fouling material is generated due to ballast deterioration, which occurs near the track surface where the loads are highest. Ballast fouling is generally thought to begin to collect toward the bottom of the ballast layer due to rainfall that can carry fouling material into the track structure and vibration from passing trains. As the ballast becomes fouled to the level of the tie bottom, tie support is reduced and track settlement rapidly increases. Highly fouled ballast permanently deforms with loading, and the tie moves laterally and longitudinally creating a tie cavity. The tie cavity is a zone ranging from a fraction of an inch to an inch around the tie perimeter where the ballast has been pushed away leaving the tie unsupported laterally and longitudinally (Figure 1.20). Poor drainage of fouled ballast often traps water in the tie cavity.

The unsupported or "hanging" tie is a particular problem for track substructure performance because the lack of vertical support produces a dynamic impact component under wheel loading resulting from the rapid vertical movement of the tie that impacts the ballast surface. This results in ballast crushing and high stresses being imparted on each layer below the tie, as well as an increased dynamic load on the superstructure and on

Figure 1.19 Ballast mud boil.

Figure 1.20 Concrete tie with fouled ballast that often results in a tie cavity.

Figure 1.21 Concrete tie showing horizontal split on side and vertical cracks.

passing vehicles. An example of such premature failure of track components is the cracked and broken concrete tie shown in Figure 1.21. Broken rail–tie fastenings are also often associated with poorly performing track substructure.

Dynamic wheel load magnitude and frequency define the stress condition that is applied to the track. The track substructure defines how the loads are distributed through the track superstructure and substructure. Because applied loads should not exceed available strength, proper performance of the track structure to reduce applied loads is necessary to ensure that the track structure is subjected to lower stresses to extend track life and improve reliability. Figure 1.22 shows how some of higher applied stresses exceed available track strength, which can be corrected by either increasing the strength of track structure or reducing peak stresses. In many

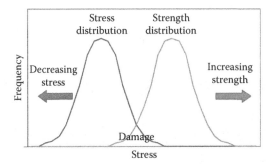

Figure 1.22 **Stress versus strength.**

cases, reducing the stress state of the track structure is considered the most efficient and economic approach.

At locations where the applied loads exceeded the strength of some track component, the load is then shed to other locations or components. For example, at the transition from strong to soft track support, a well-supported tie will see an increase in load that causes high stress in other track components. This is sometimes referred to as the "zipper" effect that can develop in concrete tie track where a single failed tie results in progressive failure of adjacent concrete ties as each tie consecutively fails and sheds load to the adjacent tie, which then successively fails.

Poor track substructure performance can also lead to track superstructure component problems, including rail problems, which tend to be more prevalent in areas of weak substructure support. Imperfections in the railhead steel coupled with the high loads due to poor substructure could result in increased occurrences of crushed railhead (Figure 1.23). Any weakness in the track structure can be exposed under the increased stress common in poor track support locations.

Figure 1.23 **Crushed rail head.**

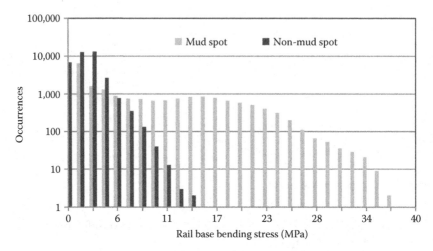

Figure 1.24 Rail bending stresses in fouled (mud spot) and clean ballast locations.

Figure 1.24 shows that track substructure can affect bending stress measured at the base of rail. Illustrated in this figure are the stress distribution results under dynamic train operations from two track locations in proximity: one with a visible mud spot and the other with no mud. As shown, rail bending stress under dynamic wheel loads is significantly greater at the mud spot than at the non-mud spot. From these measurements, it is expected that rail at the mud spot would perform worse with shorter rail life and a reduced rail fatigue life.

Figure 1.25 shows track substructure–related incidents in terms of equipment and track damage costs associated with train derailments as recorded in the Federal Railroad Administration Railroad Accident Incident Reporting System. As illustrated, from 1998 to 2013, railroads in North America incur significant loss from incidents related to track substructure.

1.6 ALTERNATIVE TRACK STRUCTURES

Ballast-less track (see Figure 1.26), also known as slab track, has rails supported on concrete slab, with rail fasteners embedded directly in the concrete or supported on concrete blocks. Located below the slab is often a subbase layer that is typically a well-graded gravel layer. The subbase is protected from surface water by the slab or another water-proof layer. Adequate drainage of subsurface infiltration must be provided so that the subbase remains stable and freely drained. Below the subbase is the subgrade consisting of either natural soil or compacted fill.

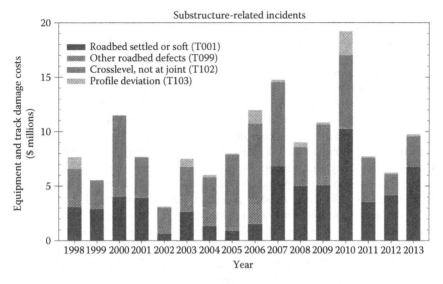

Figure 1.25 Substructure-related equipment and track damage costs. (Data from FRA Office of Safety.)

(a)

(b)

Figure 1.26 (a,b) Slab track test section. (Courtesy of Transportation Technology Center Inc., Pueblo, Colorado.)

Slab track is often used in locations with very frequent traffic where maintenance outages cannot be tolerated and in locations where maintenance of ballasted track is prohibitively expensive and logistically difficult such as the Rogers Pass Tunnel on the Canadian Pacific (Bilow and Randich 2000). Research was conducted in the early 2000s to develop a slab design that can provide the conflicting requirements of track geometry with small tolerance for error for high-speed train service, combined with the durability needed to support HAL freight traffic that would be needed for shared high-speed passenger and freight service corridors. The test of slab track has shown that slab track can perform well under HAL traffic (Li et al. 2010), although it can be more costly and may not be maintenance free.

Some of the challenges posed by traditional ballasted track structure can be reduced or eliminated by slab track such as stability against buckling, gage strength, and geometry error from ballast settlement. However, the increased cost is often not justified by the reduction in maintenance alone. Therefore, the decision to use slab track must be based on the need for the additional track structural stability and associated increase in track reliability that can be provided by slab track.

In comparison, the construction issues posed by slab track deserve special consideration because they could diminish the assumed benefits from reduced maintenance. One consideration is the inability to realign slab track once built. This necessitates the use of a more detailed site investigation to identify and repair any weak subgrade layers prior to constructing the track. This stands in contrast to traditional ballasted track that is designed to accommodate maintenance. Although the initial, up-front construction costs for slab track are significantly greater than for conventional ballasted track, some studies have found that the LCC is lower for slab track (Zoeteman and Esveld 1999; Kondapalli and Bilow 2008). However, a significant omission in both these studies is that neither appears to have considered the substructure costs associated with site investigation and subgrade preparation.

Occurrences of "flying" ballast under high-speed trains have been analyzed by société nationale des chemins de fer français (SNCF) (Saussine et al. 2011). Although airborne ballast is a highly unlikely event, it was concluded that the problem cannot be neglected. While slab track might be a solution to this type of problem, a more practical solution is to make the track surface and the underside of the train more aerodynamic. This unlikely problem has a variety of solutions, the most costly of which may be application of slab track.

A host of other alternative track structures have been developed over the years (Chipman 1930; Penny and Ingham 2003), where the main concept driving the design is improved support of the rail. For high-speed and HAL track applications, alternative track structures are being developed, such as half-frame sleeper track (Figure 1.27). These designs provide more rail–tie and ballast contact area that may improve track performance by reducing the applied load on ballast, thereby reducing the ballast settlement.

Figure 1.27 Half-frame sleeper track installed at bridge approach.

1.7 TRACK TRANSITIONS

A track transition is the location where the track structure changes abruptly. A track transition can involve a change of track superstructure (such as at grade crossing, special trackwork, tie type) or a change of track substructure (such as at bridge approach, ballasted track to slab track). In many cases, a track transition involves changes in both track superstructure and track substructure.

Track components tend to degrade faster at transitions than at other track locations. Transitions often require frequent track maintenance due to faster degradation at these locations compared to typical track. Common track transition problems include track geometry degradation (vertical surface and cross level, lateral alignment and gage), accelerated ballast breakdown, and track component failure. Furthermore, these track degradation problems may lead to poor ride quality for passenger vehicles and accelerated wear and tear on rolling stock components.

1.7.1 Bridge approach

A bridge approach is the transition from the approach embankment to the track structure on the bridge. In a short distance, the track support changes from fill embankment, to abutment wall, to bridge structure. Maintenance aimed at correcting geometry at the transitions can itself contribute to and possibly accentuate the problems associated with transitions.

Figure 1.28 Track geometry deviation from transition settlement.

Figure 1.29 Loaded profile at bridge transition.

Problems associated with bridge transitions are well recognized and have been the subject of significant research for both highway and rail infrastructure (Briaud et al. 1997; ERRI 1999; Tutumluer et al. 2012). For example, Figure 1.28 shows a common feature of profile dip at a bridge approach. Figure 1.29 shows the rail profile under load, which tends to accentuate the variation in track geometry at transitions.

1.7.2 Ballasted track to slab track

The transition from traditional ballasted track to slab track is often a zone of rough geometry. Figure 1.30 shows an example of a track transition from ballasted track to slab track. Track maintenance at this transition is compounded by the limited rail adjustment range of fasteners used on slab track.

Figure 1.30 Ballasted track to slab track transition with alignment error.

1.7.3 Special trackwork

At special trackwork locations such as switches and crossing diamonds, abrupt changes occur as a result of the change of superstructure components involving rail cross section, tie length, rail fastening, and tie plate. In some cases, the track substructure also changes to accommodate the change of track superstructure conditions. This type of track transition is especially challenging because of the variability of the rail running surface due to gaps and discontinuities associated with crossing diamonds or switches that result in extreme dynamic loading. Figure 1.31 shows several

Figure 1.31 Three types of track transition: crossing diamond, grade crossing, and turnout.

types of track transitions with two crossing diamonds, a grade crossing, and a turnout in the photograph.

1.7.4 Grade crossing

A grade crossing is a particularly important transition because this is a critical location where road and track intersect, as shown in Figure 1.31. Drainage is a significant design consideration because the only outlet are at the edges of the crossing. Poor track drainage is often a recurring problem leading to ballast degradation and accelerated track component degradation. As repair is doubly difficult given the restrictions on disturbing both the road and the track structure, it is important to design grade crossings properly and ensure that the entire zone is freely draining.

1.8 TRACK INSPECTION AND MAINTENANCE

Track inspection and maintenance are essential elements in the operation of trains over railway track. Ballasted track is designed to be maintained with open-graded ballast to facilitate maintenance work. Required maintenance is specified based on the results of track inspections. The track inspection process relies mainly on visual inspection augmented by automated inspection systems. Management systems and guidelines are used to improve the efficiency with which maintenance needs are detected and applied.

1.8.1 Inspection

Track inspection involves visual inspection with the eye of a trained inspector. However, to further improve the efficiency of track inspection, systems have been and are continuing to be developed to highlight critical zones and provide quantitative data on track conditions to the inspector. In this regard, the track inspectors' time can be prioritized and better focused on problematic track sections.

The main tool of automated track inspection is the measurement of track geometry. Track geometry data represent the position of the rails in space where relative changes in position can be detected to evaluate the roughness or smoothness of the track surface. Track settlement and other track support problems tend to be indicated by relative changes in position of the rails. Settlement trends are reflected in track geometry deterioration trends.

Track support-related measurements include track deflection or stiffness through direct measurement of the load–deflection behavior, gage restraint measurement systems to assess tie–fastener gage strength, and track lateral tie–ballast resistance from the single tie push test or lateral track panel shift. These measurements target a mechanistic element of track support to measure and infer track condition. In addition, these techniques make very good quality control assessment tools because any variations in the materials used or placement techniques will result in measurable track support variations.

For substructure inspection, track support and ballast condition measurements such as track deflection and ground-penetrating radar have been developed. An on-track cone penetrometer has also been developed for ballast and subgrade assessment. Other geophysical techniques such as spectral analysis of surface waves and resistivity techniques are being developed to assess problematic track conditions. Among these techniques are tools that could make good construction quality control inspection techniques to detect changes in material type and layer thickness to ensure compliance with construction specifications.

1.8.2 Maintenance

Track maintenance is needed to keep railway track operational. Historically, track maintenance was a labor-intensive activity. Over time, maintenance equipment has been developed that changed the methods used to maintain track. The main track maintenance operations currently used to correct or repair track substructure and support-related problems are (1) surfacing (tamping, stoneblowing, stabilizing), (2) undercutting and cleaning of ballast, and (3) drainage.

The track surfacing operation often involves a track tamper, as shown in Figure 1.32. The tamper has rolling grips on the rail that move the rail vertically and laterally to the desired position. A set of vibrating tamper tines is then inserted alongside the tie, the vibration providing for ballast particle movement, easing insertion of the tines. The tamper tines insertion on either side of the tie is followed by squeezing them together to push ballast under the tie and up against the tie bottom. Several squeezes may be used depending on how much the track was moved or if ballast conditions are problematic. The tines are then withdrawn from the ballast, and the machine moves forward to the next tie. Multiple ties can be tamped at once with some tampers capable of tamping three or more ties at a time.

Compaction of ballast is required for the stability of track, and recompaction of the ballast following tamping is needed to ensure that the track regains the stability. Most post-maintenance ballast compaction is provided by reduced speed operations, where the passing train traffic will densify the ballast. A dynamic track stabilizer (Figure 1.33) has been developed to speed up the compaction process and reduce the duration of required reduced speed operations. The stabilizer consists of a series of wheels that ride on the rail coupled with a roller grip that holds the rail laterally. The stabilizer applies a static vertical load to seat the track down into the ballast and a lateral vibration that provides for ballast rearrangement and promotes settlement of the track structure into the ballast.

Figure 1.32 Tamping machine for track surfacing and alignment.

Figure 1.33 Dynamic track stabilizer.

The track undercutter is used to remove fouled ballast from the track structure. The operation is shown in Figure 1.34, involving an excavating chain to remove old fouled ballast from the track (Figure 1.35) and passing it to a series of conveyors and through a screening plant to separate remaining useful ballast from the fouling material. The useful ballast is only a portion of the excavated ballast that has been abraded and crushed under

Figure 1.34 Track undercutting operation.

Figure 1.35 Undercutter ballast excavation chain.

traffic, but that still meets the ballast specification. The new ballast that replaces the removed, spent ballast is placed on top of the reused ballast in the more highly stressed upper ballast zone.

A large number of other maintenance operations are used routinely, but the examples discussed here serve to illustrate the variety of maintenance operations as well as available equipment. Other specialized operations include shoulder cleaning, vacuum trucks, cribbing excavators, stoneblower, and switch undercutters, among others.

1.8.3 Management

The management of railway tracks is a challenging endeavor, given the large capital investment and amount of required maintenance. Efficiently managing track performance and degradation is essential to the success of the railway industry. Making efficient use of inspection data to focus resources on critical issues and guiding routine maintenance is the central challenge. The best opportunity for success will occur when knowledgeable railway personnel can easily interpret and react to the large volume of available data. This is best illustrated by railway geotechnical problems where information from railway operations, track structure, and track geometry need to be combined with track support measurements and geotechnical information to diagnose complex problems.

While many systems have been developed to manage the data, these systems cannot replace knowledgeable interpretation of track behavior and related maintenance needs. These systems must be viewed as tools for an engineer to make good decisions more efficiently.

1.9 LAYOUT OF THE BOOK

The layout of this book is intended to build up the topics from initial discussion of the track structure through materials, mechanics, design, maintenance, and management. The first three chapters introduce the main topics of track (Chapter 1), loading (Chapter 2), and substructure (Chapter 3). Chapter 4 covers the specialized geotechnical topics of railway track mechanics. Track design from a geotechnical perspective is the focus of Chapter 5. While drainage is generally covered with other topics in many books, its importance to track geotechnical performance is so fundamental that the topic of drainage is the focus of Chapter 6. Chapter 7 covers the traditional geotechnical topic of slopes, including both cut slopes and embankment fill slopes with emphasis on slope-related railway infrastructure issues. Chapter 8 provides brief coverage of the wide variety of measurements that are used to monitor track performance and condition. Chapter 9 covers the broad topic of management, with specific focus on the management of geotechnical aspects of railway track. Case studies are the focus of Chapter 10, to provide specific examples showing the application of railway geotechnics principles.

Chapter 2

Loading

The railway track foundation is subjected to static, cyclic, and dynamic loadings. Understanding and quantifying the load environment, and how wheel loads are transmitted from the wheel–rail interface to the track foundation, are essential for the design and remediation of track and subgrade. Load characteristics and considerations can vary widely between a railway track with heavy haul freight and one with high-speed passenger operations. Transference of wheel loads to the track foundation depends on the static, cyclic, and dynamic nature of wheel loads. Furthermore, stiffness and damping properties of track substructure materials affect how wheel loads are dissipated through the depth of the track foundation.

2.1 STATIC LOADING

Static loading to the track foundation includes two components: the live load, which is the train weight, and the dead load, which is the weight of track and subgrade. In general, train weight is the dominant component of static loading, but the weight of track and subgrade becomes significant when designing and analyzing such issues as the slope stability of track built on a high embankment or subgrade problems under a large ballast layer depth.

One of the design live loads is expressed using the Cooper E series configuration adopted by the American Railway Engineering and Maintenance-of-Way Association (AREMA). For example, Figure 2.1 shows the Cooper E80 loading that is specified for railway bridges. Note that the Cooper E loading schematic was developed in 1894 using the axle loads and wheel spacing of a steam locomotive of that era.

However, design of a track foundation is often based on the actual axle load and wheel spacing. Heavy axle load (HAL) freight trains, in particular, generate significant stresses on the track and subgrade. Table 2.1 provides some data for HAL freight trains around the world. Note that the information about length of train or number of cars is important for determining

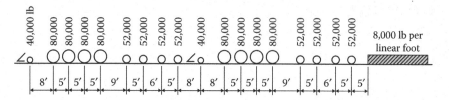

Figure 2.1 Cooper E80 loading as live design loading for railway bridges.

Table 2.1 Typical HAL freight trains around the world

Country	Axle load (tonnes)	Train length (number of cars)	Future trend (train length or axle load)
United States and Canada	33	130–140	150–170 cars
Australia	35–40	200–240	333 cars
South Africa	26–30	200–216	332 cars
Brazil	27.5–32.5	330	37 tonnes
Sweden	30	68	
China	25–27	210	30 tonnes

Table 2.2 Axle loads of high-speed trains around the world

Country	Vehicle	Axle load (tonnes)
Japan	0, 100 series	16.1
	300 series	11.3
	500, 700 series	11.1
China	Passenger car running on dedicated high-speed line	17.0
Germany	ICE 1, ICE 2, passenger car on shared line	19.0
	ICE 3, ICEM 4	16.0
England	Eurostar	17.3
France	TGV-R	17.0
United States	Acela power car	22.7
	Acela coach car	16.4

the cumulative effect of repeated loading on track foundation performance and the length of line loading to be considered for the analysis of embankment stability at large depth.

Table 2.2 includes the axle load information for high-speed passenger trains operating in various countries in the world.

Stress distribution with depth due to train weight is different from that due to weight of track and subgrade. The stress due to train weight is from concentrated point loads at wheel–rail contact points and decreases with increasing depth of track and subgrade. A stiffer track superstructure (rail and tie) leads to lower stress in ballast and subgrade. A stiffer ballast and subballast layer further reduces this stress in the subgrade. This decrease of stress with depth is a result of load-spreading capabilities of rail, tie, and granular (ballast and subballast) layers.

On the other hand, stress due to weight of track and subgrade increases as depth increases. For typical track design, this stress in the ballast and even in the subgrade is insignificant when depth is shallow. However, this dead load–induced stress begins to become more significant when the depth of concern is greater than 1 m (3 ft). In fact, it is considered an ineffective design or maintenance practice if the combined ballast and subballast thickness is >1.5 m above the subgrade (Chapter 4), whether from design or as a result of repeated ballast surfacing and tamping operations.

Calculation of track and subgrade responses including stress and deformation due to static loading is covered in Chapter 4. The following, however, gives some general considerations for determining the dead load of track and subgrade:

- Rail weight depends on the size of rail and cross section and can vary from 45 kg/m (90 lb/yard) to 75 kg/m (150 lb/yard). Lighter rail is used on transit and short/branch lines, while heavier rail is used primarily on the track for HAL freight or high-speed passenger train operations.
- Tie (sleeper) weight depends on the type of material and size of tie. For example, a typical timber tie weighs approximately 110 kg (250 lb), whereas a typical concrete tie can weigh approximately 360 kg (800 lb). In addition, tie spacing along the track should be taken into account when calculating stress due to weight of track. For example, with the assumption of 0.5 m (19.5 in.) and 0.6 m (24 in.) as the tie spacing for timber and concrete sleeper tracks, respectively, the weight distribution of tie along the track would be 220 and 600 kg/m, respectively.
- Weight of track substructure (ballast, subballast, and subgrade) depends on the material type (unit weight) and can be calculated in terms of Equation 2.1. Unit weight of track substructure material varies. For example, a typical unit weight is 1,760 kg/m^3 (110 lb/ft^3) for ballast, 1,920 kg/m^3 (120 lb/ft^3) for soil, and 2,240 kg/m^3 (140 lb/ft^3) for bedrock. Note that the reason why ballast may have lower unit weight than subgrade soil is because there are more voids in ballast.

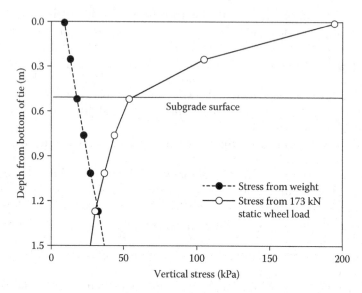

Figure 2.2 Weight and wheel load–induced stress distributions in track and subgrade.

$$w = \gamma \cdot z \tag{2.1}$$

where:

w is the weight of track substructure per unit area (kg/m²)
γ is the unit weight of material (kg/m³)
z is the depth (m)

Figure 2.2 shows an example of calculated vertical stresses due to weight of track and subgrade as well as due to a static wheel load (173 kN or 39 kips), respectively. The calculation was done for a typical track structure with concrete ties, using GEOTRACK (Chapter 4). As illustrated, stress due to dead load increases with depth, whereas the stress due to live wheel load decreases with depth.

At shallow depth, the live load stress is much greater than the dead load stress. At larger depth, however, the trend reverses. At a depth greater than 1.3 m, the stress from track and subgrade weight (dead load) becomes more significant.

2.2 CYCLIC LOADING

Cyclic loading is characterized by the following parameters: shape, duration, magnitude of loading pulse, time interval between consecutive pulses, and the total number of loading pulses.

As a train runs on a railway track, a number of loading pulses generated by moving loads are applied to the track and subgrade rapidly. These pulses are applied so quickly and repeatedly that they lead to characteristically different responses of track and subgrade than that from static sustained loading. As subgrade soil properties are affected by changes in loading rate, it is essential to determine the transient loading pulse duration. Another aspect of characterizing repeated loading pulses lies in conducting cyclic or repeated loading tests in the laboratory to evaluate material properties and quantify track and subgrade performance. A repeated loading test on ballast or subgrade soil requires the determination of loading intensity, loading pulse shape, loading pulse duration, and the number of load applications equivalent to accumulation of traffic. Finally, the analysis of track component deformation and fatigue characteristics as well as the design of track structure requires the determination of the number of load applications equivalent to the traffic.

To determine the number of repeated loading applications, a common practice to assume is that two axles under the same truck (bogie) are considered to produce one load cycle for the ballast layer and four axles under two adjacent trucks are considered to produce a single load cycle for the subgrade layer. Thus, for 1 MGT (million gross tons) of traffic, numbers of loading cycles can be determined as follows:

$$N_a = \frac{10^6}{2P_s} \tag{2.2}$$

$$N_b = \frac{N_a}{2} \tag{2.3}$$

$$N_s = \frac{N_a}{4} \tag{2.4}$$

where:
P_s is the static wheel load in tons
N_a is the number of axles per MGT
N_b is the number of load applications per MGT for ballast layer
N_s is the number of load applications per MGT for subgrade layer

These equations are intended for general use. How adjacent wheel loads interact with each other within the track structure and affect track responses depends on axle spacing, truck spacing, and depth of track under consideration. Figure 2.3 shows two examples of actual track deflection test results using a device called multidepth deflectometer. These two plots show deflection results at two different depths, as two different trains moved over the measurement location. These two trains had much different lengths between adjacent trucks on either end of the coupler. The static axle loads were

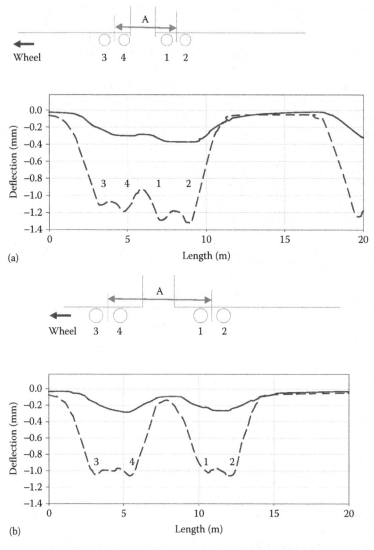

Figure 2.3 Comparison of subgrade response to moving loadings of two cars with different coupler/axle spacing: (a) short car (length between axles at coupler, $A = 2.4$ m) and (b) long car (length between axles at coupler, $A = 4.8$ m).

38 tons for the train with the short coupler length and 30 tons for the train with the long spacing between adjacent trucks.

As shown in Figure 2.3, the truck configuration with the short distance between them unloaded the subgrade only slightly between pairs of axles and so is generally considered to apply one load cycle for the four axles of the adjacent trucks near the coupler, which averages out to one subgrade load cycle per car. The long axle spacing results in nearly complete unloading between the adjacent trucks at both ends, indicating that the longer coupler length should be considered as applying one load cycle to the subgrade for two axles, which averages out to two load cycles per car.

When determining cyclic loading, it is also important to characterize the shape and duration of loading pulses applied to the ballast and subgrade materials. Two types of stress pulses are often used in a repeated load test. They are the haversine pulse and the trapezoidal pulse. A study by Li (1994) suggested that the loading pulse by a single axle can be approximately represented by a haversine pulse for the track and subgrade. However, for the two axles under the same truck or four axles under two adjacent trucks, loading pulse shape is more like a trapezoidal pulse (see Figure 2.3).

The duration of loading pulse depends on the operating speed of train and the depth of consideration and can be calculated in terms of Equation 2.5:

$$t = \frac{L}{V} \tag{2.5}$$

where:
  t is the time duration of loading pulse
  V is the train speed
  L is the influence length of an axle load or adjacent axle loads for a given depth of ballast or subgrade

Influence length is the distance between the approaching wheel that starts to have an influence for the element of concern and the leaving wheel that still has an influence on the same element. Influence length can be determined from modeling or from field measurement. In Figure 2.3, the first example has an influence length of approximately 10 m and the second example has an influence length of approximately 5 m. From a study conducted by Li (1994), an influence length can range from 2 to 12 m, depending on the depth of the element being considered and the number of adjacent axles involved. If a range of train speed from 80 to 240 km/h (50 to 150 mph) is considered, duration of a loading pulse for the track substructure can vary between 0.05 and 0.5 s.

From another study by Li (1991), the effect of loading rate on strength and stiffness properties of granular materials (ballast and subballast) is very small, unless the granular materials are saturated under undrained condition. Therefore, consideration of loading rate effect is generally meaningful only for subgrade soils.

If strength or stiffness properties of subgrade soils are to be determined by conventional laboratory test methods such as direct shear test, unconfined compression test, or an undrained triaxial test, the loading rate or loading rise time on soil specimens will generally be far lower than the rate at which an actual loading pulse is applied to the subgrade soil under the track. A conventional soil strength test generally has a range of loading rise time (defined as the time taken from zero to maximum load) from 5 to 15 min, depending on soil type and test method. Thus, the actual loading rate under the track may be much faster than that under a laboratory loading by a factor of 100–10,000. According to a study by Li (1991), a 10-fold increase in loading rate, on average, can lead to an increase in soil strength of about 10%. Therefore, subgrade soil subjected to train loading may exhibit a 20%–40% higher strength than under a laboratory loading, if other conditions are the same. This discrepancy points to the importance of running repeated load tests with proper loading pulses, as compared to a conventional laboratory test to determine subgrade soil strength and stiffness.

2.3 DYNAMIC LOADING

Broadly stated, dynamic wheel loads are either short duration forces in which the vehicle suspension plays little or no role or longer duration forces that depend on vehicle suspension response. Short duration forces are high-frequency impact loads arising from discontinuities in the wheel (flat spots, out of roundness, etc.) or rail (dipped joints, running surface discontinuities at special track locations such as turnouts, battered welds, etc.) and are a source of considerable wheel and rail damage and deterioration. Because a typical dipped joint or rail discontinuity occurs over a short track length, it produces a short duration load pulse with no time for the vehicle suspension and track foundation to react, so the corresponding dynamic load and track damage is confined locally to the wheel, rail, and tie. Longer wavelength track geometry irregularities produce a longer duration response, and so there is time for the dynamic wheel loads to be influenced by the vehicle suspension and to be transmitted into the track substructure including ballast and subgrade.

2.3.1 Dynamic load or impact load factor

Whether it is short duration/high-frequency impact load or longer duration dynamic wheel load, the load applied on the track under train operations is generally different from the static wheel load in magnitude. This dynamic wheel load can be significantly higher or lower than the static wheel load. For track design, a common method for selecting a design dynamic wheel load is simply to augment the static wheel load by a dynamic load factor or impact load factor in terms of the following equation:

$$P_d = \alpha P_s \tag{2.6}$$

where:

P_d is the dynamic wheel load
P_s is the static wheel load
α is the dynamic load factor

A range of values for dynamic load factor between 1.5 and 3.0 can be found in the literature. For example, the Code for Design of High-speed Railways (PRC 2010) recommends a value of 3.0 for design of track for speeds at or above 300 km/h and 2.5 for 250 km/h. In the same Code for Design, a lower value of 1.5 is recommended for fatigue design that involves repeated or cyclic loading.

Empirical equations have also been developed to relate this dynamic load factor to the variables that contribute to dynamic wheel load or impact load. For example, AREMA (2012) recommends the following empirical equation for calculating dynamic load factor:

$$\alpha = \left(1 + \frac{0.0052V}{D}\right) \tag{2.7}$$

Thus, Equation 2.6 becomes:

$$P_d = \left(1 + \frac{0.0052V}{D}\right)P_s \tag{2.8}$$

where:

V is the train speed (km/h)
D is the wheel diameter (m)

However, this equation is valid only for operating speeds of freight vehicles, as it was derived based on lower-speed freight operations.

2.3.2 High-frequency forces

High-frequency impact forces are often characterized by P_1 and P_2 forces (Jenkins et al. 1974). As illustrated in Figure 2.4, P_1 is a very high–frequency force occurring above 500 Hz. Although also classified as a high-frequency force, P_2 is slower and is in the range of 20–100 Hz.

Calculations of P_1 and P_2 can be done using the equations developed by Jenkins et al. (1974) and updated by British Railway Board (1993):

$$P_1 = P_0 + 2\alpha V \sqrt{\frac{K_H m_e}{1 + m_e/m_u}} \tag{2.9}$$

$$P_2 = P_0 + 2\alpha V \sqrt{\frac{m_u}{m_u + m_T}} \left[1 - \frac{\pi C_T}{4\sqrt{K_T(m_u + m_T)}}\right] \sqrt{K_T m_u} \tag{2.10}$$

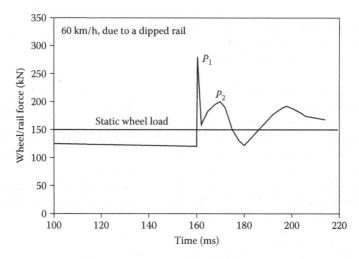

Figure 2.4 Definition of P_1 and P_2 forces.

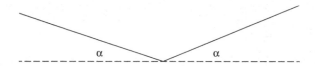

Figure 2.5 Symmetrical dip angle at joint or weld (exaggerated angle).

where:

   P_0 is the nominal static vertical wheel load (lb)
   2α is the total dip angle at a joint, weld dip, or other rail discontinuity,
      as shown in Figure 2.5 (radians)
   V is the train speed (in./s)
   m_u is the unsprung mass per wheel (lb-s²/in.)
   m_T is the equivalent track mass (lb-s²/in.) for P_2
   m_e is the effective track mass for P_1
   C_T is the equivalent track damping (lb-s/in.)
   K_T is the vertical track stiffness (lb/in.)
   K_H is the linearized wheel–rail contact stiffness

Note that all these parameters are on the per-rail basis.

   Typical values of selected parameters are $\alpha = 0.0085$ rad (0.5° on both sides of the dip), $m_T = 1.1$ lb-s²/in. for nominally stiff concrete tie track, $C_T = 670$ lb-s/in. for nominal track conditions, and $K_T = 330,000$ lb/in. for nominally stiff concrete tie track that corresponds to a track modulus of 4,000 lb/in. (assuming a track deflection of 0.10 in. under a wheel load of 33,000 lb). The effective track mass, m_e, is similar to m_T for the P_2 force, except that the magnitude of m_e is smaller (approximately 0.5 lb-s²/in.), because there is less track mass contributing to the reaction for this much shorter duration load. K_H represents a very stiff spring and is typically in the order of several hundred thousand pounds per inch.

From Equation 2.9, it is obvious that the parameters that affect P_1 force significantly are dip angle, train speed, and wheel–rail contact stiffness. From Equation 2.10, the parameters that affect P_2 force significantly are dip angle, train speed, track stiffness, track damping, and the unsprung mass.

During vehicle design, the P_2 force has been used to control the track damage produced by a new car by limiting the applied force to that produced by an existing reference car. In the analysis, both new and existing vehicles are assumed to be traveling over the same dipped joint with defined speed, static axle load, unsprung mass, and track stiffness and damping values. The three design variables of maximum speed, static axle load, and unsprung mass can be adjusted as needed for the new vehicle.

2.3.3 General equations for calculating dynamic wheel–rail forces

Equations 2.9 and 2.10 were developed for calculating high-frequency forces P_1 and P_2 associated with dipped rail joints. To calculate dynamic wheel–rail force $P(t)$ from vehicle–track interaction due to various excitations including longer wavelength track geometry irregularities, the following equation can be used (Li 1985, 1989; Li et al. 1996):

$$P(t) = \left[\frac{G_H P_0^{2/3} + y_w(t) - y_r(t)}{G_H} \right]^{3/2} \tag{2.11}$$

where:

G_H is the Hertzian contact coefficient, which depends on wheel radius and profile (Jenkins et al. 1974)

y_w and y_r are the dynamic deflections of wheel and rail, respectively, at the wheel–rail contact point

If no contact stiffness between wheel and rail is considered (i.e., assuming rigid contact between wheel and rail), the dynamic wheel–rail force can be calculated by:

$$P(t) = P_0 + m_u \left[\ddot{y}_w(t) + \ddot{\eta}(t) \right] \tag{2.12}$$

where:

$\eta(t)$ is the excitation function

m_u is the unsprung mass

2.3.4 Modeling for dynamic wheel–rail forces

As can be seen in Equations 2.11 and 2.12, dynamic track responses such as y_w and y_r need to be determined to calculate dynamic wheel–rail forces due to vehicle–track interaction. This can be accomplished using vehicle–track simulation models.

Several multibody vehicle dynamic simulation software programs are commercially available for modeling vehicle and track transient and steady-state responses, including NUCARS®, VAMPIRE®, and SIMPACK. Figure 2.6 shows the typical bodies included in a freight vehicle that can be simulated with any of these software programs.

Among those simulation programs, NUCARS has the modeling capabilities of incorporating a multilayer flexible track model (see Figure 2.7), which has the following features:

- Representation of track structure as a multibody system
- Modeling of track with various track structures such as slab track versus ballasted track, missing track components, variable ballast support, variable tie support, and so on
- Automatically integrated vehicle and track dynamic coupling system through penetrated wheel–rail contact model
- Rail movement and deflection in lateral, vertical, and roll direction under the influence of wheel–rail forces

A set of equations, generalized as Equation 2.13, is then used to obtain vehicle–track responses due to excitations. As the moving vehicle couples with

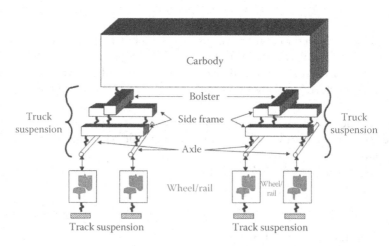

Figure 2.6 Multibody vehicle simulation model.

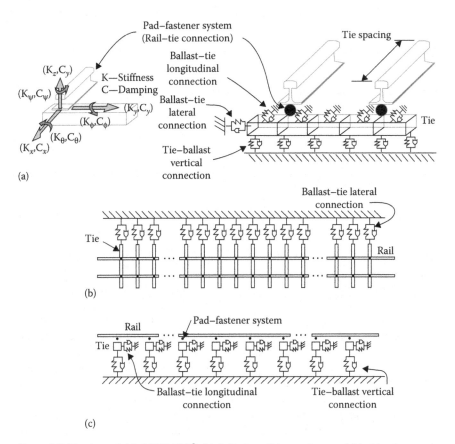

Figure 2.7 Track model in NUCARS®: (a) 3-D view, (b) top view, and (c) side view.

the track, the global mass, damping, and stiffness matrices in Equation 2.13 are wheel location dependent, that is, time dependent.

$$[m(t)]\{\ddot{y}\} + [c(t)]\{\dot{y}\} + [k(t)]\{y\} = \{p(t)\} \tag{2.13}$$

where:

$\{y\}$ is the deflections and rotations of all elements in the model

$[m]$, $[c]$, $[k]$ are matrices of mass, damping, and stiffness

$\{p\}$ is the excitation force vector related to rail or wheel irregularities or track roughness

In some cases, instead of using a longitudinally distributed track model, like the one shown in Figure 2.7, a simplified lumped parameter vehicle-track model can be used, like the one shown in Figure 2.8. Compared to a longitudinally distributed model, this lumped parameter model has far

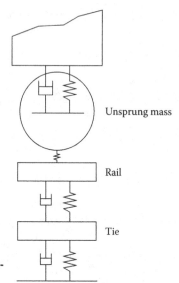

Figure 2.8 Lumped parameter vehicle–track interaction model.

fewer degrees of freedom and thus is easier to use. However, the limitation of this model is that it does not allow direct calculation of track responses at locations away from the wheel–rail contact point.

To represent an actual longitudinally distributed track structure, lumped mass and stiffness parameters for track need to be obtained based on the longitudinally distributed parameters, which are discussed next.

Figure 2.9 shows two different model systems for conversion: the distributed system (a) and (c) and the lumped parameter system (b). To obtain the equivalent lumped mass and stiffness parameters, two conversion principles can be used (Li 1989). For conversion of the mass parameter, the kinetic energy of the lumped mass under vibration is set to be equal to that of the beam under vibration. For conversion of the stiffness parameter, the

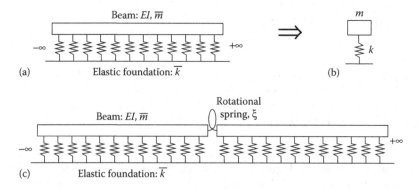

Figure 2.9 (a–c) Conversion of model parameters.

spring deflection in the lumped system under a concentrated load is set to equal to the beam deflection under the same concentrated load.

Based on these two conversion principles, the general conversion equations for mass and stiffness were derived as follows (Jenkins et al. 1974; Li 1989):

$$m = \frac{3\bar{m}}{2\beta} \tag{2.14}$$

$$k = \frac{2\bar{k}}{\beta} \tag{2.15}$$

$$\beta = \sqrt[4]{\frac{\bar{k}}{4EI}} \tag{2.16}$$

where:
m, k is the lumped mass and stiffness
$\bar{m}, \bar{k}$ is the distributed mass and stiffness
EI is the beam bending stiffness

At a rail joint, as shown in Figure 2.9c, beam bending stiffness (EI) decreases due to the existence of the joint. Consequently, the equivalent lumped mass and stiffness parameters from Equations 2.14 and 2.15 need to be modified. Using a rotational spring to represent the rail joint, the following equations were developed for the lumped parameters at the rail joint (Li 1989):

$$m = \frac{3\bar{m}}{2\beta}\mu_1 \tag{2.17}$$

$$k = \frac{2\bar{k}}{\beta}\mu_2 \tag{2.18}$$

$$\mu_1 = 0.5 + \frac{\xi}{3(\xi + \beta EI)} + \frac{\xi^2}{6(\xi + \beta EI)^2} \tag{2.19}$$

$$\mu_2 = \frac{\xi + 0.5\beta EI}{\xi + \beta EI} \tag{2.20}$$

where:
ξ is the rotational stiffness at the rail joint

From these equations, both coefficients μ_1 and μ_2 are equal to 0.5 when there is no rotational stiffness at the rail joint (i.e., when $\xi = 0$). In other words, the effective lumped mass and stiffness at the rail joint without joint bars (e.g., a broken rail) are only half of the effective lump mass and stiffness representing a continuous uniform beam.

Once the lumped model parameters are obtained, dynamic equations similar to Equation 2.13 can be established to determine vehicle–track responses due to dynamic excitations. Then, Equation 2.11 or 2.12 can be used to predict dynamic wheel–rail forces. Again, depending on whether rigid contact or Hertzian contact is used between wheel and rail, prediction of wheel–rail force may use a different equation.

2.3.5 Stiffness and damping parameters in track modeling

Applications of dynamic simulation models depend, to a great degree, on the proper characterization of the system parameters such as stiffness and damping properties. To obtain accurate simulation results, stiffness and damping values must be known. A vehicle–track model without proper parameter inputs will have limited practical uses.

The definition of a stiffness parameter is generally based on the linearization of a load versus deformation relationship. Because most track and subgrade components behave in a nonlinear and hysteretic way, a stiffness value needs to be defined for a given magnitude of deformation. To analyze dynamic track responses of small deformation magnitudes, a linearized stiffness parameter is often satisfactory. However, when a large track deformation (including both elastic and permanent components) is of interest, a more accurate characterization of a load–deformation relationship requires the use of a piece-wise linear representation.

Damping can be expressed by either a viscous damping coefficient, c, or a loss factor, η. These two parameters can be related to each other by the following equation:

$$c = \frac{k\eta}{\omega} \tag{2.21}$$

where:
   k is the stiffness
   ω is the angular frequency

Table 2.3 lists typical ranges of track stiffness and damping values (lower bound, nominal, and upper bound values) based on a study by Li et al. (1999).

2.3.6 Dynamic track modeling results

Based on the NUCARS track model (Figure 2.7), this section presents some modeling results to illustrate the effects of track stiffness and damping on dynamic track responses. First, Figure 2.10 shows the vertical receptance (deflection divided by dynamic input force) result for a track with nominal track parameters. The dynamic force (swept sine) is applied at the middle of

Table 2.3 Range of track stiffness and damping values for dynamic track modeling

Parameters	Rail–tie connection (pad)	Tie–foundation connection (ballast)
k_z (MN/m)	50-300-1000	20-100-500
c_z (kNs/m)	5-50-100	30-100-250
k_y (MN/m)	25-100-250	25-70-200
c_y (kNs/m)	5-10-30	20-30-80

Note: y, the lateral direction; z, the vertical direction.

1 MN/m = 5.7 kips/in., 1 kNs/m = 5.7 lb/in./s.

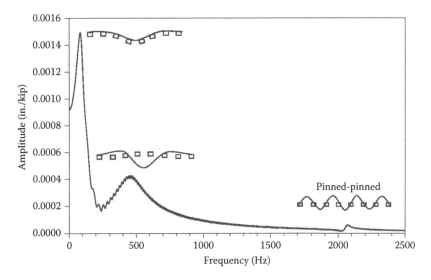

Figure 2.10 Vertical dynamic track response under a vertical dynamic force.

the span between the two adjacent ties, where the corresponding deflection is used for receptance calculation.

As shown in Figure 2.10, the receptance of this track contains two major peaks. The first peak at lower frequency corresponds to the mode where rails and ties move in phase relative to the ballast. The second higher frequency peak is associated with the mode of the system where the rail and ties move in anti-phase. Furthermore, another peak associated with the rail "pinned-pinned" mode can be seen. In this mode, the displaced shape of the rail has a wavelength of approximately two support spacings, with nodes close to the tie supports. This mode is not affected by the characteristics of the pad/fasteners and ballast material (Li et al. 1999).

Figures 2.11 and 2.12 show the effects of stiffness and damping parameters. The vertical receptance results (amplitudes) were calculated for the

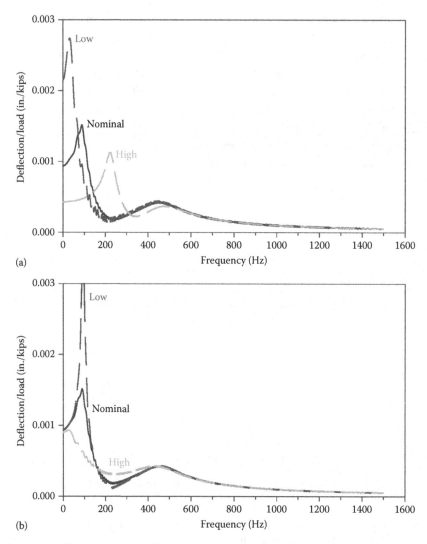

Figure 2.11 Effects of ballast stiffness (a) and damping (b) on vertical track response.

nominal, lower bound, and upper bound values of each track parameter listed in Table 2.3.

Figure 2.11 illustrates the effects of ballast (tie–foundation connection) stiffness and damping on the magnitude of track receptance. As shown, the ballast characteristics mainly affect the first fundamental mode. An increase in ballast stiffness increases the fundamental frequency and decreases the magnitude of rail displacement. The fundamental frequency is between 50 and 210 Hz, depending on the ballast stiffness. On the other

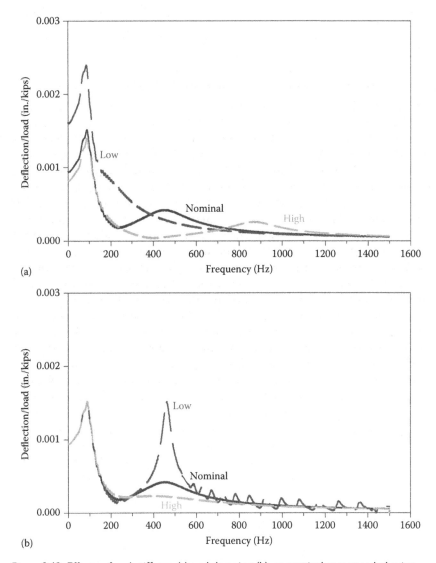

Figure 2.12 Effects of pad stiffness (a) and damping (b) on vertical response behavior.

hand, an increase in ballast damping leads to a decrease in magnitude of track deformation at the fundamental frequency.

In contrast, the pad characteristics mainly affect the second mode. As shown in Figure 2.12, an increase in pad stiffness leads to an increase in the frequency corresponding to the second mode. Moreover, an increase in pad damping leads to a decrease of deformation corresponding only to the second mode.

Nevertheless, damping from ballast or pad can help attenuate dynamic track deflection, although it works effectively in a range of different frequencies.

2.3.7 Measured dynamic wheel–rail forces

Wayside (trackside) load station and onboard instrumented wheel sets (IWS) can be used to measure dynamic wheel loads resulting from dynamic vehicle–track interaction. A wayside load station measures wheel–rail forces generated at a selected track location and is ideally suited to identify vehicles or wheels in a train that perform poorly and causes large wheel impact forces. An IWS measures wheel–rail forces under a selected truck (bogie) along the track and is ideally suited to investigate vehicle–track interaction due to track geometry irregularities and rail discontinuities.

Figure 2.13 shows the distributions of maximum wheel loads recorded using a wayside measurement method (strain gages on the rail) from heavy haul freight trains that passed over a short span of concrete bridge and its approach located in the western United States. As shown, a large number of large wheel loads, at least twice the static wheel load of 160 kN (36 kips), were recorded. Furthermore, the number of large wheel loads measured on the bridge was greater than in the bridge approach, because the track on the bridge was stiffer and had less damping. As a result, mud pumping and cracked concrete ties occurred at this bridge approach (Li et al. 2010).

Depending on the sampling rate, both wayside load station and IWS can measure wheel–rail forces of a wide range of frequencies. However, there is

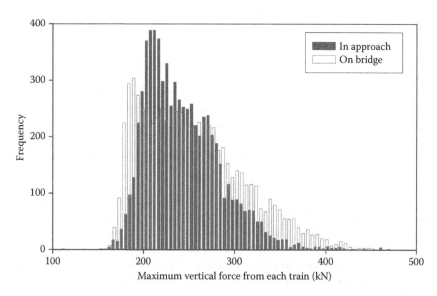

Figure 2.13 Measured vertical wheel loads under heavy axle load freight trains.

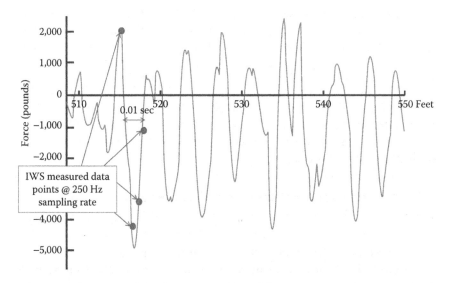

Figure 2.14 Dynamic load waveform and digital representation from IWS.

a limitation of IWS in measuring high-frequency impact forces, especially the P_1 force. This is because IWS samples at a rate lower than 500 Hz.

Figure 2.14 shows the dynamic wheel load augment (+ and − the static vertical wheel load) that was predicted by modeling for a wheel traversing a track with slight bumps and dips. Note that the dynamic wheel load pulse alternately adds to and then subtracts from the static vertical wheel load. Suppose that the largest force is the load of interest. Note that this load pulse duration is 10 ms (typical for a P_2 force) and that when this waveform is digitized by sampling with the IWS at a rate of 250 Hz, 2.5 data samples are obtained over this 10 ms (0.01 s × 250 data samples/s = 2.5 samples). Most of the downward load pulse is then characterized by the four data points that are evenly spaced along the x-axis, as shown in Figure 2.14. By visual inspection, this sampling rate is sufficient to capture the peak load with little error, even if the positions of the actual sampling points shift forward or backward from their shown location.

The error in estimating the peak load for a chosen sampling rate can be determined from the following equation (SAE Fatigue Handbook 1997):

$$PK_{err} = 2\sin^2\left(\frac{\pi f_d}{2 f_s}\right) \tag{2.22}$$

where:

    f_s is the data sampling frequency
    f_d is the frequency of the data or the frequency of impact force to be
       measured

Because the example waveform in Figure 2.14 has a 3-ft cycle at a speed of 60 mph, its bounce frequency is approximately 30 Hz (88 ft/s/3 ft per cycle). This bounce frequency and a 250 Hz sampling frequency results in a peak value error of approximately 7%. Therefore, the maximum dynamic load predicted by analysis as shown by the solid line should be reasonably well approximated by the IWS measurement at the indicated sampling rate. Provided the sampling rate is high enough, the IWS can be used for measurement of dynamic wheel loads such as the P_2 force. However, the IWS with a sampling rate of 250 Hz is not sufficient to measure higher frequency impact loads from the P_1 force.

2.3.8 Critical speed of high-speed passenger trains

There is a growing amount of evidence that high-speed passenger trains can induce resonant ground vibrations where the vertical track deflection under loading becomes dynamically amplified as trains approach a certain "critical speed." The surface waves induced in the soil by train motion radiate outward from the train, oscillating as the waves travel along the soil surface. This wave travels fast in rock or very stiff soil but is slowed by the presence of soft soil with high water content, making it easier for a train to catch up to its own wave. If train speed approaches the ground wave propagation velocity, there will be a buildup and amplification of deformation waves that will then form a shock front, similar to the Mach effect when an airplane approaches the speed of sound.

"Critical speed" is the speed at which trains induce a large resonant response in the track with soft ground conditions resulting in excessive vertical vibrations. Perhaps the best known and most well-documented case is that at Ledsgard, Sweden (Holm et al. 2002). This critical speed could be defined as that corresponding to the peak vertical displacement such as the one identified in the Ledsgard study occurring at 235 km/h, as shown in Figure 2.15. However, because the condition leading up to the peak amplitude vibration may itself be undesirable, it may be more appropriate to define critical speed as any train speed that produces track vibration with amplitude significantly greater than the elastic deformation that occurs under static loading. Therefore, it is more accurate to speak of a range of speeds that produce this condition rather than a singular speed.

Note that not only is the track downward deflection significantly increased at these speeds, but the track uplift is also introduced, amplifying the vibration further. This particular type of propagating wave causing these large vibrations is known as the Rayleigh wave, which travels along the ground surface and produces a strong vertical component of oscillation. The Rayleigh wave velocity is reduced in soft and wet soil conditions.

Train loading causes an elastic vertical deformation in the soil that has a concentric pattern around the load for train speeds up to approximately

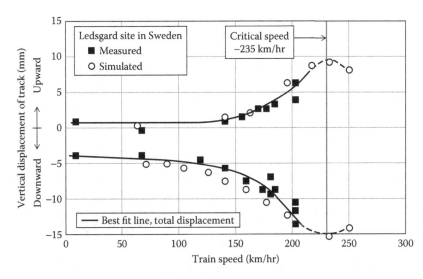

Figure 2.15 Measured and predicted vertical track displacement versus train speed with soft soil ground. (Modified from Holm, G., et al., *Mitigation of Track and Ground Vibration by High Speed Trains at Ledsgard, Sweden*, Swedish Deep Stabilization Research Centre, Linkoping, Sweden, 2002.)

25% of the Rayleigh wave velocity. This concentric pattern is shown in the top portion of Figure 2.16 where the iso-vertical deformation contour lines are shown from above looking down. The sizes of the vertical arrows indicate the relative magnitude of static vertical wheel loads for this trainset.

With increasing train speed the iso-vertical deformation lines grow laterally and become compressed in the direction of travel as shown in the lower two portions of Figure 2.16. Also, although not apparent from the two-dimensional plots, the vertical deformation amplitude increases exponentially with train speed.

Because testing track response at train speeds near the maximum vibration amplitude is risky, researchers used analysis to simulate track vibration for speeds above the highest speed tested (Figure 2.15). From the simulations, the critical speed at maximum vertical oscillation was estimated to be 235 km/h. The analysis also showed that the largest vertical track displacement is predicted to shift from being under the heaviest vehicles for lower speeds to being under the lighter vehicles in the middle of the trainset as critical speed is approached.

Obviously, train speeds should stay well below that corresponding to the peak resonance, but what is a safe speed reduction below that corresponding to peak resonance? The Swedish practices define the critical speed as that corresponding to the maximum track deflection (assuming this could be reliably estimated from analysis) and limit train speed to no more than

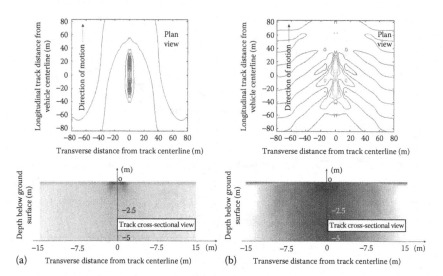

Figure 2.16 Ground deformation patterns under (a) vehicle at "nominal" speed and (b) critical speed. (Courtesy of Hai Huang.)

two-thirds of this. Where improvement of soft soils is needed to allow the desired train speed, the Swedish researchers specify that remediation must increase the soil strength such that the critical speed is increased by a factor of 1.5–2.0 beyond the intended maximum train speed (Chrismer 2011).

It should be noted that "critical speed" does not become a practical problem on the vast majority of soils until very high-speed trains running on the track with soft soil conditions and is not a problem for the vast majority of the lengths of existing railway lines globally where passenger trains run at 200 km/h or more.

2.4 LOAD TRANSFER IN TRACK FOUNDATION

Whether it is static, cyclic, or dynamic, loading is transmitted to the track and subgrade, resulting in stresses and strains in the track foundation. This transference of load, stress, and strain to the track foundation depends to a great degree on the strength and stiffness properties of track substructure. With adequate strength and stiffness of track foundation materials, stresses and strains due to various types of loadings should not cause track and subgrade failure or excessive deformation.

The following presents some fundamental discussions regarding stresses and strains in the track foundation, as well as strength and stiffness properties that directly affect the transference of loading to the track foundation.

2.4.1 Stresses and strains in track foundation

Figure 2.17 illustrates the stress state of three different elements in the track foundation under a moving wheel load. Only stresses in the vertical and longitudinal directions are shown. In general, the three-dimensional stress state for an element in the track foundation can be expressed as

$$\{\sigma_{ij}\} = \{\sigma_x, \sigma_y, \sigma_z, \sigma_{xy}, \sigma_{yz}, \sigma_{zx}\} \tag{2.23}$$

where:

σ_x, σ_y, σ_z are the normal stresses in the longitudinal, lateral, and vertical directions

σ_{xy}, σ_{yz}, σ_{zx} are the shear stresses in xy, yz, and zx planes

Figure 2.17 shows the normal and shear stresses in the zx plane, as such $\sigma_{zx} = \tau$.

A normal stress is called a principal stress when the shear stress is equal to zero. As shown in Figure 2.17, for the element directly under the wheel load, both vertical and lateral normal stresses are principal stresses. The larger principal stress is referred to as the major principal stress and is denoted as σ_1. The smaller principal stress is referred to as the minor principal stress and is denoted as σ_3.

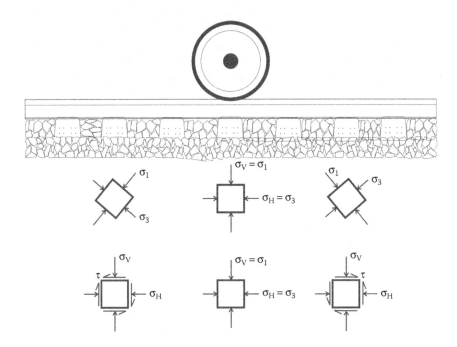

Figure 2.17 Stresses in track foundation.

Bulk stress, as defined in Equation 2.24, is a stress parameter generally used for the granular layer (ballast and subballast) analysis, while deviator stress, as defined in Equation 2.25, is a stress parameter generally used for the analysis of subgrade consisting of cohesive soils.

$$\theta = \sigma_1 + \sigma_2 + \sigma_3 \tag{2.24}$$

$$\sigma_d = \sigma_1 - \sigma_3 \tag{2.25}$$

where:
θ is the bulk stress
σ_d is the deviator stress

The ratio of horizontal to vertical stress is expressed by a factor called the coefficient of lateral stress or lateral stress ratio and is often denoted by the symbol K:

$$K = \frac{\sigma_x}{\sigma_z} \tag{2.26}$$

Where there is no lateral strain, this ratio is called the coefficient of lateral stress at rest or lateral stress ratio at rest and is often denoted using the symbol K_0. Typically, lateral stress ratio at rest is between 0.4 and 0.5, although it is possible that the horizontal stress exceeds the vertical stress, and in that case, this coefficient is greater than 1.0. This is especially true when residual stress in the ballast is high and resulting lateral stress ratio can be significantly higher than 1.0 (see Chapter 3). How to determine these stresses through modeling is discussed in Chapter 4. After stresses are determined, strains can be obtained by

$$\varepsilon_x = \frac{1}{E}\left[\sigma_x - \nu(\sigma_y + \sigma_z)\right] \tag{2.27}$$

$$\varepsilon_y = \frac{1}{E}\left[\sigma_y - \nu(\sigma_x + \sigma_z)\right] \tag{2.28}$$

$$\varepsilon_z = \frac{1}{E}\left[\sigma_z - \nu(\sigma_x + \sigma_z)\right] \tag{2.29}$$

$$\gamma_{xy} = \frac{\tau_{xy}}{G} \tag{2.30}$$

$$\gamma_{yz} = \frac{\tau_{yz}}{G} \tag{2.31}$$

$$\gamma_{zx} = \frac{\tau_{zx}}{G} \tag{2.32}$$

where:

ε$_x$, ε$_y$, ε$_z$ is the normal elastic strain in the longitudinal, lateral, and vertical directions

γ$_{xy}$, γ$_{yz}$, γ$_{zx}$ is the elastic shear strain in xy, yz, and zx planes

E is the elastic modulus

ν is the Poisson ratio

G is the shear modulus

The relationship between E, ν, and G is defined by

$$G = \frac{E}{2(1+\nu)} \tag{2.33}$$

Under repeated wheel load applications, however, cumulative plastic strain may become a deformation parameter that is more critical than elastic strain. Figure 2.18 illustrates the difference between resilient (elastic) deformation and plastic deformation from repeated load applications under a cyclic loading condition. Elastic modulus, when defined in terms of resilient deformation and deviator stress, is referred to as resilient modulus. Resilient modulus and plastic deformation are two main parameters critical to the

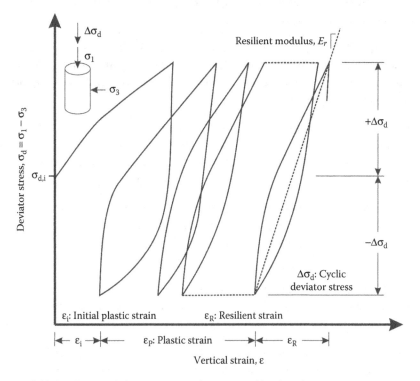

Figure 2.18 Resilient and plastic strain under repeated load applications.

analysis and design of track foundation subjected to repeated wheel load applications, as discussed in Chapter 4.

2.4.2 Load transmission in track

Vehicle–track interaction determines magnitudes of wheel–rail interaction forces. Depending on the spectrum of frequency components, some of higher frequency force components (such as P_1 force discussed previously) will dissipate quickly when they are transmitted from the wheel–rail interface downward to the track substructure layer. Static or dynamic wheel loads of lower frequencies, however, are spread to the underling track substructure layers over larger area with increasing depth.

This spreading of load over larger area at greater depth, to a large degree, depends on the stiffness characteristic of each layer of the track structure, including the rail, ties, ballast, and subgrade. Chapter 4 includes discussions and models that can be used to compute how the static or quasi-dynamic wheel loads are transmitted from the wheel–rail interface to the underling track substructure layers both along and across the track. The following gives two examples using GEOTRACK (Chapter 4) to illustrate how a static wheel load is transmitted from the wheel–rail interface to the rail–tie interface, to the tie–ballast interface, and to various depths in the ballast and subgrade layers.

One example is for a typical concrete tie track, and the other is for a typical wood tie track, as shown in Figure 2.19. In these two examples, a single 40-kip wheel load is applied on the rail directly above a tie, and this figure shows how this load is transmitted downward to different depths both along and across the track.

The top plots show the results at various depths under the rail along the track longitudinally. The left plot is for the concrete tie track and the right plot is for the wood tie track. As illustrated, for a 40-kip single wheel load on the concrete tie track, 12.4 kips (31%) is transmitted to the rail–tie interface directly under the load, and 9 kips (23%) is transmitted to the two adjacent concrete ties. The entire wheel load is spread only to the fourth tie, because the rail–tie force is essentially zero on the fifth tie.

On the other hand, for the wood tie track, 16.6 kips (42%) is transmitted to the rail–tie interface directly under the 40-kip wheel load, which is spread over a shorter distance along the track, roughly to the third tie, as the plot on the top right shows.

At the ballast surface, maximum vertical stress is generated directly under the wheel load, although the wood tie track has higher maximum vertical stress than the concrete tie track, because the former does not spread the load over a distance along the track as much as the latter does. The stress contours illustrate how vertical stress is distributed at various depths along the track. As shown, directly under the wheel load, stress is still somewhat significant (e.g., 5 psi) at a depth of approximately 4 ft in the subgrade (from

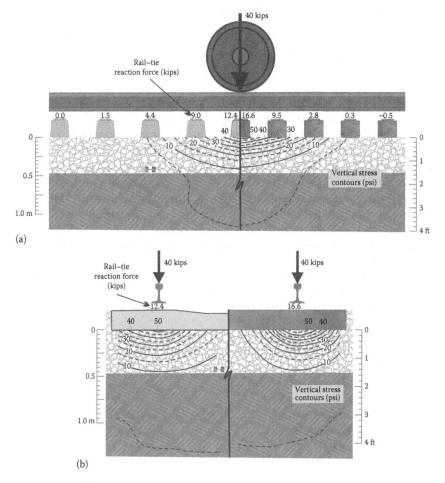

Figure 2.19 Load transmission from wheel–rail interface downward along (a) and across (b) concrete and wood tie tracks.

bottom of the tie), but this stress reduces quickly and becomes insignificant further away from the wheel load.

The bottom two plots show how the 40-kip single wheel load is transmitted downward across half the track with respect to the center of the track, for both the concrete and wood tie tracks. Again, higher vertical stress (over 50 psi) is generated directly under the rail for the wood tie track compared to the concrete tie track. Distribution of vertical stress across the track at various depths in the ballast and subgrade can be seen to be more uniform for the concrete track than for the wood tie track. It is also worth noting that toward the end of tie, the concrete tie generates higher vertical stress on the ballast surface than the wood tie does.

Note that the load transmission patterns, as shown in Figure 2.19, are based on an assumption of a typical concrete and wood tie track with a nominal ballast layer thickness of 18 in. with modulus of 60,000 psi and a subgrade modulus of 8,000 psi.

2.4.3 Strength properties

Strength of track foundation materials is what prevents failure or excessive deformation of track foundation under various types of loadings. Strength properties of track foundation materials, however, can be defined in many different ways, depending on what loading condition is considered. Soil mechanics textbooks cover the details of how soil strength should be determined for a given loading condition (e.g., Lambe and Whitman 1969).

Fundamentally, the stress state of track foundation materials, whether ballast or cohesive soil, can be described in terms of the Mohr's circle, as shown in Figure 2.20. The Mohr's circle at failure stresses defines the Mohr–Coulomb failure envelop that describes soil shear strength.

The Mohr–Coulomb failure envelop is tangent to the Mohr's circle at failure and is used to describe the stresses at failure (or shear strength) based on the equation as follows:

$$\tau_f = c + \sigma \tan \phi \tag{2.34}$$

where:

τ_f is the shear strength of soil
σ is the applied normal stress
ϕ is the soil friction angle
c is the soil cohesion (psi)

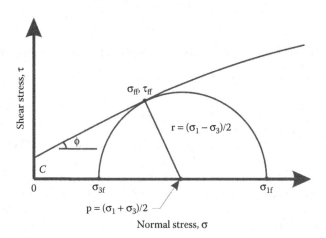

Figure 2.20 Mohr's circle and Mohr–Coulomb failure envelope.

For granular materials such as clean ballast, soil cohesion is typically zero, and shear strength comes from friction angle between ballast particles.

From Equation 2.34, it is obvious that the shear strength of any track foundation material depends directly on the normal stress applied, friction angle, and cohesion.

2.4.4 Total stress, effective stress, and pore water pressure

Water in the track foundation, when not free flowing or drained, will greatly affect the load transference and strength properties of track foundation materials. Basically, if water is present, it will share with the soil skeleton the normal stress applied to a given element but will not resist any shear stress. In other words, for a given element in the track foundation, the total normal stress applied is equal to the sum of pore water pressure and the effective stress resisted by soil skeleton in terms of the following equation:

$$\sigma' = \sigma - u \tag{2.35}$$

where:
σ is the total stress
σ' is the effective stress
u is the pore water pressure

The effective stress controls certain aspects of soil behavior, notably soil strength. In fact, shear strength of soil as described in terms of the Mohr–Coulomb equation can be rewritten in terms of the following equation:

$$\tau_f = c' + \sigma' \tan \phi' \tag{2.36}$$

As indicated by Equations 2.35 and 2.36, higher water pressure reduces effective stress, which therefore reduces the soil shear strength.

Pore water pressure is generated not only by the hydrostatic water pressure but also by the external loading applied. When there is no time for water to drain under wheel loadings, excess pore water pressure is generated. The phrase "excess pore water pressure" is used because it is above hydrostatic pressure. This excess water pressure further reduces soil strength and should be avoided by providing good drainage in the track foundation.

Because ballast is generally well drained and does not develop excess pore pressure due to wheel loadings, it is often taken for granted that the total stress is equal to the effective stress. This assumption can be increasingly violated as ballast ages, deteriorates, and becomes more fouled.

2.5 MOVING LOAD AND PRINCIPAL STRESS ROTATION

As a train approaches, passes over, and moves away from a point below the track, a load or stress pulse acts on the element of soil at that point. The stress state on this soil element will change as shown in Figure 2.17. Part (b) of this figure illustrates the effect of a moving load, whereby the principal stresses rotate to resist the applied load. In this way, shear stresses develop on the horizontal and vertical planes. This phenomenon is referred to as the "rotation of principal stress" due to a moving wheel load.

In Figure 2.17, the shear stress reverses direction as the load transitions from approaching to moving away from a soil element location. This incremental change in shear stress orientation may be problematic, especially under certain loading situations where shear stress reversal drives dynamic soil behavior and, in some instances, can lead to failure.

A static or quasi-static analysis is commonly used for most stress–strain response analyses of the track substructure. In some cases, the dynamic load of the train is applied. However, the substructure response can be different for a moving load, and some track responses or failure modes must be analyzed using a moving load–based model (O'Reilly and Brown 1991; Grabe 2002; Brown 2007), for example, the tie cavity and adjacent mud boil illustrated in Figure 2.21.

The state diagram (Figure 2.22), compiled based on trends described by Castro and Poulos (1977), Selig and Waters (1994) describes soil behavior based on drained and undrained repeated loading conditions. The line separating the loose and dense states represents the points at which the critical void ratio might be attained in a drained test or flow might occur along the steady-state line in undrained tests. The relatively well-known failure mode of liquefaction occurs in loose sand subjected to repeated loading

(a)   (b)

Figure 2.21 Tie cavity and adjacent mud boil: (a) tie cavity and (b) mud boil.

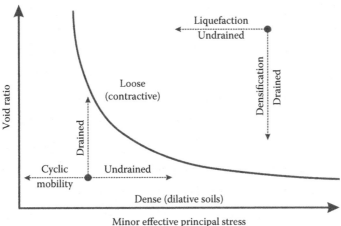

Figure 2.22 State diagram for drained and undrained loading.

that generates increased pore pressure, resulting in reduced effective confining pressure and, ultimately, flow. Although some cases of loose sands can be found in transportation structures such as pavement or track, the occurrence can be limited with good construction practices and compaction (densification).

The dynamic loading–related behavior of dense soils is represented by the modes below and to the left of the solid line in Figure 2.22. As volume must remain constant for undrained conditions, pore pressure can be generated. In this zone, negative pore pressure often develops in monotonic loading of dilative soils that would tend to follow the undrained line to the right. However, there is a tendency for all soils to decrease in volume under repeated loading due to particle rearrangement (Selig and Waters 1994). Cyclic mobility refers to the accumulation of large strains due to increased pore pressure when the soil tends to decrease in volume under repeated or dynamic load. Cyclic mobility can occur in both loose and dense soils (Kramer 1996), but this type of behavior would be mainly relevant for dense soils in railroad applications. The increased pore pressure results in reduced effective confining stress and increased soil deformation (Selig and Waters 1994).

The rate of development of increased pore pressure and the resulting deformation trends depend on the applied loading conditions for a given soil. The most rapid pore pressure development occurs when the applied cyclic loads result in shear stress reversal. In general, shear stress reversal will occur when the applied cyclic deviator stress exceeds the static deviator stress (Figure 2.18). Figure 2.23 presents a diagram that categorizes failure according to the occurrence of shear stress reversal. With no shear stress reversal, the soil deformation trends followed typical behavior illustrated for static strain governed failure in Figure 2.24. In this case, failure is

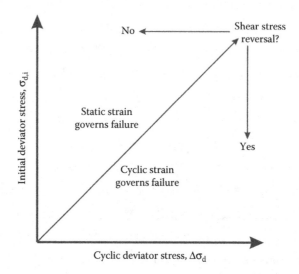

Figure 2.23 Failure mode related to shear stress reversal. (Based on Selig, E.T. and Waters, J.M., *Track Geotechnology and Substructure Management*, Thomas Telford, London, 1994.)

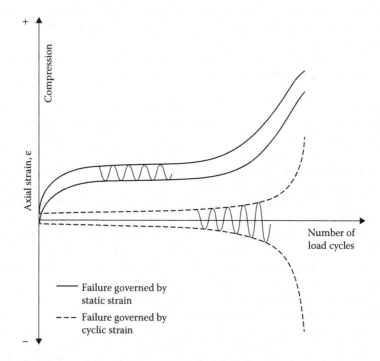

Figure 2.24 Static strain and cyclic strain governed failure trends in repeated load test. (Based on Selig, E.T. and Waters, J.M., *Track Geotechnology and Substructure Management*, Thomas Telford, London, 1994.)

defined by the accumulation of plastic stain over many cycles as the sample tends to compress to a shorter unloaded static length. When shear stress reversal occurred, the pore pressure developed more rapidly and deformation followed the trend for cyclic strain governed failure in Figure 2.24. The test results for the samples where failure was governed by cyclic strain had a more significant increase in pore pressure and more rapid accumulation of strain than samples where static strain governed failure (Selig and Chang 1981; Selig and Waters 1994).

Relating the results to track, samples not subject to shear stress reversal can continue to deform relatively slowly so that the deformation might be remedied by periodic maintenance and predicted according to traditional cyclic load behavior models. Also, the relatively progressive failure mechanism can result in redistribution of applied stress and changes in void ratio or in-situ stress conditions that can stabilize the deformation. For samples subject to shear stress reversal, the failure may occur rapidly. This possibility of potentially unsafe conditions coupled with the reported occurrence of rapid track failure in practice (Read and Li 1995) indicate that further effort to more fully describe and characterize the potential for soil failure under cyclic loading in track structures is needed.

Chapter 3

Substructure

The track substructure includes the ballast, subballast, and subgrade layers that support the track superstructure of rails and ties. Track substructure behavior has a significant influence on track superstructure stability and performance as well as vehicle dynamics. The main function of the track substructure is to support the applied train loads uniformly and without permanent deformation that might affect the track geometry. This chapter focuses on the properties and roles of track substructure components of *ballasted* railway track structure, although other track structure types are briefly discussed.

3.1 BALLAST

Railway ballast is the crushed stone that forms the top layer of the substructure, in which the tie (sleeper) is embedded and supported. Mainline ballast material is usually large, uniformly graded crushed stone. Although crushed stone is used for a variety of engineering purposes, as railroad ballast it is subjected to a uniquely severe combination of loading stresses and environmental exposure. In particular, the upper portion of ballast that is directly below the tie is the zone that must endure the highest stresses from traffic loads and from the surfacing operation, leading to more rapid ballast deterioration in this area (Chrismer 1993). There are many quarries that supply crushed rock for other types of construction, but the rock's suitability as ballast must consider the distinctive demands of the railroad environment. This should be kept in mind when evaluating new ballast suppliers, because good experience of other industries with a given source of rock may not translate into its success as railway track ballast.

This chapter's treatment of ballast covers its material considerations, but the subject is also discussed in Chapter 4 with regard to ballast layer behavior and mechanics, and ballast life is discussed in Chapter 9.

3.1.1 Ballast functions

The acceptability of material for use as ballast must be judged by its ability to perform its intended functions. The ballast functions described by Hay (1982) and Selig and Waters (1994) include the following:

1. Resist applied loads (vertical, lateral, and longitudinal) to maintain track position.
2. Provide positive and rapid drainage of water.
3. Accommodate track surfacing and alignment maintenance.
4. Provide resilience needed to dissipate large and dynamic loads.
5. Provide adequate void space for storage of ballast-fouling material without interfering with ballast particle contact.

Additionally, Selig and Waters (1994) mention several secondary requirements for ballast: resist frost action, inhibit growth of vegetation, reduce propagation of airborne noise, and provide electrical resistance between rails.

The track cross section in Figure 3.1 shows the interface of the ballast and tie and several important ballast zones. As noted by Selig and Waters (1994), the ballast layer can be subdivided into these zones including the upper ballast layer, which is the zone where track maintenance disturbs the ballast and influences ballast performance, and the lower ballast, which is usually not disturbed by tamping maintenance. Additionally, this zone of upper ballast that supports the tie directly below the rail seat area is termed the *tamping zone* and is often the most heavily loaded ballast zone. The ballast below the crib should be noted as the subcrib ballast. The importance

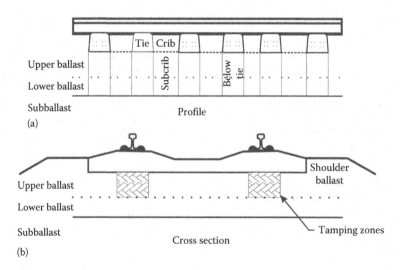

Figure 3.1 (a,b) Ballast zones.

of distinguishing between crib and subcrib zones arises from consideration of the zones affected by the applied train load and track maintenance, in particular, track surfacing. The tamper used in track surfacing lifts the track structure and has tines that are inserted in the ballast to force the rock particles under raised ties in the heavily loaded tamping zone.

3.1.2 Parent rock characterization

The type of parent rock that the ballast is crushed from will affect the derived ballast particle size, shape, and angularity of individual particles, which consequently affects the frictional resistance to sliding between two particles in contact. Some rock has cleavage planes and bedding layers that set up preferred failure planes that can control particle shape. An understanding of parent rock characteristics provides insight into ballast performance.

The characterization of crushed rock ballast begins with evaluating the strength and inherent structure of the parent rock mass including cleavage and bedding planes. The rock strength can be assessed by testing the unconfined compressive strength of an extracted rock core or by testing the ballast. Natural variations within a quarry should be evaluated to determine if they could adversely impact ballast shape, size, or performance. Once a potentially suitable parent rock mass has been identified, a detailed evaluation of the crushed rock aggregate should be undertaken. The following sections discuss some of the more common tests for ballast aggregates.

3.1.3 Aggregate characterization

The performance and longevity of ballast particles is dependent on the material properties and loading conditions. Aggregate characterization is important for judging its performance as ballast, whether one is determining suitable quarry sources for new ballast or is assessing the performance of existing ballast in track.

3.1.3.1 Grain size distribution

The grain size distribution (gradation) of ballast is the most common specification for new ballast and a common technique for the assessment of worn ballast. The grain size distribution of the aggregate mass allows for reliable correlation to strength, deformation, and drainage characteristics. Mainline ballast specifications typically require a relatively narrow range of particle sizes, which maximizes interparticle void volume. This large void volume facilitates drainage and provides for substantial storage of ballast-fouling material. When narrowly graded ballast is adequately compacted, it performs well and provides superb drainage and fouling material storage. More broadly graded ballast will provide increased strength and resistance to deformation due to the denser packing arrangement of the particles, but broadly graded ballast can be expected to have a lower void volume than narrowly graded ballast. Although a more broadly graded ballast can provide superior resistance to deformation and might provide the desired drainage and fouling material storage capacity, the final challenge then becomes material handling and transport because the particles will tend to segregate during transportation to the site. The as-placed ballast gradation is critical to ballast performance and often most difficult to assess. Keeping in mind sample size requirements from the American Society of Testing Materials (ASTM), it is also important to follow sampling guidelines to ensure that a representative sample is obtained when evaluating field placement.

The grain size distribution test (ASTM D6913) consists of placing a sample in the top of a stack of sieves arranged in the order of decreasing opening size from top to bottom (Figure 3.2). The particles in the top sieve falls through the stack as it is shaken, and particles are retained on the sieves through which they are too large to pass. The weight retained on each sieve is recorded and expressed as a percentage of the original weight. The percent passing each sieve size is typically plotted on a semilog chart of sieve size versus percent passing, which is the grain size distribution plot (Figure 3.3).

The allowable grain size range of two common ballast gradations, according to the American Railway Engineering and Maintenance-of-Way Association (AREMA) manual of recommended practices, is shown in Figure 3.3, and the percent passing certain sieve sizes for a few gradations are presented in Table 3.1.

Particle size is an important factor in ballast performance because the amount of void space between particles increases with particle size, and therefore the amount of void volume for storage of fouling fines increases.

Ballast drying   Sample preparation   Ballast sieving

Figure 3.2 Ballast gradation analysis. (Courtesy of Transportation Technology Center Inc., Pueblo, Colorado.)

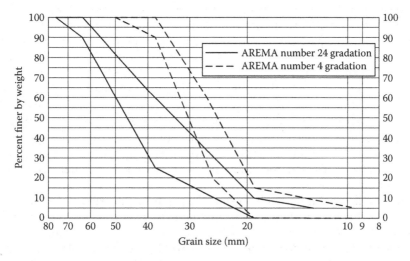

Figure 3.3 AREMA #24 and #4 ballast gradation particle size ranges.

Table 3.1 AREMA recommended ballast gradations 3, 4, and 24

Sieve designation (ASTM E11-09)	Sieve opening size (mm)	AREMA ballast gradation Percent passing		
		4	4A	24
3"	75.0	–	–	100
2 1/2"	63.0	–	100	90–100
2"	50.0	100	90–100	–
1 1/2"	37.5	90–100	60–90	25–60
1"	25.0	20–55	Oct-35	–
3/4"	19.0	0–15	0–10	0–10
1/2"	12.5	–	–	0–5
3/8"	9.5	0–5	0–3	–

However, the effect of particle size alone on strength is difficult to quantify because the associated effects of void ratio, particle shape, and texture, among other variables, also tend to affect the aggregate strength. This requirement for increased void volume must be balanced with the difficultly in working with and compacting larger particles. The compaction force required to densify ballast increases with increasing particle size.

As ballast deteriorates and becomes increasingly fouled, it is important to characterize the amount of fine-grained material to determine its expected performance and remaining life. The fine material that passes a #200 sieve is silt and clay sized, which is particularly damaging to the ballast permeability and strength in the presence of water. Dry sieving of a fouled ballast sample might result in incorrect assessment of the percentage of these fine particles if a portion of fine particles agglomerate to form a larger mass retained on the #200 sieve leading to the incorrect designation of this material as coarse grained. The presence of agglomerated fine-grained particles can be difficult to distinguish from small coarse-grained particles while conducting the test. If an incorrect assessment is made of the amount of fine particles contained in the ballast, it would cause incorrect inferences about future performance and ballast life. The wet sieve approach (ASTM C117) is very effective at separating the fine fraction by using water to disperse the finer particles so they pass through the openings of the #200 sieve.

Ballast deterioration tends to shift the grain size distribution to smaller sizes (to the right in Figure 3.3). To the extent that ballast is randomly crushed to a smaller size under traffic, it might seem that the grain size plot would shift fairly uniformly to smaller particle sizes with ballast degradation. However, it is more common for the grain size distribution of fouled ballast to develop a distinct "tail" in the distribution curve with perhaps 10%–20% or more particles finer than coarse sand indicating that ballast deterioration can be dominated by a wear mechanism that generates a predominance of small particles.

3.1.3.2 Shape, angularity, texture

Particle shape, angularity, and surface texture are important elements needed to assess potential strength characteristics of an assembly of particles. Shape, angularity, and surface texture are critical elements that affect ballast performance since they affect ballast interlocking, which layer contributes to ballast strength and deformation behavior. Well-proportioned particles that tend to be somewhat cubical are required for stable and strong ballast, whereas elongated, flaky, rounded, or smooth shapes should be avoided. Selig and Waters (1994) provide a detailed discussion of ballast particle shape, angularity, and surface texture requirements and assessment criteria.

Flaky or elongated particles have a specific dimension that is significantly greater than other dimensions, and these tend to set up preferential planes of weakness when present in large proportion and clustered together. Flat and elongated (F&E) ratio is defined as the ratio of the maximum dimension of an aggregate particle to the minimum one, and AREMA specifications currently allow a maximum of 5% by weight of total ballast having greater than 3 to 1 F&E ratio. An alignment of an abundance of flat or elongated ballast particles may lead to particle breakage, settlement, and possibly the development of slip failure surfaces and shear movement.

Surface texture contributes greatly to the interlocking of ballast particles that is required to develop a stable ballast layer. Particles with rough surface texture tend to interlock better than smooth. The roughness refers

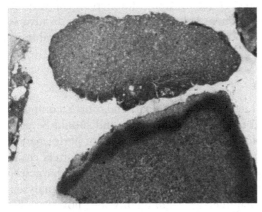

Figure 3.4 Surface texture roughness observation from
thin section.

to the surface texture of the particle with the most basic assessment of roughness being by touch where marble particles feel smooth and granite particles are rough. Surface texture is the main contributor to surface friction with rough particles having a higher friction factor and thus developing a higher friction force at the same normal force. An image showing the surface texture of several particles is illustrated in Figure 3.4.

Angularity is a measure of the sharpness of the edges and corners of an individual particle. Rounded particles tend to roll past one another under load, whereas angular particles interlock to resist the applied load. The angularity contributes to the ability of the particles to resist shear. A combination of both angularity and surface roughness are required for ballast particles to interlock properly and form a stable layer.

Assessment of both angularity and texture characterization typically rely on comparison with standardized charts where angularity is expressed on a qualitative rating scale of angular, subangular, subrounded, and rounded. Surface texture is described as rough, subrough, subsmooth, or smooth (see Figure 9.18 for ballast particle shape indices).

Improved techniques for measuring and analyzing ballast shape and roughness are available. Fractal analysis is an effective technique for quantifying both the angularity and the surface texture of particles (Hyslip and Vallejo 1997). In addition, Moaveni et al. (2013) describes an image-processing and analytical technique to better quantify particle shape. The number of fractured surfaces can also be used as an indicator of aggregate roughness (ASTM D5821) and is specified for a variety of coarse aggregate applications such as concrete and asphalt, including as a ballast specification (Selig and Waters 1994).

Particle shape, angularity, and surface texture most notably affect the mass shear strength of the ballast material, although it has been noted (Selig and Waters 1994) that increased angularity and roughness are related to increased ballast wear and breakage because the loaded contact point between particles cause local stress concentrations and increased abrasion. However, the benefit of increased shear strength outweighs the disadvantage of increased breakage/wear potential, and therefore, it is desirable to have rough, angular ballast.

3.1.3.3 Petrographic analysis

Petrographic analysis is a visual technique to evaluate the source, composition, and nature of the material making up the sample and should be performed on representative samples to provide an indication of performance and weathering/breakdown-related potential of the material. Watters et al. (1987) note that the performance of ballast in the field depends on characteristics of the parent rock, which can be determined by petrographic analysis. Petrographic analysis can be a reliable aggregate quality test; however, the results are strongly dependent on the competence of the petrographer (Selig and Waters 1994). An investigation by Wnek et al. (2013) demonstrated the challenge in identifying small changes in petrographic characteristics that substantially influenced ballast performance. Petrographic examination involves visual assessment of specimens under microscopic examination including preparing "thin sections," as depicted in Figure 3.5, among other tests. Thin sections are slices of rock polished to a thickness of 30 μm and examined under a polarizing microscope to determine mineral modes, grain size, grain morphology, and textural fabric.

Petrographic examination applies to both parent rock assessment and ballast performance assessment. When evaluating the parent rock material, petrographic analysis should be performed on the ballast material to evaluate the bulk composition, texture, and presence of secondary minerals

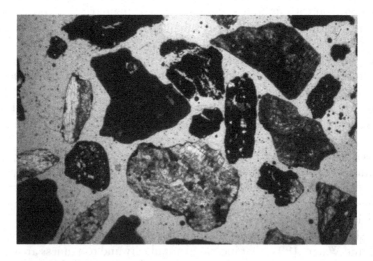

Figure 3.5 Example of granite/syenite ballast by thin section.

Table 3.2 Rock properties related to ballast performance

Rock property	Deterioration problem
Mineralogical	
• Rocks made from soft minerals like clay shale	Rapid wear and crushing
• Minerals that deteriorate rapidly or weather	Weathering or deterioration of rock
Textural	
• Poorly consolidated	Rapid abrasion
• High porosity	Susceptible to freeze-thaw deterioration
Shape and surface characteristics	
• Smooth surface	Poor stability
• Unsuitable shape	Poor stability, breakage of long flat particles

Source: Watters et al. (1987).

that may affect performance. The petrographic analysis can indicate other potential mechanisms of weakness, as shown in Table 3.2.

3.1.3.4 Crushing and abrasion resistance

Crushing and abrasion of ballast are often leading causes of ballast degradation that can cause track settlement and drive track geometry maintenance costs. The tests considered in this section can be used to identify ballast that may be prone to crushing and abrasion and include Los Angeles Abrasion (LAA), Mill Abrasion (MA), bulk specific gravity, absorption, sulfate soundness, angularity, surface texture, and petrographic analysis. Whenever possible, standard testing methods such as the ASTM, British Standard (BS), or other equivalent standard specification methods should be used to ensure both the quality of results and the ability to readily duplicate the testing methods and compare results from other labs or ballast sources.

3.1.3.4.1 Los Angeles Abrasion

The LAA test provides a measure of crushing resistance, which is used to evaluate ballast particle strength and characteristics to resist the breakage under the tie. The LAA test involves rotating 10 kg (22 lb) of ballast with 12 steel balls (5 kg total weight) for 1,000 revolutions in a steel drum (Figure 3.6). The rotation of the drum at 30–33 rpm results in impact between the steel balls and the ballast particles as they tumble in the cylinder, crushing the ballast. After 1,000 revolutions, the material is removed from the drum, and the sample is wash sieved on a 4.25 mm (#12) sieve. The LAA value is the amount of material passing the 4.25 mm sieve generated

Figure 3.6 Los Angeles Abrasion apparatus. (Courtesy of University of Illinois Urbana-Champaign, Champaign, Illinois.)

by the test as a percentage of the original sample weight. A small LAA value is desired for ballast as it indicates increased resistance to crushing. The version of the test most suited to ballast is ASTM C535, "Standard Test Method for Resistance to Degradation of Large-Size Coarse Aggregate by Abrasion and Impact in the Los Angeles Machine." The LAA test is also designated as BS EN 1097-2 and BS EN 13450 Annex C.

BS 812-110 and 812-112 are other aggregate tests of the crushing resistance. In BS 812-110, a standard cup is filled with aggregate and a static load of 390 kN is gradually applied over a period of 10 min. The crushing value is taken as the percentage of fines produced during the test to the original sample weight. In BS 812-112, a standard cup is filled with aggregate, onto which a standard weight is dropped from a given distance. The aggregate impact value is the weight of smaller particles generated from impact as a percentage of the original sample weight. Taken together, BS 812-110 and BS 812-112 reasonably simulate the mechanisms of ballast crushing due to monotonically increasing load and of sudden impact loading. The standard test method is straightforward and reasonably fast, which are desirable characteristics. However, careful sample preparation is critical for accurate and reliable results.

Although the LAA and the BS-812 test conditions represent valid ballast degradation modes, an additional test is needed to measure ballast degradation due to abrasion by attrition through particle on particle grinding, which is a common ballast degradation mechanism.

Figure 3.7 Mill Abrasion apparatus. (Courtesy of University of Massachusetts, Amherst, Massachusetts.)

3.1.3.4.2 Mill Abrasion

The MA test provides a measure of abrasion resistance of ballast and is an indicator of ballast hardness. The MA test concept is based on the micro-Deval abrasion test (ASTM D6928) for aggregate crushing and abrasion but has the advantage of testing separately for abrasion resistance without crushing, where crushing of particles is better measured in the LAA test. The MA is a wet abrasion test and consists of revolving 3 kg (6.6 lb) of material with a specified amount of water about the longitudinal axis of a 229 mm (9 in.) outside diameter porcelain jar (Figure 3.7) at 33 rpm for 10,000 revolutions. The rotation of the porcelain jar causes the ballast particles to tumble and roll over each other, resulting in wear without significant particle crushing as in the LAA test. The MA value is the amount of material passing the 0.075 mm (#200) sieve generated by the test as a percentage of the original sample weight. Low MA values are desired for ballast material. There is currently no ASTM standard for MA testing; however, Selig and Boucher (1990) discuss the test in detail and present recommended procedures.

Alternatively, ballast aggregate may be evaluated for similar abrasion characteristics with the micro-Deval abrasion test as described by ASTM D6928, "Standard Test Method for Resistance of Coarse Aggregate to Degradation by Abrasion in the Micro-Deval Apparatus" (alternatively specified in BS EN 1097-1). The micro-Deval abrasion test procedure results in some particle crushing that overlaps the LAA test specification. A better practice is to separate ballast particle wear and crushing characteristics in a ballast specification by using both the LAA value and the MA value separately (Klassen et al. 1987).

3.1.3.5 Bulk-specific gravity, absorption, sulfate soundness

Bulk-specific gravity and absorption tests provide indicators of strength and breakage potential based on the concept that higher strength is associated with higher density aggregate. Bulk-specific gravity is typically determined using Archimedes principle based on water displacement. Absorption is an indication of the rock porosity that relates to its strength and freeze-thaw resistance. Absorption is defined as the weight of water absorbed by the particles divided by the dry particle weight, expressed as a percentage. Bulk-specific gravity and absorption tests are specified in ASTM C127, "Standard Test Method for Specific Gravity and Absorption of Coarse Aggregate."

Sulfate soundness is a measure of a stone aggregate's freeze-thaw resistance, which is important to assess where ballast is exposed to freezing temperatures and precipitation. The sulfate soundness test is a form of accelerated weathering of the particles and is conducted by subjecting stone aggregate repeatedly to saturated solutions of either sodium sulfate or magnesium sulfate. These solutions carry chemicals into the void space of aggregate particles, which then form crystals that exert internal expansion pressure to simulate the deterioration mechanism of freezing water. The weight of smaller particles generated from breakdown during the test is measured, and this weight is expressed as a percentage of the initial specimen weight. Sodium sulfate tests should be performed in accordance with ASTM C88, "Standard Test Method for Soundness of Aggregates by Use of Sodium or Magnesium Sulfate." The results of the sulfate soundness test should be interpreted with an understanding of the inherent variability of these tests and the challenges associated with trying to simulate aggregate deterioration under freeze-thaw conditions. The sulfate soundness test results depend on the type of solution used during the test, and variability between labs using the sodium sulfate solution has been high leading to concern that the test may not be reliable (Selig and Waters 1994). The magnesium sulfate solution appears to provide less variability in results than the sodium sulfate solution (New York State DOT Geotechnical Test Method 21).

The sulfate soundness test using magnesium sulfate solution provides a reasonable indication of freeze-thaw variability. Other indications of freeze-thaw resistance, such as particle density and strength, must also be considered when evaluating results from sulfate soundness. When possible, field experience with the freeze-thaw resistance of ballast must be considered directly and weighed with laboratory test results.

3.1.3.6 Ballast aggregate specifications

Table 3.3 presents specifications for the ballast aggregate characterization tests. Also included in this table are specifications recommended by AREMA (2013) and the Canadian Pacific Railroad (Klassen et al. 1987).

Table 3.3 Tests and specifications for characterization of ballast material

Test	Procedure	AREMA	CPR
Los Angeles Abrasion	ASTM C535	35% max.	45% max.
Mill Abrasion	Selig and Waters (1994)	–	9% max.
Bulk specific gravity	ASTM C127	2.60 min.	2.60 min.
Absorption	ASTM C127	1.0% max.	0.5% max.
Sulfate soundness	ASTM C88	5.0% max.	1.0% max. mainline

3.1.4 Ballast fouling

Ballast fouling is the process by which the voids between particles become filled with fouling material. The fouling material is derived from ballast deterioration from repeated loading and tamping and from sources external to the ballast. Although a variety of potential sources of fouling material are possible and fouling sources are often site-specific, the most common source is breakdown of the ballast itself. A definitive study by Selig and Waters (1994) reported that 76% of sites studied had ballast breakdown as the major source of fouling. The far less prevalent second largest source of fouling material, which was the predominant source of fouling at 13% of sites, was upward migration of material from a lower granular layer. In these cases, the lower layer was often old roadbed material that was mixed with the ballast at the interface of the layers. The old roadbed materials observed to foul ballast were typically granular rather than fine grained and did not infiltrate very far into the ballast. Ballast contamination from the subgrade is rare. The fact that fouled ballast often appears as mud leads to the common misconception that subgrade has pumped up into the ballast. The appearance of mud in the ballast typically results from the abrasion of ballast under traffic in the presence of water.

Although the most common source of fouling material is abraded or crushed ballast, other sources can include

1. Contaminants shipped with the ballast or material mixed with the ballast while it is handled or installed
2. Material dropped or spilled on the track from lading, or from traction sanding
3. Windblown material
4. Soil penetrating the ballast from below
5. Tie or other deteriorated track materials

Although any of these possible sources of fines could be the dominant contributor to fouled ballast at a specific site, it appears reasonable to assume that fouling material is generated from ballast breakdown unless there is reason to believe that another cause is evident or likely. Color can be a clue to the fouling source, although color can often be misleading because the presence of only a small amount of fouling from certain single sources can control the color of fouling material that is predominately from other sources. For example, in locations where concrete tie wear is a problem, fouling material can take on a light gray color even though the abraded concrete generally makes up only a small percentage of the fouling material. On lines with significant coal traffic, coal dust from passing cars falls onto the track often giving a dark black color to the fouled material even if the coal fines are a minor fouling component. In locations where subgrade has penetrated the ballast, the fouling material may take on the color of the subgrade although the color of the subgrade may not be distinct. Ballast contaminants such as coal, subgrade, and other natural materials are often not distinct in color from other ballast contaminants, making visual clues unreliable. Any definitive assessment of the source of ballast fouling will need a test to assess the source and nature of the fouling material.

Ballast performance is much worse when fouling materials have plasticity compared to granular, nonplastic fouling material, making the characterization (plasticity, grain size, etc.) of the fouling material important for determination of track performance and maintenance requirements. ASTM D2488 Standard Practice for Description of Soils (Visual Manual Method) can provide clues about the plasticity of ballast-fouling material that can help to identify the likely source.

One measure of the amount of fouling material is the percentage of material passing the #4 sieve (opening size of 4.76 mm or smaller). Typical ballast gradation specifications limit the amount of material smaller than this size as delivered; therefore, the amount of material passing the #4 sieve is often related to the degree of fouling. A practical definition of fouling material is the presence of material that is of smaller particle size than is allowed according to the original specification for ballast. Selig and Waters (1994) define fouling material as the particles passing the #4 sieve, which is subdivided into coarse fouling, the particles between the #4 sieve and the #200 sieve, and fine fouling, the material passing the #200 sieve. Ballast performance has been observed to be significantly worse with fine-fouling material in the ballast compared to coarse-fouling material (Selig and Waters 1994).

Selig and Waters (1994) proposed the fouling index to quantify the degree of ballast fouling:

$$FI \ (\text{fouling index}) = P_4 + P_{200} \tag{3.1}$$

where:

P_4 is the percent passing the #4 sieve (4.76 mm)

P_{200} is the percent passing the #200 sieve (0.074 mm)

This accounts for the material passing the #200 sieve twice to accentuate the effect of the finer material due to its very large influence on permeability. The *FI* has been observed to correlate well with ballast performance problems.

As the ballast begins to become fouled, the initial fouling material generated is expected to be coarse particles due to breakage of asperities as the ballast compacts. Although the coarse broken particles may reduce ballast strength slightly, they are not expected to be very damaging to ballast performance as they fall into the voids between ballast particles and will not reduce ballast permeability appreciably. Coarse broken particles may even improve ballast performance slightly, if they inhibit ballast movement. This process, along with particle interlocking, is thought to be responsible for the more stiff response (similar to a more broadly graded material) of ballast after a significant number of load cycles because the coarse breakage may, under the right conditions, inhibit ballast particle movement by in effect wedging the ballast into position (Selig and Waters 1994; Indraratna and Salim 2006).

However, as load cycles accumulate, ballast wear continues and produces progressively smaller particles. As these finer particles accumulate, they will tend to inhibit ballast performance by retaining water and reducing drainage, creating the appearance of mud in the presence of water. The behavior of fouled ballast is further degraded if the fouling material behaves plastically (Han and Selig 1997). In general, fouling material in the ballast contributes to some degree of ballast drainage problems, settlement, and increased track maintenance.

The effect of fouled ballast on track performance is dramatic. Sussmann et al. (2001a) studied the effects of track condition and stiffness (Sussmann 2007) on track geometry profile deterioration rates and found that the geometry deteriorated most rapidly at sites where fouled ballast and drainage problems existed. A detailed geotechnical investigation of the track found that sites with poor drainage, fouled ballast, and subgrade failures had the most advanced deterioration. The fouled ballast sites had the highest rates of track geometry degradation when there was also a lack of drainage that retained water in the fouled ballast.

Efforts to develop a quick, reliable, and robust method to quantify the fouling condition of ballast relative to in-service performance have been pursued for decades. The fouling index by Selig and Waters (1994) provides a robust method of discriminating ballast conditions, but the heavy reliance on the percent passing the #200 sieve make this impractical to implement on a large scale due to the difficulty of reliably sieving material on such a very fine mesh in the field. Sussmann et al. (2012) developed a comparison of the fouling index to the criteria based on the amount of material passing the 3/4 in. sieve as developed by Canadian National (CN). Table 3.4 shows the comparative results where standard ballast gradations at varying fouling levels published in Selig and Waters (1994) were equated to CN criteria based on the percent passing the 3/4 in. sieve.

Table 3.4 Fouling index comparison with CN criteria

Category	Fouling index (Selig and Waters 1994)	% Passing 3/4 in. sieve (CN)
Clean	<1	–
Moderately clean	1 to <10	–
Moderately fouled	10 to <20	25–40
Fouled	20 to <40	40–50
Highly fouled	>40	>50

Source: Sussmann, T.R. et al., *J. Transport. Res. Board*, 2289, 87–94, 2012.

Notably, the CN comparison does not have corresponding data for clean to moderately clean ballast mainly because the criteria were established only for ballast nearing its end of life. Moderately clean ballast corresponds to less than 25% passing the 3/4 in. sieve. For the moderately fouled, fouled, and highly fouled conditions, a clear relationship between the breakpoints in the fouling index and the CN criteria were found. The benefit of the CN criteria is that the use of a single sieve with large openings lends itself to reasonably accurate testing in the field while reducing the need to transport samples to the laboratory for gradation analysis.

An alternative fouling index was developed by Spoornet (Vorster 2013) based on the percent passing (P) the 19, 6.7, 1.18, and 0.15 mm sieves:

$$FI_{Spoornet} = 0.4P_{19} + 0.3P_{6.7} + 0.2P_{1.18} + 0.1P_{0.15} \tag{3.2}$$

The use of four sieves provides a better indication of the deteriorated grain size distribution compared with that of new ballast and may be an improved indicator of ballast performance. However, this formulation for fouling index is moving away from the simplified form needed for field assessment. In addition, the small value of the 0.1 multiplier for the finest particles should be investigated because this appears counter to the argument that the finer material affects ballast performance most significantly.

Ballast cleaning intervention limits have been developed, and Table 3.5 compares several ballast-fouling intervention levels. Two main criteria are

Table 3.5 Ballast-fouling intervention level comparison

Agency or source	Criteria/size mm (in.)	Limit (%)	Rationale for intervention
Selig and Waters	Fouling index (FI)	40	Highly fouled
		30	Drainage reduction
Spoornet	$FI_{Spoornet}$	80	Gradation change
Canadian National (Ruel)	19 mm (3/4 in.)	40	Highly fouled
UIC	14 mm (1/2 in.)	30	N/A
ERRI	22.4 mm (7/8 in.)	30	N/A

used: either a fouling index or single sieve size percent passing criteria. The fouling index approach requires laboratory testing to develop the required grain size distribution, while the single sieve approach lends itself more to field application.

The single sieve approaches for ballast-fouling intervention are needed because these methods provide the opportunity for more direct comparison of ballast condition to the surroundings. Understanding the variability of track conditions and the relationship of these variations to the local surroundings and site constraints can provide the basis for improved specification of field work operations. However, the goal for the industry is a continuous measure of ballast fouling along the track to provide a map of fouling condition. This can be provided by ground-penetrating radar, as described in Chapter 8. Typically, a combination of field mapping of ballast-fouling variations and field spot and laboratory testing to more definitively document fouling conditions will provide the best insight to the cause and source of ballast fouling as well as the required extents of remedial action.

Obtaining ballast samples from selected locations can provide useful information when interpreting ballast-fouling grain size variations. To assess the ballast general condition, it is often desirable to have samples obtained from both the loaded zone beneath the rail seat area of the ties and from between ties. Sample selection for assessment of in-track ballast must be done with care due to variability along the track that is dependent on ballast placement, density, and influences of subgrade and track structure variations, as well as traffic and maintenance. Recommended ballast sampling locations from Klassen et al. (1987) are shown in Figure 3.8. Subballast and subgrade samples should also be obtained, and these excavations can also provide useful information on the interface boundary location and profile.

3.1.5 Ballast layer behavior

Characterizing the aggregate material using the previous tests is often accompanied by assessment of the stress, strain, and strength behavior of ballast to provide insight into performance of the ballast as a layer. Assessment of ballast layer performance is a critical aspect of track structural design and performance evaluation.

3.1.5.1 Ballast layer strength parameters

Ballast deforms a small amount under each load cycle. Typically, this deformation is mainly elastic, but there is a small component of plastic deformation. Therefore, ballast layer performance is usually best defined in terms of a limiting deformation criterion. This is different from statically loaded geotechnical structures where strength is defined in terms of material failure and not deformation.

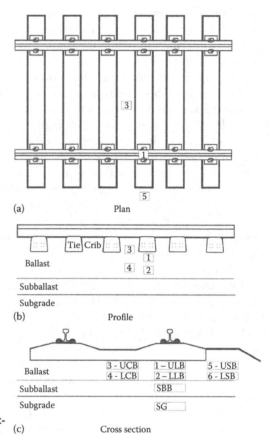

Figure 3.8 (a–c) Ballast sampling location recommendations.

Ballast deformation can be due to settlement and particle rearrangement, ballast fracture/crushing, and ballast wear/fatigue. These different deformation modes combine by varying degrees to develop the overall ballast layer deformation. Further development of these criteria could be based on laboratory tests where the particle gradation changes are accurately measured and linked to repeated load performance of ballast under a variety of test scenarios. This would allow for the better definition of failure, better life-cycle cost prediction, and could lead to improved screening tests to aid in ballast selection. Modeling techniques to replicate and extend the use of the results from these tests are being developed using discrete element modeling and the finite element method.

When compared with an ultimate strength parameter that simply provides an allowable static load not to be exceeded, the utility of the deformation-based strength criteria to develop material properties and performance models becomes clear (Sussmann and Hyslip 2010). Deformation-based criteria represent in-service failure modes, whereas an ultimate strength parameter

Table 3.6 Ballast properties

Property	Clean
Friction angle, ϕ	40°–55°
K_0	1–10
Resilient modulus, M_R	140–550 MPa (20–80 ksi)
Poisson's ratio	0.3

is not directly related or proportional to common track failure mechanism under normal operating loads. Therefore, the links between ultimate strength criteria and screening tests for ballast quality are limited.

Although ballast performance is best defined by the repeated load behavior, the ultimate strength of ballast can be related to a friction angle based on the Mohr–Coulomb failure envelop friction angle, ϕ, as described in Section 2.4.3. Clean ballast may have friction angles in the range of 40°–55° (Table 3.6).

The value K_0 is the coefficient of lateral earth pressure, which is the ratio of lateral stress divided by vertical stress in the ballast. Under repeated loading of the ballast layer, the lateral stress tends to accumulate and remain even after the vertical load is removed. The magnitude of residual stress in ballast relates to the cohesion intercept on the y-axis in a Mohr–Coulomb stress diagram. For ballast, the cohesion intercept is often taken as zero because it is a noncohesive material. However, when properly densified to the normal compacted ballast state, the behavior can include a cohesive component due to large interparticle compressive forces induced from repeated loading. This residual stress (Selig and Waters 1994) that develops in the ballast can produce a K_0 as large as 10 or greater.

The material properties of degraded, fouled ballast are expected to reflect a reduced interparticle frictional resistance and lower layer resilient modulus than the values for clean ballast shown in Table 3.6. However, there is a lack of available information to indicate these degraded property values for fouled ballast.

One major point of distinction is whether the fouled ballast is wet or dry. Wet, fouled ballast tends to have reduced strength and increased deformation under load when compared with clean ballast. Conversely dry, fouled ballast can have increased strength and stiffness; however, this may not translate into reduced, repeated load deformation or settlement when compared with clean ballast. This dry, fouled ballast behavior is caused by specifics of the sample related to particle size, gradation, initial density, plasticity, and cementitious characteristics of the fouling material.

After being disturbed by tamping, fouled wet ballast is expected to resettle faster than clean ballast. Although it is true that the settlement rate of ballast is influenced by the fouling material, it should not be assumed that the ballast particles lose contact with each other with increasing fouling, nor that the deformational characteristics of the fouled ballast become those of the fine material in the interparticle voids of the ballast. After tamping and recompaction of fouled ballast under traffic loading, the ballast particles will be forced back into contact, thus diminishing the effect of the fouling material on the mechanical behavior of the ballast structure. If the point contacts between ballast particles can be re-established by traffic compaction, the layer becomes increasingly resistant to deformation with further repeated loading. Under these conditions, the settlement trend of fouled ballast is expected to increase at a decreasing rate and become more stable with deformation rather than increase at an increasing rate as would, for example, a plastic clay that becomes progressively weaker with repeated load cycles.

Understanding the strength and variability of clean and fouled ballast is important for developing new repair strategies, improved maintenance procedures, and assessment of track load capacity and track life for maintenance planning.

3.1.5.2 Ballast stress–strain behavior

The bulk density or unit weight of ballast affects strength, with increased density leading to increased strength. Selig and Waters (1994), Lambe and Whitman (1969), and others clearly show that both the angle of internal friction and the deviator stress at peak $(\sigma_{1f}-\sigma_{3f})$ increase with increasing density (decreasing void ratio). As with any granular material, the strength increase is attributable to volume change behavior, that is, dilatancy, where denser material at the same confining pressure will have higher peak strength than a less dense material at that same confining pressure. The less dense material tends to force particles into a tighter packing when loaded, where a dense material cannot allow particles into a closer spacing without requiring the particles to dilate or pass over one another, resulting in an increase in volume to accommodate deformation. Much of this behavior is stress dependent and a result of the nonlinearity of the failure envelop.

Loose ballast tends to have a unit weight of 90–100 lb/ft³ (1.44–1.6 Mg/m³), while that for compacted ballast is approximately 110 lb/ft³ (1.76 Mg/m³) (Table 3.7). In Table 3.7, fouling index refers to the amount of ballast contamination. While this is a small change in unit weight, the associated engineering behavior changes distinctly. As reported in Selig and Waters (1994), Knutson

Table 3.7 Unit weight/volume relationships of ballast

Ballast condition	FI	Unit weight (lb/ft³)	Void ratio	Void volume (%)
Loose	0	95	0.68	41
Compact	0	110	0.53	35
Moderately fouled	20	125	0.35	26
Heavily fouled	40	135	0.25	20

Source: Sussmann, T.R. et al., J. Transport. Res. Board, 2289, 87–94, 2012.

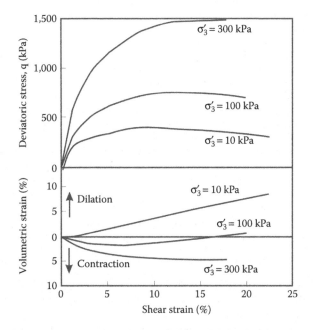

Figure 3.9 Ballast behavior in triaxial test. (Idealized behavior based on data from Indraratna, B. and Salim, W., *Mechanics of Ballasted Rail Tracks, A Geotechnical Perspective*, Taylor & Francis, London, 2005; Selig, E.T. and Waters, J.M., *Track Geotechnology and Substructure Management*, Thomas Telford, London, 1994.)

and Thompson (1978) noted large permanent deformation associated with loose ballast that had a unit weight of 90 lb/ft³ (1.44 Mg/m³) and almost negligible axial strain comparatively for ballast that had unit weight of 107 lb/ft³ (1.71 Mg/m³), confirming the trends of large differences in ballast behavior with compaction.

In order for ballast to densify, the stress state must be adequate to force the ballast into a contractive state. As shown in data reported by Indraratna and Salim (2005), the stress state has a controlling influence on the change in ballast behavior from dilative at low-confining pressure to contractive at high-confining pressure (Figure 3.9).

Compaction from repeated cycles of traffic loading is required in the field to densify ballast after maintenance to ensure track stability. The increase in stiffness and strength observed to develop after a large number of train passes is attributed to increased packing density of the aggregate mass. In Figure 3.10, the data from Suiker (2002) show the general trend of the effect of increasing packing density on ballast deformation response. Ballast densifies if the applied loads are adequately large to force the particles together. At low applied stress ratio ($n = \sigma_1/\sigma_3$), the data curve labeled as "2" in Figure 3.10 shifts from the behavior of the virgin, as placed material, but does not change significantly. However, with higher loads, the behavior after 1 million load cycles changes

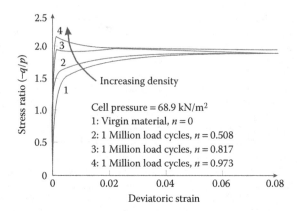

Figure 3.10 Influence of cyclic loading on ballast densification and performance (n = cyclic stress ratio = σ_1/σ_3). (Idealized behavior based on data from Suiker, 2002.)

dramatically. The data curves labeled as "3" and "4" in Figure 3.10 show an increase in stiffness and peak strength. Suiker (2002) attributed this increase in peak strength to an increase in density.

The strong relationship between ballast unit weight and strength has been amply demonstrated (Hay 1982; Selig and Waters 1994, among others), but a practical ballast density measurement device or technique has not been developed. Part of the challenge in assessing ballast density arises from the large particle size and the required sample sizes to obtain a representative measurement. Non-nuclear density gages may provide the needed data to better quantify the influence of ballast density and associated strength changes on ballast performance and to develop associated quality control criteria.

Ballast grain size distribution has an effect on ballast deformation characteristics. The effect of grain size distribution on strength can be attributed, at least in part, to void ratio changes and changes in the particle interlocking behavior, which is known to affect the propensity for dilation/contraction. Roner (1985) found that differences in "parallel gradations" (i.e., which have a fixed gradation curve distribution shape but are scaled uniformly up or down in grain size) have essentially the same friction angle when compared at the same void ratio. This finding highlights the need to compact ballast (reduce the void ratio) to increase ballast layer strength.

3.1.5.3 Ballast deformation

In addition to strength parameters, ballast characterization must include a definition of the deformation behavior. Resilient deformation and resilient modulus are important parameters required for analysis of track performance and load distribution throughout the track structure. While Young's

Modulus describes stiffness as the slope of the stress–strain diagram, ballast stiffness under repeated loading is described by the resilient modulus, M_r, where the slope of the stress–strain diagram is taken after the material has stiffened under an initial series of load cycles.

The deformational behavior observed in the static triaxial compression test can be characterized by the tangent or secant modulus, taken as either tangent to the stress–strain curve in the early portion of shear or as the slope of the line through zero. One problem with these definitions is that the strain experienced during shear is composed of both an elastic and plastic component, and there is no way of distinguishing between the two in monotonic loading. The second problem is that the modulus changes with the number of loads, with the most significant change after the first load cycle. This is attributed to densification and rearrangement of the ballast particles, sometimes referred to as shakedown. For these reasons, cyclic load tests are generally more desirable for evaluation of repeated load strain behavior and characterization of modulus at various stages of ballast life. After an appropriate number of load cycles, the differences between secant and tangent moduli often become insignificant. However, accurate assessment of ballast deformation and resulting track settlement can only be obtained based on knowledge of these changes derived from resilient modulus testing.

Other properties of ballast include Poisson's ratio, which is rarely evaluated and often assumed to be 0.3, and K_0, which is also seldom measured but is often assumed to be 1, although under certain conditions it can be considerably greater for ballast as indicated in Table 3.6. These material properties coupled with an accurate assessment of ballast layer thickness are the common parameters used to describe ballast behavior in analytical models like GEOTRACK (described in Chapter 4).

3.1.5.4 Ballast layer residual stress

To provide the needed stress reduction to the layers below it, ballast must be stiff compared to lower layers. Figure 3.11 shows the ballast layer subdivided into three zones. In the upper zone, the load from the tie is applied to the ballast in the most typical mode of ballast loading: common triaxial compression. In the middle of the layer, the vertical load has reduced because of load distribution of the track structure, while the confining pressure increases. The lower layer shows the incremental tension that develops in the ballast under each successive wheel load. Failure is predicted in layered elastic models when the incremental tension exceeds the confining pressure. As the confining pressure in ballast is often low, very little incremental tensile stress can actually develop without failure, which often takes the form of particle spreading and rearrangement. Conceptually, the incremental tension predicted by the models in the bottom of the ballast layer results from the softer underlying layer tending to deform more than the

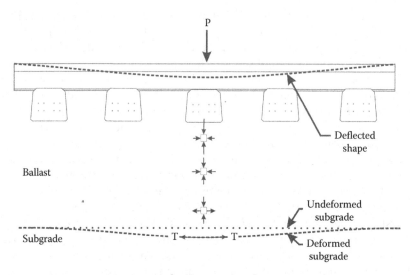

Figure 3.11 Ballast layer loading resulting in the tendency for cyclic tensile (T) strain in a material with no tensile capacity.

upper stiffer layer as described for pavements in Garber and Hoel (2015). The deformed interface that develops at the bottom of the ballast layer is longer than the undeformed interface, indicating conceptually the tendency for ballast to spread near the subgrade interface.

The tendency for tensile stress to develop in the bottom of the ballast layer occurs during the initial process of ballast compaction or shakedown. When the incremental tension exceeds the confining pressure in this bottom ballast layer, particle spreading and rearrangement will occur in this zone. As the particles spread under tension, ballast particles from above fall into the spaces created. When the tension is released the particles that have fallen are wedged into position very tightly. Under repeated load, conceptually this becomes a ratcheting type of process where the incremental tension from each wheel load spreads the ballast, and particles above fill in the created space. When the load from the wheel is removed, the particle is held tightly in position. This occurs until the combination of the compressive stress that interlocks the particles and the confining stress exceeds the incremental tensile stress from loading. At this point, the incremental stress changes under load are still tensile, but the compressive stress in the layer, due to the interlocked stress between particles, is greater (Selig 1987). In this case, the total stress is still compressive even though the incremental stress may be tensile, thereby reducing the magnitude of the compressive stress. This high-interlocked stress (Uzan 1985) is most likely the mechanism by which large K_0 conditions develop, as has been observed in ballast tests (Selig and Waters 1994).

The interlocked stresses (Barksdale and Alba 1993) that develop have sometimes been termed *residual stresses*. The development of residual stress is only possible if the ballast is supported by an elastic layer that can deform under load and then return to its original position. If either plastic deformation occurs or the supporting layer does not return to its original position, residual stress cannot develop in the ballast, although ballast spreading may continue. Ballast residual stress is the main mechanism by which the ballast interlocks and compacts and is sometimes incorrectly referred to as consolidation in lateral stability-related research documents.

The residual stress in the layer that develops under this process of rearrangement is unique and represents a significant stress level. In bound layers such as cement or asphalt where particles cannot rearrange to accommodate this incremental tensile stress, the strength of the material must be greater than the incremental tensile stress. In a similar manner, use of reinforced layers like geogrid or geoweb must also be designed with an understanding of this behavior. Reinforced layers can perform adequately, but compaction will be critical so that the particles will not be able to rearrange freely to build up residual stress because this process might damage the reinforcing material.

3.1.5.5 Ballast strength and deformation tests

Assessment of ballast performance is an important aspect of track structural design and performance evaluation. However, testing of ballast performance is often challenging due to difficulties in obtaining representative samples and preparing and performing tests that accurately represent field conditions.

3.1.5.5.1 Direct shear test

The shear strength of ballast, obtained from a strength test such as the direct shear test (Figure 3.12) or the triaxial compression test, is typically

Figure 3.12 Large direct shear apparatus. (Courtesy of University of Illinois Urbana-Champaign, Champaign, Illinois.)

characterized by the stresses at failure and reported as the friction angle, ϕ, as described in Section 2.4.3. A shear strength assessment provides a good indication of the ballast friction angle and, as such, may be a suitable screening test, but ballast rarely fails with a well-defined shear failure plane, and so this test does not represent a failure mechanism that is common for ballast.

3.1.5.5.2 Triaxial test

The triaxial test is often more suitable for ballast characterization. The evaluation of ballast strength with a triaxial test can provide both the friction angle and Young's modulus (stiffness). Repeated load (cyclic) triaxial tests provide the resilient modulus and the friction angle. The resilient modulus is an important parameter in ballast assessment because the ballast often stiffens with increasing load cycles (Figure 2.18). The stiffness attained by the ballast after it stabilizes following a large number of repeated load cycles is a more meaningful parameter than is the initial stiffness.

Because ballast is generally well-drained and typically does not develop excess pore pressure during compressive testing, it is often assumed that the total stress is equal to the effective stress. Although it is possible that excess pore water pressure may develop as ballast becomes increasingly fouled, the amount of fouling corresponding to this condition is likely to be impractical. Tests of fouled, wet ballast should identify whether effective stress or total stress analyses was being used.

Triaxial testing of ballast requires a large test apparatus as shown in Figure 3.13. Typically, the diameter of the aggregate sample being tested should be at least 2.5–3 times greater than the largest individual particle size according to ASTM standards. The triaxial test can be used in monotonic loading to determine the ultimate shear strength of ballast. However, as the load on ballast does not monotonically increase until failure, a better use of the triaxial test is for repetitive load testing, where the samples are tested for particle breakage and permanent deformation after a large number of load cycles. This test method most closely resembles the load environment and failure mechanisms of ballast in track. Use of the repeated load triaxial test is recommended for ballast evaluations that include life-cycle cost–related ballast performance parameters of strength, stiffness, and deformation.

3.1.5.5.3 Ballast box test

Over the years, various boxes have been created for lab tests to simulate the tie–ballast interface. Investigations have been conducted into the influence of the type of box, material, and associated measurements to ensure that the influence of significant parameters can be measured within the

Figure 3.13 Large (0.3 m, 12 in. diameter) triaxial test apparatus. (Courtesy of University of Illinois Urbana-Champaign, Champaign, Illinois.)

limitations of the device (Bennett et al. 2011). The ballast box design must consider the stiffness and frictional resistance of the sides to ensure reasonable fidelity with field performance observations. A bigger box is generally desirable to limit the influence of the edges of the box, but there is also the concern that bigger boxes require more sample preparation, setup, and applied load. The ballast box provides a way to overcome the limitations of more traditional lab tests such as the triaxial cell, which does not consider the tie–ballast interaction. This interaction is best assessed with a physical simulation like a box test (Figure 3.14) coupled with analysis and modeling.

The influence of fouling on the mechanical behavior of the ballast can be estimated from a triaxial test. However, this test cannot assess the influence the settlement has on the track. One common track maintenance question concerns the influence of hanging ties on track settlement and how this affects required maintenance. Ballast settlement routinely causes track settlement and track geometry anomalies, but sometimes it results in the ballast settling away from the bottom of certain ties leaving these ties hanging. Selig and Waters (1994) present data to indicate that ballast settlement increases slightly with increasing fouling index. However, when a gap is introduced between the tie and fouled ballast to model a hanging tie, the ballast settlement increases by a factor of approximately 4 or more at the same fouling index (Figure 3.15).

The ballast box tests provide evidence that hanging ties damage track and cause more settlement than well-supported ties, but the real benefit of

Figure 3.14 Ballast box test loaded and prepared for a test.

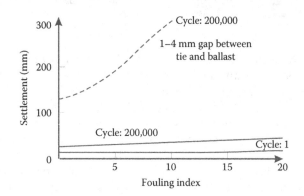

Figure 3.15 Ballast box evaluation of ballast fouling. (Adapted from Selig, E.T. and Waters, J.M., *Track Geotechnology and Substructure Management*, Thomas Telford, London, 1994.)

this test is that it provides the ability to quantify the damage. This quantification can support planning and justification for maintenance to limit damage from hanging ties, and it points to the value of instrumenting the tie section for load to evaluate if the tie is structurally suitable for the applied loading. Although this type of test opens up many possibilities, care must be exercised because these are not standard tests and success is dependent on understanding the problem and experience with this type of testing.

Further testing by Han (1997) developed detailed relationships between various ballast parameters and expected performance. This analytical tool was based on significant testing of ballast to develop data on ballast settlement for many conditions, an example of which is shown in Figure 3.16.

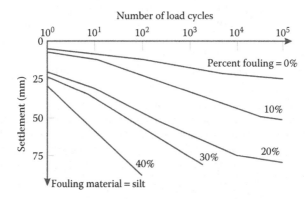

Figure 3.16 Influence of fouling level on ballast settlement rate. (Idealized from data presented by Han, X. and Selig, E., Effects of fouling on ballast settlement. *Proceedings of the 6th International Heavy Haul Railway Conference*, Cape Town, South Africa, pp. 257–268, 1997.)

3.1.6 Ballast compaction

Compaction of ballast, sometimes colloquially referred to as consolidation, typically results from traffic loading but can also be provided in a more controlled way by application of the dynamic track stabilizer to help ensure lateral track stability after maintenance that disturbs the ballast. As compaction changes the unit weight or density of the material, it is common practice to use a measure of unit weight to judge the level of compaction of most geomaterials. However, measurement of ballast unit weight is challenging due to the large-sized aggregate (Yoo et al. 1978). Nuclear density probes are sometimes used, although the disturbance produced by the probes can influence the measurement. In-place tests of unit weight using the water replacement technique (Panuccio et al. 1978) is probably the most reliable method, although this is very time consuming and results in a small number of tests. Ballast settlement potential can be very sensitive to even small changes in its density due to compaction, degrading ballast condition, or track maintenance. The small range of expected results coupled with the dramatic changes in performance demonstrates the need for accurate data to predict ballast settlement.

Density testing was developed as a measure of the soil physical state indicative of its mechanical properties. The measured density of soil is often a simplified way to estimate its mechanical strength or stiffness because density is usually easier to measure than tests that measure strength or stiffness directly. As ballast density tests can be more difficult in ballast than field mechanical strength or stiffness tests, several mechanical tests have shown promise for field ballast assessment including plate load tests used on the Channel Tunnel Rail Link (O'Riordan and Phear 2001). On this project, the plate load test data were correlated with unit weight to provide the more common density terminology for specification and discussion, while using the stiffness criteria for field assessment. Further discussion of the plate load test is provided in Chapters 8 and 10.

The process of ballast compaction involves an interlocking of the particles into an increasingly tighter arrangement with increasing number of repeated load cycles. When this process works properly, the compacted ballast layer becomes strong enough to support the applied loads without significant plastic deformation while providing resiliency so that the ballast absorbs some of the applied energy. The compaction process is necessary to provide interlocking of ballast particles and increased ballast layer stiffness to substantially reduce the stress transmitted to the lower layers.

The interlocking process relies in part on a ratcheting mechanism that involves deformation of the ballast layer under load with the tendency of the applied load to spread the bottom of the ballast layer. This tendency for temporary spreading reduces the ballast interparticle contact force allowing the particles above to slip into a tighter packing with the particles below (Selig et al. 1986; Selig and Waters 1994). As the load is removed, the tendency for spreading is removed and the interparticle contact force returns at an even larger magnitude, provided the ballast below is confined and not allowed to deform plastically. Without adequate confinement, the bottom ballast will tend to spread and this interlocking process cannot occur. With too much confinement, the interparticle force will be too high and cannot be reduced enough to allow the particles to slide past one another into a tighter packing. This is one of the basic mechanisms likely resulting in the optimization of track stiffness for track performance longevity (Hunt 1999). After a period of time, the ballast compaction process reaches equilibrium as the tendency for spreading of the ballast will not reduce the interparticle contact force enough to allow further interlocking. As long as the loading stays within the range of applied loads during the compaction process, the ballast will be stable. However, any significantly higher loads would lead to further compaction.

It is important that the compacted condition of the ballast be restored following maintenance disturbance to minimize settlement to provide track lateral strength and buckling resistance. Compaction under traffic loading can provide this, but a more controlled and quicker compaction can be provided using a track stabilizer after tamping. The track stabilizer is designed specifically to provide quick on-track ballast compaction (Figures 1.33 and 3.17). The amount of compaction provided by the stabilizer is not generally known and is difficult to estimate. The stabilizer has a head (Figure 3.17) that grips the track laterally, applies a downward force, and provides a 25–42 Hz horizontal vibration to rearrange and compact the ballast. The amplitude and frequency of the vibration coupled with the magnitude of the downward force determine the amount of compaction that can be achieved, and this varies somewhat along the track due to variable support

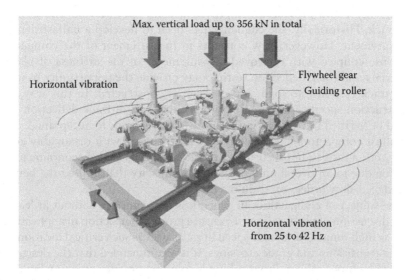

Figure 3.17 Dynamic track stabilizer head. (Courtesy of Rainer Wenty, Plasser & Thuerer, Vienna, Austria.)

and track condition. Although a track stabilizer can only provide a small percentage of the needed ballast compaction, the compaction it does provide is generally more uniform and controlled compared to ballast compaction from traffic (Sussmann et al. 2003b).

As stated, the ballast compaction achieved using the dynamic track stabilizer is expected to vary depending on the track and field conditions. Lateral track strength test results on concrete tie track have shown increased lateral resistance following track stabilization (Sussmann et al. 2003b). However, the stabilizer might not be optimized for the particular track and substructure characteristics under test, and the performance of the stabilizer should be verified with field test data to ensure good results.

Obviously, providing a controlled density of ballast in track is beneficial. Techniques are being developed to provide monitoring of the compaction process (intelligentcompaction.com) for highways and pavements, with a minimum of field testing needed for verification. Intelligent compaction provides a direct measurement of the soil stiffness based on the response of the vibratory compactor. This effort recognizes the benefits of uniformity of support and of properties. Typical compactors for aggregate and subgrade compaction use a single drum roller (Chang et al. 2011), which can be modified to monitor compactor response to assess the compaction of aggregate. For newly constructed track, the criteria for proper placement of ballast has been assumed to be ballast density where a minimum density could be specified. By correlating the stiffness of support and its effect on the vibration frequency of vibratory rollers, intelligent compaction technology could provide an improved density specification and control for track. Historically, the challenge has been to develop a ballast density measurement. However, new concepts in measurement of the compactor response coupled with improved measurements of the stiffness of placed materials and track deflection testing may provide the opportunity to measure track response to load as each layer is constructed. In this way, the construction specifications can be written in terms of track deflection and could be verified by a deflection test of each layer prior to acceptance.

Ballast compactors have been tried over the years with reasonably good compaction results, but their use in those early trials lacked economic justification due to an overabundance of track capacity that limited the benefit of reduced slow orders on train operations.

The subject of compaction, in general, is not complete without at least a brief discussion of proof rolling. To ensure proper compaction in applications such as high tonnage and high-speed lines, as well as such critical locations as bridge transitions and grade crossings, it is recommended that the design be verified by proof rolling. Proof rolling is common practice on many building project sites and a similar application on track to compact the old exposed roadbed on which the new track will be constructed to reveal weak spots that require over excavation and replacement, reinforcement or drainage. If these weak spots are adequately addressed, this will result in a more uniform platform for track construction. It is recommended that the proof rolling concept be employed to ensure adequate track support from lower layers in construction and maintenance operations that remove the track structure and expose the lower layers.

3.1.7 Used ballast

Used ballast typically refers to the reclaimed material returned to track during the undercutting cleaning operation. As the finer material is removed, only the larger particles are returned to track during undercutting/ cleaning. Used ballast is typically smaller in size than the specification for new ballast, and the wear and abrasion under traffic that reduced the particle size usually reduces the particle angularity, especially when water is present. Although newly broken ballast from track would be angular, continued wear under traffic is expected to develop worn ballast that is not as angular as new ballast, and this reduced particle angularity may affect its deformational behavior somewhat under loading.

Generally, the amount of old ballast with particle sizes large enough to be returned to track constitutes less than half of the ballast. Therefore, the majority of the ballast will be new. If the used ballast could be mixed with new ballast, the distribution of a small proportion of less angular used ballast particles into a new ballast layer would not be expected to be particularly detrimental to performance. Moaveni et al. (2013) describe application of a method to assess ballast in the field so that properly graded and shaped particles are placed to ensure ballast performance. However, ballast-handling procedures during undercutting do not allow for this type of used and new ballast mixing during placement. Rather, the used ballast is placed in track as a distinct layer below the new ballast and above the subballast with new ballast placed on top.

As a distinct layer, Indraratna and Salim (2005) found that the peak principal stress ratio was reduced about 60% from the level measured for new ballast at low-confining pressure of 10 kPa. The difference between new and used ballast was found to be negligible at a confining pressure of 200 kPa (Indraratna and Salim 2005). Therefore, with adequate confinement a used ballast layer can perform well in track. Means to confine the used ballast might be provided mechanically by soil reinforcement, but a substantial amount of confinement will be provided simply by placing the used ballast as deep in the track section as possible and confining the used ballast to the center of the track to avoid the low-confinement conditions in the shoulder region.

3.2 SUBBALLAST

Subballast is the granular layer that is located below the ballast and above the subgrade, which has been either placed as a specific layer or evolved in-place from the particle wear, densification, and settlement of old ballast layers due to decades of loading and track maintenance. The latter condition is very typical of railway lines that have been active for a long time and where the old roadbed acts essentially as a subballast layer.

As a structural layer, subballast reduces stress to the subgrade, similar to ballast, by an amount dependent on its resilient modulus (stiffness) and thickness. Subballast stiffness generally controls the load-spreading ability of the subballast layer and depends largely on its compacted density, which in turn is controlled by its gradation. The typically well-graded subballast allows a high relative density and stiffness, but the narrowly graded ballast is stiffer due to the dominating effect of interlock of the larger particles.

Subballast must not plastically deform over many load cycles. This requires subballast to (1) be well drained and avoid positive pore water pressure under repeated load and (2) have durable, angular particles that interlock and resist abrasion.

Under well-drained conditions, the strength and settlement characteristics of the subballast will be governed primarily by the density of the material, provided that durable and angular gravel is used. Although broadly graded subballast has the advantage of achieving a high-compacted density, strength, and layer stiffness, an overabundance of fine particles will inhibit drainage and potentially lead to excessive settlement under loading if saturated (see Chapter 6). For the subballast layer to be stable under loading, its gradation should be designed to be well drained as the first priority, and secondarily optimized for density and ease of compaction after considering strength and stiffness requirements.

The strength and settlement characteristics of subballast are affected by the amount of moisture in the aggregate. In an unsaturated condition, the presence of a limited amount of water acts to "lubricate" the particles and increase both elastic and permanent deformation, even without developing pore pressure (Thom and Brown 1987). High fines content is undesirable because fines tend to both attract and retain moisture, which causes problems due to low permeability and the possibility of retaining water within the subballast layer and in contact with the subgrade.

3.2.1 Subballast functions

The subballast is a select crushed stone or gravel and sand mixture that is used to cover the natural subgrade soil or fill material (on an embankment) to provide a solid and well-draining layer under the track. As an intermediate gradation generally between the fine subgrade and coarse ballast, subballast must be designed to separate these layers to prevent intermixing.

The primary functions of subballast are to (1) provide drainage out of the track, (2) help reduce applied stress to the subgrade, (3) provide separation between the ballast and subgrade, and (4) help to provide frost protection to the subgrade in colder climates.

3.2.2 Subballast characterization

The most important parameter for characterization and selection of subballast is the grain size distribution of the material because of the influence of gradation on the separation and drainage functions. The most significant characteristic of the subballast gradation is the fines content, which is the amount of material in the silt to clay size of <0.075 mm. When subballast has a fines content of approximately 10% or more and is in the presence of water, it becomes susceptible to excessive deformation under cyclic loading, and it tends to hold/retain water through capillary tension. Subballast with low fines content will typically have good drainage characteristics, but might not provide the needed separation between subgrade and ballast.

Small increases in fines content can dramatically change the drainage characteristics of subballast. The permeability and water retention characteristics of granular aggregate are strongly influenced by the size of the individual void spaces. Smaller average void space results in an increased tendency to retain water in the material due to lower permeability. A US Federal Highway Administration study (Tandon et al. 1996) of the water retention capacity of subbase found that higher fines content leads to higher subgrade saturation levels. It was noted that small amounts of fine material (passing the #200 sieve, 0.075 mm) in the road base material drastically affected the amount of water retained. In two comparable densely graded base samples, the water retention doubled from 41% to 82% saturation with an increase in fines from 1.6% to 5%.

The AREMA manual specification for subballast limits the percent passing the #200 sieve to 5%. However, subballast with 5% fines may still remain nearly saturated (Tandon et al. 1996). To provide free-draining subballast, the data from Tandon et al. suggest that 2%–3% or less of fines will be required.

The drainage requirements must be weighed against the filtration/ separation criteria presented in Section 6.4.2. The conflicting drainage and separation requirements must be balanced. This becomes especially critical on fine-grained subgrade where drainage is equally important to separation criteria. The fine sand portion of the subballast is needed to provide the required separation of the subgrade from penetrating the ballast, but the fines portion should be as low as possible to provide proper drainage characteristics.

A subballast gradation that conforms to both the drainage and separation functions will generally have good strength characteristics, provided that the material has been compacted properly and has adequate angularity and interparticle friction. The typical range of subballast properties for friction and other parameters are shown in Table 3.8.

Due to exposure to a harsh environment, the subballast must be comprised of material with high particle durability and low environmental reactivity. The durability and reactivity are functions of material type and are measured as described in the section on ballast. The AREMA manual recommends performing LAA (ASTM C131) and sodium sulfate soundness (ASTM C88) tests on subballast but does not specify values for acceptance. Reasonable limiting values for subballast acceptance of LAA and sodium sulfate soundness tests are <50% and <5%, respectively. Subballast material must not react with conditions in the environment that produce cementing, expansion, or deterioration. ASTM D2940 and D124 help to ensure that subbase material for highways will not break up when subjected to freeze-thaw or wetting-drying cycles and that the subbase has a LAA value no greater than 50. These are also desirable characteristics for railroad subballast.

Table 3.8 Subballast properties

Property	Value
Friction angle, ϕ	25°–40°
K_0	0.4–1
Resilient modulus, M_R	55–105 MPa (8–15) ksi
Poisson's ratio	0.3–0.4
Permeability (minimum)	100 m/day

3.2.3 Subballast performance

The effect of moisture on performance of granular material will be described in subsequent sections as a softening of the material, one that provides a mechanism for increased deformation. Both elastic and plastic deformation are influenced, but moisture increases the plastic deformation more dramatically.

In service, when subballast becomes saturated and subjected to cyclic loading, increased pore pressure can develop that can lead to severe loss of strength and high amounts of deformation, through mechanisms of liquefaction or cyclic mobility. This type of saturated failure of subballast was documented at the Association of American Railroads (AAR) test facility in Pueblo, Colorado (see Read and Li 1995 and Chapter 6). Subballast should have low fines content and reasonable thickness not exceeding approximately 12 in. (300 mm) to allow for quick drainage and to prevent pore pressure development and subsequent settlement. However, as stated, the separation function requires the subballast to have a certain amount of fines to protect ballast from penetration of finer soil below.

The gradation and unit weight (in-place density) of subballast have the biggest influence on its performance. An investigation by Thom and Brown (1987), which included a range of gradations from uniform gravel down to a well-graded soil containing nearly equal parts of silt, sand, and gravel showed that the compacted unit weight had more of an effect on strength than did gradation. Gradation was found to have more of an effect than density on permeability. Therefore, selection of subballast gradation is critical to track drainage considerations, and so gradation should be considered as comparable in importance to the density–strength relationships required for structural performance of the subballast layer to ensure proper long-term track performance.

3.2.4 Subballast drainage

A clean and free-draining ballast layer underlain by well-draining subballast is necessary for good substructure stability and performance. If the ballast layer is properly maintained to be free-draining, then water that enters the track from above by precipitation is drained out of the track by the ballast and subballast layers (Tandon et al. 1996). Subballast functions to provide drainage of water from downward infiltration from the ballast, lateral infiltration from side of track, and upward migration from the subgrade. In most cases, the predominant source of water to the track substructure is from downward infiltrating water from the ballast.

Granular subballast layers do not function well as water-shedding layers until they saturate. Heyns (2000) found that even for relatively impermeable subballast (permeability of 0.2 to 0.6 m/day) the subballast layer does not shed water across the subballast/ballast interface, unless the rainfall rate is sufficiently high. This water-shedding off the top of the subballast layer only occurs when the rainfall intensity is in the order of 25–50 mm/h for typical subballast gradations. For low-permeability subballast (e.g., subballast with 20% fines), only in extreme rainfall events (>70 mm/h) does water shed off the top of the ballast/subballast interface. The testing also concluded that the slope of subballast/subgrade interface does not play a significant role in avoiding subballast saturation. Rather, the interface slope is a significant factor for estimating the time for drainage of the subballast once the rainfall stops.

As found by both Tandon et al. (1996) and Heyns (2001), it is more common for subballast with fines to saturate during rainfall with the fines holding the moisture longer after the rainfall event, which would lead to more subgrade saturation. To avoid subgrade saturation, an impervious seal of the subballast material surface would provide the best method to shed water over the surface. Alternatively, well-draining subballast with very little fines could be used to minimize the time required to drain the subballast layer.

As water tends to infiltrate the subballast layer rather than being shed laterally at its surface, it is necessary to ensure that the subballast material is as permeable as possible considering the constraints imposed on the grain size distribution by the separation criteria. At a minimum, this would require establishing and maintaining a free-draining end of the subballast layer immediately at the toe of the ballast shoulder or establishing a zone of free-draining material outside the edge of tie. Further details on track drainage design considerations are included in Chapter 6.

3.2.5 Separation

Subballast provides an essential function of separating the ballast and subgrade. The grain size distribution for subballast must prevent intermixing of ballast and subballast, and the migration of subgrade into the ballast. As depicted in Figure 3.18, the soil particle at location "A" is carried with water into the subballast above. If the subballast grain size distribution is properly sized for the subgrade, the soil particles will become trapped in the lower portion of the subballast and will not travel into the material. The filtration and separation criteria are described in Chapter 6.

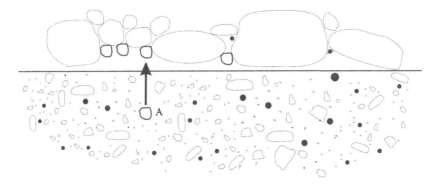

Figure 3.18 Soil migration and separation/filtration.

Although subgrade fouling of ballast is not the most common source of fouling because an adequate separating subballast layer is usually present, one of the cases where this does occur is through subgrade attrition where ballast is placed directly over a hard but erodible subgrade material such as a hard clay or soft rock. Attrition is the wearing away of a hard subgrade surface by ballast grinding on it in the presence of water that produces muddy fines that are subsequently squeezed up into the ballast by cyclic train loading. Subballast, if present, acts to cushion the hard ballast particles from the erodible subgrade surface and thus preclude subgrade attrition.

For conditions of water seepage through soil, the soil gradation requirements for separation, clogging, and filtration can be determined based on the separation criteria outlined by Terzaghi, and more fully developed by the US Army Corps of Engineers and the US Bureau of Reclamation. Cedergren (1989) provides thorough coverage of this topic. The specific guidelines and their application, including application to geosynthetics, are further covered in Chapter 6.

3.3 SUBGRADE

Subgrade, the foundation on which everything above depends for support, is often the most variable and potentially the weakest of track components. The inaccessibility of subgrade makes it challenging to assess its condition, diagnose a problem, and prescribe or implement a remedy with confidence. Changes in soil type, properties, and physical state can all occur over short distances of tens of meters making identification of problematic conditions a particular challenge due to the long distances involved. Understanding subgrade behavior provides the basis for the investigative tools and remedial methods that have been developed.

The term *subgrade* is sometimes used to denote the elevation grade line of the track ready to receive subballast during construction. This tends to oversimplify the nature of subgrade and minimize the distinction of the various subgrade elements that often are defined as the formation and fill. In a cut, the subgrade is usually exposed natural soil or rock, which is often considered the formation layer. The formation layer is most commonly defined as the natural soil layer supporting the track. In a fill, the subgrade surface is usually the top of the embankment constructed of locally available fill on top of the formation layer. At grade, subgrade is usually natural soil and a thin layer of placed fill to smooth out local variations in elevation. If the subgrade is granular, then track stability is usually assured and drainage becomes the main concern, although the subgrade will not be overly sensitive to moisture variations. If the subgrade is fine grained, then questions of stability can arise requiring some classification of the subgrade and drainage becomes even more of a concern, especially for moisture-sensitive clays.

Unlike buildings or other structures with a limited footprint, track right-of-way, like all transportation infrastructure, is a low-cost per foot structure that has not historically justified a detailed site investigation. The maintainable nature of track has been used to argue against the need for a site investigation, and for the approach of relying on train traffic to eventually find the weak subgrade zones for maintenance forces to repair, using a "build now, fix later" approach.

However, this "fix as needed" approach to subgrade has its limitations. Some desired subgrade features may be very expensive to restore once they have deteriorated because of the costs of reduced track availability during maintenance and cost of the work. For example, the drainage cross-slope is needed to avoid trapped water on the subgrade and to rapidly drain the subballast, but re-establishing the cross-slope once subgrade has settled and deformed requires removal of the track structure and taking the track out of service for a period of time. In most mainline situations, track time for maintenance work windows and train delay time due to slow orders are expensive opportunity cost items. The subgrade repair and maintenance costs should be considered in light of present-day opportunity costs of maintenance, repair, and rehabilitation work items and train delay scenarios to develop an economically justifiable site investigation budget. Remote sensing technology currently provides a means to cost-effectively identify zones of subgrade variations that can be targeted with limited site investigation and reconnaissance to characterize the subgrade, develop appropriate track designs, and reduce the likelihood that subgrade deformation will affect track performance.

3.3.1 Subgrade functions

To function as a stable foundation layer, the subgrade must be structurally sound and not sensitive to environmental damage. Structural considerations for the subgrade include ensuring the subgrade is

1. Stable under self-weight such that it does not deteriorate through consolidation settlement or massive track instability, sometimes referred to as massive shear
2. Stable under train loading and does not strain plastically to form a progressive shear failure known as a subgrade squeeze and does not deform and produce ballast pockets

The track must be designed to also protect the subgrade against environmental problems to minimize the effects of frost heave and shrink–swell-related volume change. Finally, the track must be designed to protect hard but erodible subgrade from attrition loss due to sharp-edged ballast particles grinding on it in the presence of water Selig and Waters (1994).

3.3.2 Soil types

The subgrade generally consists of any naturally occurring soil including the coarse soil types of sand and gravel and the fine-grained soil types of silt and clay. Construction over previously developed areas may lead to construction debris and other by-products of development in the subgrade. Natural soils include both organic and inorganic soil masses, with the general recommendation that organic soils be removed and replaced. Although track has been built over existing organic subgrade deposits like peat, this is not a common or desirable subgrade material.

Subgrade soil characterization should include determination of soil type, grain size distribution, physical state, and mechanistic properties of strength and stiffness. The primary governing issue for subgrade is the soil type. This is followed by determination of the physical state of the soil including the density and moisture content. Finally, specific questions about the physical properties and engineering characteristics should be investigated.

Coarse-grained subgrade (sands and gravels) with little (<5%) clay or silt tends to be least problematic. With adequate drainage and reasonable surface compaction, most sand and gravel subgrade tends to perform well. However, coarse-grained subgrade is not trouble free and is subject to unique failure mechanisms including liquefaction and cyclic mobility. The grain size distribution, unit weight, water table elevation, and moisture content tend to be the main indicators of potential problems with liquefaction or cyclic mobility, although specific evaluation of engineering properties may be needed for design or remediation.

Clays and silts tend to be the focus of most subgrade improvement work because these types of soils are subject to the common subgrade failure modes of ballast pockets and subgrade squeeze. But clay- and silt-sized particles do not need to be the predominant constituent in a soil for these failures to occur. Even mostly coarse-grained subgrade with as little as 10% of clay or silt can take on the behavior of a fine-grained soil and be subject to these failure modes. Distinct subgrade behavior variations occur between silt and clay, making soil classification the first question when dealing with fine-grained subgrade. Once the soil type is identified, the physical state needs to be characterized because moisture content and degree of saturation, density, plasticity, and soil fabric are all important elements that might identify specific problems or highlight unique solutions. The depth of the water table and engineering characterization then become the next questions needed to identify problems or develop solutions or designs.

Organic soils such as peats are often the most problematic subgrade soils. Organic soils are identified by traditional soil characterization tests such as moisture content (possibly exceeding 100%) and grain size distribution. Specific testing for organic content and the state of decay of the organic material are also common. Peat is routinely recommended to be removed from the track subgrade, although this is not always practical for deep deposits or existing track built on organic soils. Methods to mitigate the impact of organic soils and peat include densification to strengthen the deposit and limit settlement, reinforcement of layers to reduce the applied stress to an acceptable level, or confinement to reduce the mobility of the peat.

An indicator of subgrade performance is the amount of pressure it can withstand without deforming substantially. Although the AREMA design manual advocates a uniform minimum pressure for all North American subgrade, the properties of subgrade can vary dramatically and can provide widely varying levels of track support. Instead of considering the track as a structure that should provide equal support to the train for varying levels of subgrade support, the implicit assumption of the AREMA design procedure is that if the track is built to standards that protect against most problematic subgrade conditions, then the remaining problem zones can be repaired later as their locations become apparent by deforming under traffic loading ("build now, fix later"). However, once the track is in place and pressure to continue operations limits the track outage time, the options for repair of subgrade problems are often limited and costly.

The bearing pressure of subgrade soil is one parameter to consider when evaluating the variability of subgrade support conditions and when identifying subgrade that should be corrected prior to track construction. Existing problematic conditions are seldom known prior to construction without a detailed inspection of the track structure or local geologic conditions. Although conducting these tests during the planning stages for new track adds to the project cost, they provide more detailed understanding of conditions that can lead to a design with lower life cycle track cost by minimizing the expenditures over time.

3.3.3 Common problems

Unstable subgrade tends to mainly consist of fine-grained soil that often corresponds to lower strength and low permeability relative to coarse-grained subgrade. In general, more finely grained soil and/or greater plasticity soil tend to be found in poor performance zones. Figure 3.19 illustrates the

Figure 3.19 Track settlement zone due to weak subgrade.

signs of a poorly performing location with a fine-grained embankment that has settled under traffic.

Moisture content of the soil has a profound influence on subgrade performance. Most nonorganic soil types, even fine-grained soil, could function as stable subgrade if maintained at low moisture content. However, it can be very difficult in practice to limit the access that water has to moisture-sensitive soils. Sources of subgrade moisture include surface water infiltrating through the ballast and subballast, water in drainage ditches to the side of track that may infiltrate, and groundwater that may migrate up into the track. Yoder and Witczak (1975) indicate that groundwater can be a major subgrade concern if the water table is within approximately 6.1 m (20 ft) of the ground surface.

Stresses from loading can cause subgrade deformation under either the static weight of overburden or the repeated train dynamic loading, and one of these sources of loading will typically control the amount of deformation. The static weight of overburden material on a soft, deformable subgrade (Sluz et al. 2003) can cause consolidation settlement and, possibly, massive shear failure. Repeated traffic loading can be characterized by the magnitude of wheel loads, and their corresponding number of load cycles. A simple case is a bi-modal wheel load distribution consisting of unit trains of empty freight cars and unit trains of loaded freight cars. However, the actual wheel load distribution is typically more complex and should consider the range of static wheel loads including their dynamic augment as described in Chapter 2. The subgrade behavior under static load can be much different than under repeated loading because

1. Fine-grained soils will exhibit lower strength under repeated loading than under a sustained load of the same magnitude.
2. Small, plastic deformations under individual wheel loads can accumulate to form substantial track settlement.

It is necessary to characterize subgrade stability under both static and repeated loading cases.

Although far less common, unstable conditions can arise from coarse-grained soil that is saturated. In such a case, increased settlement of coarse-grained soils due to cyclic mobility or liquefaction can result from the increase in pore water pressure under repeated loading. In some anecdotal cases, this has been observed in the granular subballast layer as well.

Soil temperature is a concern due to cyclic freeze and thaw damage. Under certain combinations of temperature, soil suction, soil permeability, and availability of water, an ice lens will form when the water freezes, causing ground heave. When the soil thaws again excess water from the ice lens will be trapped in the subgrade and weaken the soil.

Table 3.9 summarizes subgrade problems in terms of loading, environmental conditions, and their causes and features. Several of these problems are discussed in more detail next.

The loading that drives massive shear failure is the substructure self-weight, and the weights from the train and track superstructure. The resistance to massive shear is from the shear resistance of the substructure layers. This type of failure often occurs shortly after track construction or

Table 3.9 Characteristics of track subgrade problems

Cause	Type	Factors	Features
Dead load	Massive shear failure	• Weight of train, track, and subgrade • Inadequate soil strength	• High embankment and cut slope • Often triggered by increase in water content
Dead load	Consolidation settlement	• Embankment weight • Saturated fine-grained soils	• Increased static soil stress compared to before construction
Live load	Subgrade attrition	• Repeated loading of hard subgrade by ballast • Contact between ballast and subgrade • Clay-rich rocks or soils • Water	• Muddy ballast • Inadequate subballast
Live load	Cyclic mobility/ liquefaction	• Repeated loading • Saturated silt and fine sand	• Large displacement • More severe with vibration • Can happen in subballast
Live load	Progressive shear failure	• Repeated overstressing • Fine-grained soils • High water content	• Squeezing near subgrade surface • Heaves in crib and/or shoulder • Depression under ties
Live load	Cumulative plastic deformation	• Repeated loading • Soft or loose soils	• Differential subgrade settlement • Ballast pockets
Environment	Frost action (heave and softening)	• Periodic freezing temperature • Free water • Frost-susceptible soils	• Occur in winter/ spring period • Rough track surface
Environment	Swelling/ shrinkage	• Highly plastic soils • Changing moisture content	• Rough track surface
Environment	Slope erosion	• Running surface and subsurface water • Wind	• Soil washed or blown away
Environment	Soil collapse	• Water inundation of loose soil deposits	• Ground settlement

after heavy rainfall or flooding and is characterized by an abrupt loss of track alignment and surface.

When the subgrade is constructed as part of an embankment, the stress on the embankment foundation will increase due to the added weight from the embankment. If the embankment soil is not well drained, excess pore pressure will develop reducing the effective stress that can compromise stability, potentially causing a massive shear failure. The excess pore pressure dissipation and resulting settlement are analyzed in basic consolidation theory. In coarse-grained soils, the dissipation of excess pore pressure is rapid, and so most settlement is rapid and will occur during the construction of the embankment and track. In fine-grained soils, the dissipation of pore pressure is slow, resulting in settlement after construction.

Liquefaction and cyclic mobility are dynamic loading–related soil deformation mechanisms. These types of failures are the result of a loss of shear strength under repeated loading due to increased pore water pressure. Liquefaction occurs when loose, saturated cohesionless soils (particularly coarse silts and fine to medium sands) are dynamically loaded or vibrated. Under vibrated loading, soil particles tend to compact, transferring intergranular stress into pore water pressure. Liquefaction occurs when the pore water pressure exceeds the total stress in the material, resulting in zero effective stress and complete loss of frictional shear resistance. Liquefaction can lead to rapid settlement and deterioration of track geometry. While liquefaction primarily occurs in loose samples, cyclic mobility occurs in either loose or dense soils. Cyclic mobility tends to be driven by loading and has been associated with shear stress reversal (Selig and Chang 1981). Liquefaction and cyclic mobility may also occur in the subballast layer if the subballast has silt to fine sand particles, and the water in the subballast cannot drain adequately.

Progressive shear failure is the plastic flow of the soil caused by overstressing at the subgrade surface under repeated loading. The subgrade soil gradually squeezes out from under the tie and then upward at the tie end, following the path of least resistance as shown in Figure 3.20. This is primarily a problem with fine-grained soils, particularly those with high clay content. The addition of more ballast above the subgrade squeeze zone results in an increase in ballast depth and a corresponding tendency for a reduction in stress at the subgrade level, which might improve subgrade stability. However, the depression in the subgrade surface will trap water that can further soften the subgrade.

The formation of ballast pockets is an extreme case of cumulative plastic deformation. In the AREMA manual from 1921, the term *water pocket* is used to describe ballast pockets because the depression was filled with water. Although progressive shear failure is accompanied by progressive shear deformation near the subgrade surface, cumulative plastic deformation is classified as a separate type of subgrade problem because it includes not only the vertical component of progressive shear deformation

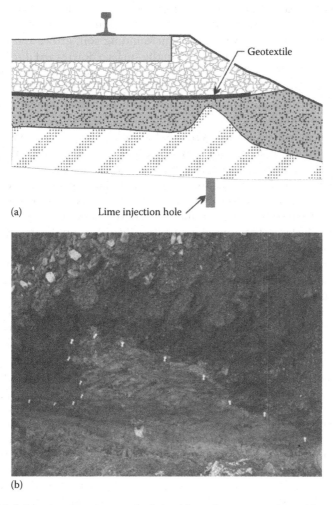

Figure 3.20 Subgrade squeeze at end of tie: (a) track cross section with a subgrade squeeze and (b) subgrade squeeze and lime injection hole.

but also the vertical deformation caused by progressive compaction and consolidation over a considerable depth of subgrade.

Frost heave occurs in the presence of frost-susceptible soil, free water, and freezing soil temperatures. Frost-susceptible soils are those that are sufficiently fine grained to have large capillary rise but are sufficiently pervious to allow an adequate flow of water to feed ice lens growth. These soils include silts, silty sands, and low-plasticity clays. Granular soils with very few fines are generally not frost susceptible. Frost action has two components: heave during freezing and softening during thawing. The frost heave develops from ice lenses that form within the soil, expand, and lift the ground surface. Rocks present near the surface will cool more rapidly than the adjacent ground, potentially causing an ice lens directly below the rock that can draw water from surrounding soil, grow, and force the rock upward. When the ice melts from the surface downward during thawing periods, excess water is released in the soil that causes a significant reduction of subgrade soil strength. During this period of thaw softening, the subgrade is susceptible to plastic deformation and shear failure, as described previously.

Volume change susceptible soils that shrink when dry and swell when wet move the track structure. Clay is the soil type predominantly susceptible to volume change associated with moisture content changes. These soils are mostly found in arid areas and contain large amounts of clay minerals. The volume change varies along the track and may lead to settlement (shrink) in some locations and heave (swell) in others, affecting track geometry. The largest volume change typically occurs on the initial swell of partially saturated soils.

Slope erosion occurs with running surface and subsurface water, but sometimes wind can also erode material at the subgrade slope and toe. Surface erosion may not immediately affect track operation. However, if not repaired and allowed to progress, erosion can undermine the track, lead to sinkholes, or otherwise impact track operation and stability. On an embankment, erosion may undermine the ballast shoulders and the ties. For a cut slope, this may lead to blocking drainage ditches, as eroded material flows down toward the track structure and may even be carried into the track structure in severe erosion situations.

Soil collapse is not common but can occur in soils that are most often windblown and deposited into a dispersed structure that is characterized by low compressibility when dry but suddenly softens when saturated. Loess deposits are notorious for this type of problem and, due to the stiff nature of the deposit, can often withstand very steep cut slopes, which are needed to reduce the soil's exposure to water. Loess is typically windblown silt with some clay that binds the particles together in a dispersed and open structure.

3.3.4 Subgrade improvement methods

Subgrade problems vary widely, and each situation may require a unique solution. Subgrade repair and improvement is typically most successful if the problem is properly understood, and the devised solution addresses the problem. Properly understanding the problem involves understanding local geology, site history, previous construction methods and trends, local traffic trends, track maintenance history, track condition measurements, and site hydrology and hydrogeology followed by conducting a detailed site reconnaissance to identify the problem. Based on this level of understanding, a site investigation can be planned to further characterize the problem and design the repair. As subsequent sections of the book consider repair and stabilization in detail along with specifics on site conditions, this section will focus on a few improvement methods that hold promise. Subgrade improvement techniques include grouting, load transfer, and soil densification–based techniques, and each of these will be discussed next.

Techniques of compaction grouting and displacement grouting tend to be practical options for existing lines, while permeation grouting tends not to be useful when the problem soil is fine grained and impermeable. Grouting techniques that rely on compaction or displacement for grout injection are likely to work better for repair of existing railroad subgrade problems. Compaction grouting relies on a stiff concrete mix that is compacted into a bulb in the ground compacting the neighboring soil and providing concrete structural support. Displacement grouting works in a similar manner, but the concrete is less stiff so it can be pumped under pressure while displacing adjacent soil. The grouting program should be designed to provide the needed confinement to stabilize the track structure. This can often be accomplished by identifying the path of least resistance for the failure, and reinforcing that section of the embankment so that other grouting locations do not accelerate the failure. Then, grouting in other planes can occur to retain the subgrade.

Load transfer involves installing a structural element in the track to transmit load through a problematic layer so that it reacts on a lower, more stable layer. Load transfer–based mechanisms include the compaction grouting method noted earlier, where the concrete column that remains is suitable for transferring load. Load transfer structural elements can be concrete or other cementitious material, steel, wood, and so on. Soil nails, concrete columns, and helical piles are likely candidates for load transfer. The main requirement is that the structural element be installed with the track structure in place, leading to the use of particularly small elements that fit between the ties.

Soil densification–based improvement methods include stone columns and other displacement-based techniques to compact the soil and fill the void with a structural element that can help transfer the load. Soil densification–based techniques, again like the compaction grouting method noted earlier, have the distinct advantage of compacting the surrounding soil during installation, resulting in improved soil behavior. Stone columns are another method for creating a structural element while compacting the surrounding soil and can be constructed by simply compacting stone in bulbs below the surface like the method of compaction grouting. Stone columns are trademarked as a rammed aggregate pier® or geopier®.

3.4 OTHER SUBSTRUCTURE MATERIALS

3.4.1 Ballast treatment methods

A variety of treatment methods have been proposed for ballast over the years ranging from ballast glue and other pour-in-place material used to lock the ballast in place to the use of geosynthetics to provide tensile reinforcement and geotextile fabric as a separator to keep fines in lower layers from migrating into the ballast. These materials are used to address the recognized need for ballast stability and confinement. However, claims regarding some of these products and treatments have not always been supported by evidence (Sussmann and Selig 1997). Therefore, the application of alternative ballast materials and stabilizing materials or additives should be viewed skeptically. For example, pouring anything into the void space of ballast that will block at least a portion of the voids will reduce the drainage and void storage capability of ballast at the least and in the worst case may attract and hold water.

The desire to stabilize ballast usually follows from the recognition that the ballast section is not properly compacted, because compact ballast is generally very stable. The main reason for poor ballast compaction is that ballast density is generally not controlled and compaction under traffic often accumulates slowly, even if aided by a dynamic track stabilizer following tamping or ballast placement.

Geotextiles have been proposed to provide separation from lower layers containing fine material that may contaminate the ballast. However, ballast has been found to abrade or puncture geotextiles, which removes any protection against separation and confinement (Figure 3.21). But even when geotextiles remain intact and do not puncture or tear, they often fail to prevent the upward migration of soils with an abundance of silt and clay particles from below, because such small particles readily pass through the geotextile and into the ballast. This indicates that the soil particles are finer than the holes in the geotextile, which should be designed using the filter separation criteria described in Chapter 6. Even if the geotextile does separate the fine material, problems can still develop as the geotextile may become caked with fine material and inhibit drainage. Lastly, consideration

Figure 3.21 Abraded geotextile.

should be given to how the geotextile may be removed during track maintenance or rehabilitation. Geotextiles have been particularly problematic because the material binds up on the undercutter chain, teeth, and sprockets.

The innovative use of new materials to improve track performance should be encouraged because the industry needs every economic advantage. However, healthy skepticism is required to ensure that the solution does not create problems beyond those being corrected. There are cases where installing geosynthetics and associated ballast treatments in track are appropriate; however, the application must fit within product capabilities and not inhibit future maintenance. It is prudent to place any geosynthetics below the maintained ballast layer at least 250 mm below the tie bottom, unless installing the material higher in the track structure will not inhibit track maintenance or ballast performance during or after its expected life. The material must not degrade ballast or substructure performance if destroyed by maintenance. For instance, some pour-in-place glues and binders confine ballast well, but if tamping is required and destroys the material matrix, the resulting small compressible (at least relative to hard rock ballast) pieces must not inhibit ballast interlocking, or create a track maintenance problem.

3.4.2 Alternative subballast materials

A variety of materials have been used in place of traditional granular subballast. From concrete slab to asphalt layers, the intent is to place a material that is stiffer than subballast to provide improved track structure support. The track may already be underlain by old roadbed consisting of materials that provide good structural support. Sometimes this strong, stiff old roadbed layer is called hardpan, which is a very stable structural layer below the track that has compacted over time. In some cases, the hardpan is a result of decades of compaction that have produced a very dense layer that may be nearly as resistant to penetration as concrete. In other cases, this hardpan layer was intentionally placed, consisting of brick or cobble layers as an earlier attempt to support the track. In few instances, the placement of a layer of cementitious slag has been reported.

Alternative subballast materials that are sometimes considered for contemporary track designs include asphalt layers, reinforced soil layers, cement-stabilized layers, and impermeable membranes. Each of these alternatives has a variety of characteristics and requires knowledgeable designers. For example, asphalt can create a strong and stiff layer like a pavement or it can create a resilient layer for vibration attenuation, when properly specified and placed. Proper application of asphalt requires specific expertise in asphalt technology to ensure proper mix design, track section design, asphalt transportation, and placement techniques. Soil-reinforced layers can range from the use of high-strength geotextiles, to cellular geoweb, and to flat geogrid (Figure 3.22). The type of geotextile (woven or nonwoven)

(a)   (b)

Figure 3.22 Geoweb and geogrid examples: (a) geoweb and (b) geogrid. (Courtesy of Transportation Technology Center Inc., Pueblo, Colorado.)

will control the opening size that affects the filter characteristics, and also will affect the strength. Most geoweb is made from high-density polyethylene (HDPE), which is commonly used for plastic pipe, drainage infiltrators, and many other products. One problem with the use of HDPE in a structural layer is that HDPE tends to creep when loaded, which can result in strain over time, loss of structural support, and settlement. New materials are being used to limit the creep properties of HDPE geoweb, but these should be evaluated for durability and resistance to wear and abrasion.

Selecting an alternative subballast material may be a necessity when other solutions appear unlikely to create the required track structure and to adequately balance competing functional requirements. To ensure that alternative subballast materials are applied successfully requires that they address specific problematic soil conditions that the material is known to correct. Particularly, moisture-sensitive clay subgrade has successfully been treated using a glass fiber-reinforced geomembrane to seal the subballast and subgrade from water infiltration. Cement-stabilized soil layers have been used to successfully support track over soft subgrade. Soil reinforcement has been used to stabilize weak track structure on a steep embankment. These successes developed from understanding the problem and its failure mechanism, identifying a cost-effective solution, developing a conservative design, and following up to make sure the design is properly implemented.

3.4.3 Hot mix asphalt

The use of hot mix asphalt (HMA) as a ballast underlayment is the most common application of HMA in the track. As an underlayment, HMA is placed directly on top of the subgrade soil, or on top of a granular subbase, with ballast then placed on top of the HMA. HMA underlayment has been used in mainline track, road crossings, tunnel and bridge approaches, and special trackwork such as turnouts and crossing diamonds (Chrismer et al. 1996). HMA has also been placed in the track as an overlayment with ties placed directly on the asphalt layer with no ballast. However, this application is not used in North American freight track applications due to the inability to surface and realign the track.

The main areas in which HMA underlayment can improve conventional track substructure performance are to provide drainage out of the track, reduce stresses on the subgrade, increase track stiffness, and improve constructability of new track.

Subgrade stress levels will be reduced beneath an HMA layer compared to an equal thickness of granular material (subballast) due to the increased resilient modulus that can be provided by HMA. This benefit is useful in situations where the thickness of the granular layer is limited due to overhead height or excavation depth restrictions that result from a variety of construction constraints. Without these restrictions, the desired subgrade stress reduction can often be accomplished with traditional granular materials.

The HMA also provides an improved degree of confinement to the track because HMA is much better than ballast at resisting spreading due to its higher resistance to tension and shear. The increased confinement will lead to a stiffer ballast layer response, if properly compacted. In some cases, the use of HMA can also result in track stiffness that is too high, causing accelerated rates of ballast deterioration and track component degradation (e.g., concrete tie cracking). The stiff HMA also typically results in less damping in the substructure (Singh et al. 2004) so dynamic loads may be more acute. HMA with a modified binder mix (e.g., rubberized HMA) can potentially improve the damping characteristics of the substructure to reduce high-impact stresses (Buonanno et al. 2000; Singh et al. 2004).

An HMA underlayment can also provide a smooth, rut-free surface for track laying operations, even in wet conditions, which can accelerate construction schedules. However, in locations with this problem, the construction of the HMA itself will need a reasonable work platform underneath it, requiring the placement of a granular working base in most cases.

It should not be assumed that traditional ballast and subballast materials cannot provide the advantages associated with HMA. For example, HMA may provide a stable work platform, but granular material placed over a soft soil can provide the same. Also, increased track stiffness can be provided by a stronger, stiffer, more confined ballast or subballast layer and does not require HMA. HMA has been applied at a number of locations throughout the industry partly due to widespread availability of both material and contractors who are knowledgeable in highway HMA applications. However, the use of HMA in track is different than in highway applications, and so careful consideration must be given to the design details, contractor selection, and material specifications.

HMA design and property considerations involve two main areas of concern: mix design/quality control and structural design of the track cross section.

Mix design is covered thoroughly in many textbooks, design manuals, and industry standards (e.g., Huang 1991; Asphalt manual 2012). These references describe the process of developing a mix design from desired properties through proper selection of aggregate and amount and viscosity of asphalt binder. Applications of HMA in track have ranged from providing an alternative structural layer to sealing the surface of problematic subgrade. However, the use of asphalt to seal water from penetrating into poorly performing subgrade cannot be successful if the layer is not properly designed for structural performance. In almost any HMA track installation, asphalt layer strength will be a primary concern, which will require use of a large angular aggregate relative to the common highway gradations. From that point forward, the mix design will vary depending on the desired characteristics including resilience, strength, and deformation properties. See Chapter 5 for discussion of design trade-offs and considerations as well as references on various designs for track installation.

Any installation of asphalt should include basic quality control parameters to ensure conformance with design specifications. Temperature, density, and strength testing should be required for any installation, in addition to other required quality control parameters specific to the design or application. The logistics of trucking the material and maintaining the asphalt temperature is a concern due to the remoteness of many construction sites. Achieving the desired compacted density can be somewhat challenging to a contractor not familiar with railroad HMA applications because compaction is often difficult on track, especially over a soft subgrade. Control of density is essential because it will control the strength of the layer, and strength is critical to long-term performance. Finally, samples from the placed layer should be obtained, tested, and compared with the mix design and specification as well as the expectations of the track design engineer.

Often the most difficult information to estimate is the performance of HMA in distributing applied stress and resisting deformation. Use of a geotechnical track analysis software programs like KENTRACK or GEOTRACK is the main tool available for this analysis. The only way to verify the design from these types of analysis is to obtain field samples to test for resilient modulus and other mechanistic design parameters and then conduct a field test to evaluate the deformation of each individual layer, or the overall track system. More detailed design considerations and material properties are discussed in Chapter 5.

3.4.4 Concrete and cementitious material

The use of concrete in track has a long history from use of concrete beams to support the rails (Chipman 1930) to the use of contemporary concrete cross-ties. Along the way, major problems with the widespread use of concrete have been encountered including alkali-silica-related concrete tie failures of the 1980s. Continued vigilance of the concrete product quality is required to ensure that track integrity and stability are not compromised. All concrete quality control, design, and maintenance-related specifications should be heeded.

In addition to the requirement that good-quality concrete be used in track, the track structure must be stable for the use of any concrete structure (Li et al. 2010). Concrete requires solid support, whether for a concrete tie or slab. Concrete is unforgiving of poor support conditions, and excessive settlement could lead to cracks and rapid structural deterioration. One well-documented case of advanced track design and failure of concrete ties and a cement-stabilized subgrade layer is the Kansas Test Track (Cooper et al. 1979).

The Kansas Test Track was a 1.5-mile-long experiment in the design and construction of advanced track technology. A wide variety of advanced track support structures were installed to provide different levels of track stiffness in the test zone with a plan for evaluation over multiple years. Signs of distress became apparent after only 2.5 months of operation and the track was closed to traffic permanently within 6 months. The poor track performance resulted from poor drainage of the top of the embankment onto which these various track support structures were installed. This poor drainage resulted in softening of the embankment, which compromised the performance of the track support structures (Cooper et al. 1979). This highlights the importance of the drainage and geotechnical aspects of track support, especially for concrete structures that cannot withstand large track deformations without cracking and structural deterioration.

3.4.5 Water

The water trapped in and draining through track is a drainage concern as already described, especially considering the potential effects on subballast and subgrade performance. In many locations at a depth of perhaps only one or two ballast particles below the surface, moisture can be noted that is retained in the track from previous wetting. Deeper, perhaps near the bottom of tie, the presence of water is common when ballast is fouled, except in the most arid of conditions. Water is often a part of the track, and even in well-drained situations, inundation during a rainfall event is likely to cause water to be carried through the track structure.

Although water is typically thought to be chemically neutral, it may have high or low pH and/or carry chemical and organics that can cause damage. Water is a solvent that tends to take soluble elements, organics, and chemicals into solution and transport them along with the soil particles as described in the separation/filtration discussion of the subballast section. Salt is an element that most readily dissolves and is abundant in many areas of the world. Drainage water can carry dissolved salts and sulfates that can be detrimental to various track components and affect grounding of track signals or development of stray currents. Salts attack steel and metals and cause rust through oxidation-based corrosion. Sulfates attack concrete elements or anything containing cement, including drainage pipe. Precautions to deal with these potential problems include special coatings for steel or cast iron pipe and use of chemical resistant concrete. To anticipate and manage these problems one must evaluate the site conditions to determine their likelihood, understand the site-specific nature of this problem, and recognize signs indicating that it could or may be occurring, and apply the available means to deal with these conditions.

3.5 TRACK TRANSITIONS

A track transition is a change in track structure, such as encountered by a vehicle moving from a conventional ballasted track on a soil subgrade to a bridge, slab track, or grade crossing. A transition can either involve a change of track superstructure (such as at grade crossing and turnout) or a change in track support (bridge, slab, tunnel, etc.). In most cases, a track transition involves changes in both track superstructure as well as track substructure. Mechanistically, a transition is often characterized as a large change in vertical track stiffness, although this is not always the case. The problems at transitions vary, but the most common feature is differential track settlement, such as a dip in the vertical rail profile shown in Figure 3.23.

Typical problems associated with transitions and the settlement-induced dips are rapid track geometry degradation, mud pumping, poor ride quality, difficulty in conducting effective track maintenance, and damage to track and vehicle components from dynamic wheel loads. There are also problems transitioning from low track stiffness to high track stiffness, such as when

Figure 3.23 Dip at bridge transition.

trains travel onto a bridge where the common problems include excessive vibration and breakage of track components on the bridge.

A site investigation is often required to determine if the recurring dip that follows maintenance is related to improper maintenance or from instability of the track structure or substructure. Maintenance procedures that work well in open track away from the transition may require modification to be effective in the transition. Many misconceptions exist about the source of these problems and their appropriate repair. A full discussion of track transition–related problems, causes, and potential remedial measures is provided in Chapters 5 and 9.

Chapter 4

Mechanics

This chapter presents models and methodologies that have been developed to characterize and predict behaviors and responses of track and subgrade components under various load environments. The first section of this chapter describes models that can be used for analysis of stress, strain, and deformation in track and subgrade under wheel loads. The second section is focused on the resilient behavior or, more specifically, on the estimation of resilient modulus of granular ballast materials and fine-grained subgrade soils. The third section of this chapter discusses the prediction of plastic or permanent cumulative deformation for both granular ballast and fine-grained subgrade soils, as a result of repeated wheel load applications. Plenty of examples are given in this chapter to illustrate how track and subgrade responses, resilient modulus, and cumulative deformation can be analyzed and predicted. Particular attention is given to illustrate what factors significantly affect track and subgrade behaviors and responses. Understanding of the fundamental mechanics as described in this chapter is essential for design, construction, and maintenance of railway track and subgrade.

4.1 TRACK AND SUBGRADE MODELS

An analytical model that incorporates major track structural components and foundation characteristics is important for assessment of track performance and design of track foundations. In general, four types of analytical models are available for railway track and subgrade, including the beam on elastic foundation (BOEF) model, the elastic half-space model, multilayer models, and finite element models (FEM).

4.1.1 BOEF model

BOEF model is a classic model that was developed and advanced by many pioneer engineers and researchers (e.g., Winkler, Timoshenko, and Talbot). The theoretical formulation of the BOEF model is based on the assumption

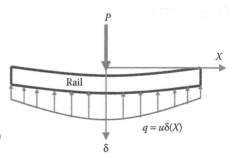

Figure 4.1 Beam on elastic foundation
model.

that each rail acts like a continuous beam resting on an elastic foundation, as shown in Figure 4.1. The elastic foundation is represented by track modulus, u, which is defined as the supporting force per unit length of rail per unit vertical deflection of the rail. As defined in Figure 4.1, track modulus includes the effects of fasteners, ties, ballast, subballast, and subgrade, that is, everything under the beam (rail).

The differential equation for this model is as follows:

$$EI\frac{d^4\delta}{dx^4} + u\delta = 0 \qquad (4.1)$$

where:
  E is the rail modulus of elasticity
  I is the rail moment of inertia
  δ is the rail deflection
  x is the distance from the applied load P
  u is the track modulus

From this equation, track deflection, rail-bending moment, shear force, and vertical foundation reaction force can be derived as follows:

$$\delta = \frac{P}{\left(64EIu^3\right)^{1/4}} e^{-\beta x}(\cos\beta x + \sin\beta x) \qquad (4.2)$$

$$M = P\left(\frac{EI}{64u}\right)^{1/4} e^{-\beta x}(\cos\beta x - \sin\beta x) \qquad (4.3)$$

$$S = -\frac{P}{2}e^{-\beta x}\cos\beta x \qquad (4.4)$$

$$q = P\left(\frac{u}{64EI}\right)^{1/4} e^{-\beta x}(\cos\beta x + \sin\beta x) \qquad (4.5)$$

where:

$$\beta = \left(\frac{u}{4EI}\right)^{1/4} \tag{4.6}$$

The maximum deflection and bending moment occur at the point of load application, where $x = 0$. In other words, the following equations can be derived to calculate maximum deflection, bending moment, shear force, and foundation reaction force:

$$\delta_0 = \frac{P}{(64EIu^3)^{1/4}} \tag{4.7}$$

$$M_0 = P\left(\frac{EI}{64u}\right)^{1/4} \tag{4.8}$$

$$S_0 = -\frac{P}{2} \tag{4.9}$$

$$q_0 = P\left(\frac{u}{64EI}\right)^{1/4} = -u\delta_0 \tag{4.10}$$

4.1.2 Track modulus

Because of oversimplification of ballast, subballast, and subgrade layers as one elastic foundation, as shown in Figure 4.1, the BOEF model is no longer considered adequate for analysis of a track structure. However, track modulus, originally defined from this model, has become a parameter that is used extensively to quantify track foundation support, the load-carrying capacity, and the overall stiffness of ballast, subballast, and subgrade layers.

From Figure 4.1, track modulus is defined as

$$u = -\frac{q}{\delta} \tag{4.11}$$

On the other hand, track stiffness, based on the BOEF model, is defined as

$$k = \frac{P}{\delta_0} \tag{4.12}$$

In other words, track stiffness is simply the force applied on the rail divided by the rail deflection directly under the force. From a practical point of view, track stiffness can be directly measured, if the corresponding rail deflection under a given load can be measured.

From Equations 4.10 through 4.12, track modulus can be calculated from track stiffness as follows:

$$u = \frac{k^{4/3}}{(64EI)^{1/3}} \tag{4.13}$$

Because of the definition of track stiffness by Equation 4.12 and calculation of track modulus from track stiffness in terms of Equation 4.13, track stiffness and track modulus can be calculated from any analytical model or track measurement, as long as one can calculate or measure track deflection under an applied wheel force.

From their definitions, track stiffness includes the effect of rail flexural rigidity, EI, whereas track modulus does not. In other words, track modulus is a parameter that characterizes overall stiffness of track foundation directly under the rail.

Section 4.1.5 will include discussion of how track stiffness and track modulus are affected by individual properties of track and subgrade components, as well as how track modulus can be used to characterize overall load-carrying capacity and deformation, and thus overall performance of the track foundation.

4.1.3 Half-space model

The elastic half-space model assumes a homogeneous, isotropic, and elastic media for the track substructure. This model can be used for the calculation of stresses, strains, and deflections in the track substructure, which, however, cannot be obtained from the BOEF model. In the elastic half-space model, the surface load is applied either through the contact pressure between the tie and the ballast surface or through a concentrated wheel load. The solutions of stresses and strains due to a point load at the surface have been given by Boussinesq and can be found in many geotechnical textbooks. For example, the vertical stress at any depth below the surface can be determined as follows (see Figure 4.2):

$$\sigma_z = \frac{3}{2\pi \left[1 + (r/z)^2 \right]^{5/2}} \frac{P}{z^2} \tag{4.14}$$

where:
   r is the distance radially from the point load
   z is the depth
   P is the point load

Solutions for a uniformly distributed circular or rectangular load can be obtained by integration of the solution due to point load.

Many geotechnical textbooks give detailed equations and charts for calculating stresses, strains, and deflections in the soil foundation due to various shapes of surface loads. However, the elastic half-space model is

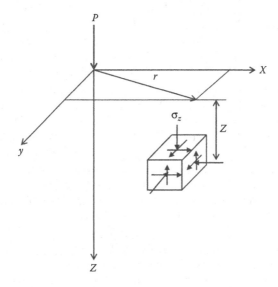

Figure 4.2 Stresses in an elastic medium to a point load.

no longer considered adequate for track and subgrade, especially with the development of multilayer computer models and FEM that can characterize and represent actual track and subgrade conditions more accurately than the elastic half-space model.

4.1.4 GEOTRACK

Several multilayer computer models have been developed for the analysis of track and subgrade response under wheel loads, including GEOTRACK (Chang et al. 1980) and KENTRACK (Huang et al. 1984). A major improvement of a multilayer computer model over the BOEF or elastic half-space model is the incorporation of all major components of the track superstructure and substructure, that is, rail, tie, fastener, ballast, subballast, and subgrade. Another advantage is that a multilayer computer model can take into account nonlinear material properties, such as resilient modulus of track substructure layers.

The main disadvantage of a multilayer model is that track responses, such as stress, strain, and deflection, determined from such a model are not found by solving simple equations like those developed under the BOEF or half-space elastic space models. That is why these multilayer models are also referred to as multilayer "computer" models, because a computer is needed to solve complicated equations to determine the solutions for track and subgrade response under given wheel loads, and track and subgrade conditions.

GEOTRACK (owned by the Transportation Technology Center, Inc.) is a three-dimensional (3D), multilayer model for determining the responses of the track and subgrade as a function of (1) wheel loads, (2) properties of rails and ties, (3) properties of ballast, subballast, and underlying soil layers, and (4) dimensions including tie spacing and layer thicknesses. The model provides predicted rail seat loads, tie–ballast reactions, rail and tie deflections, and rail and tie bending moments. The model also provides vertical displacements and complete 3D stresses at selected locations in ballast, subballast, and subgrade. In addition, the model calculates track modulus.

Figure 4.3 shows the components of track and subgrade contained in the model. The rails are represented as linear elastic beams with support at each tie. The rails span 11 ties and are free to rotate at the ends and at each tie. Each connection between the rails and ties is represented by a linear spring. The ties are represented as linear elastic beams supported at ten equally spaced locations on the underlying ballast. The individual tie–ballast reactions are applied to the ballast surface over circular areas, whose sizes are related to the tie dimensions.

The ballast, subballast, and underlying subgrade are represented as a set of up to five elastic layers of infinite horizontal extent. The bottom layer extends to infinite depth. Each layer has a separate resilient modulus and Poisson's ratio. No slip is permitted at the layer interface.

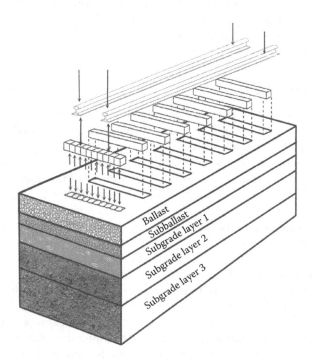

Figure 4.3 Forces and component in GEOTRACK model.

The model only allows vertical components of wheel loads. For a single load, the wheel load is placed directly over the center tie of the 11-tie track structure. For multiple loads, the single load solution is shifted to each additional load location and combined algebraically with the center tie single load solution.

In general, the wheel spacings will not be integer multiples of the tie spacing, and thus the wheels will not all be located over all the ties. However, as an approximation, either the wheels are located at the ties nearest to the actual locations or a wheel load is subdivided into two parts that are assigned to adjacent ties. Because the tie spacings are small relative to the wheel spacings, and the rails are sufficiently stiff, the error caused by this approximation is not of practical importance.

Each layer in the GEOTRACK model is characterized as an elastic material with a resilient modulus. The magnitudes of resilient moduli may be defined in several ways for use in the GEOTRACK model. A linear relationship between the log of resilient modulus and the log of bulk stress is commonly used to characterize granular materials (Section 4.2). A linear relationship between the log of resilient modulus and the log of deviator stress, on the other hand, is most appropriate to represent resilient modulus for fine-grained soils (Section 4.2). In addition to these stress-state–dependent modulus formulations, any layer may be assigned a constant modulus, independent of the modulus formulation used for the other layers when GEOTRACK is used for parametric study.

To determine the stress state, the static tie and rail weights are converted to surcharge pressure acting at the surface of the ballast layer. The geostatic vertical stress is determined by summing the products of layer unit weights and depths from the bottom of the tie to a selected point within each layer, usually in the middle. The geostatic vertical stress is then added to the surcharge pressure. The vertical stress multiplied by the coefficient of lateral earth pressure in each layer gives the unloaded horizontal stress for the layer. These unloaded stresses are then added to the incremental stresses generated by wheel load to get the stresses in the loaded state.

When stress-state–dependent modulus is used in the model, these aforementioned stresses are then used to calculate the resilient modulus. The layer modulus is updated after each successive iteration of stress calculation, and the calculations are repeated until convergence is achieved between the calculated stresses and the corresponding resilient modulus. Generally, three iterations are sufficient for reasonable modulus convergence.

The stress state obtained from the GEOTRACK model is 3D and is defined by the stress tensor as follows:

$$\{\sigma_{ij}\} = \{\sigma_x, \sigma_y, \sigma_z, \sigma_{xy}, \sigma_{yz}, \sigma_{zx}\} \tag{4.15}$$

where:

σ_x, σ_y, σ_z are the normal stress in lateral, longitudinal, and vertical direction

σ_{xy}, σ_{yz}, σ_{zx} are the shear stress in xy, yz, and zx planes

In addition, the GEOTRACK model can calculate the equivalent stress state for a triaxial test. The stress state for a triaxial test is represented by an axial stress (vertical), σ_a, and a radial (confining) stress, σ_r. This calculation of equivalent triaxial stress state is based on the assumption that the octahedral shear and normal stresses in the field are equal to those in a triaxial test.

The octahedral normal stress, σ_{oct}, is a function of the first invariant of stresses, I_1, and the octahedral shear stress, τ_{oct}, is a function of the first and second invariants, I_1 and I_2, as follows:

$$I_1 = \sigma_x + \sigma_y + \sigma_z \tag{4.16}$$

$$I_2 = \sigma_x\sigma_y + \sigma_x\sigma_z + \sigma_y\sigma_z - \sigma_{xy}^2 - \sigma_{xz}^2 - \sigma_{yz}^2 \tag{4.17}$$

From the GEOTRACK model, the octahedral stresses are

$$\sigma_{oct(field)} = \frac{I_1}{3} \tag{4.18}$$

$$\tau_{oct(field)} = \left[-\frac{2}{3}\left(I_2 - \frac{I_1^2}{3} \right) \right]^{1/2} \tag{4.19}$$

For a triaxial test, the octahedral stresses are

$$\sigma_{oct(triaxial)} = \frac{\sigma_a + 2\sigma_r}{3} \tag{4.20}$$

$$\tau_{oct(triaxial)} = \frac{\sqrt{2}}{3}\left(\sigma_a - \sigma_r \right) \tag{4.21}$$

Set

$$\sigma_{oct(field)} = \sigma_{oct(triaxial)}$$

and

$$\tau_{oct(field)} = \tau_{oct(triaxial)}$$

Thus,

$$\sigma_a = \sigma_{oct(field)} + \sqrt{2}\,\tau_{oct(field)} \tag{4.22}$$

$$\sigma_r = \sigma_{oct(field)} + \frac{1}{\sqrt{2}}\,\tau_{oct(field)} \tag{4.23}$$

In the case of a triaxial test, deviator stress is thus defined as

$$\sigma_d = \sigma_a - \sigma_r \tag{4.24}$$

Table 4.1 Input and output parameters in GEOTRACK

Input	Output
• Rail—gage, weight, Young's modulus, moment of inertia, and cross-sectional area	• Rail—deflections and rail seat load at all tie locations, maximum bending moment (maximum bending stress)
• Fastener/pad—vertical stiffness	• Ties—deflections and tie–ballast reactions at ten segments along each tie, bending moments at tie center and at rail–tie seats
• Number of ties: up to 17	
• Ties—tie spacing, length, height, width, Young's modulus, and moment of inertia	
• Substructure layers (ballast, subballast, subgrade): up to five	• Substructure layers—vertical deflections, complete 3D stresses, and strains at any location (x, y, z)
• Substructure layers: modulus, Poisson's ratio, layer thickness, unit weight, and coefficient of lateral earth pressure	• Track modulus
• Wheel loads: magnitudes and spacings of up to four wheel loads	

Table 4.1 is a summary of input and output parameters for the GEOTRACK model. This model can be used for a number of track types, such as wood tie track, concrete tie track, slab track, track with standard gage, wide gage, or narrow gage, for applications including

- Analyze the effects of loads on track component performance.
- Analyze the effects of rail and tie properties on track behavior and performance.
- Analyze the effects of pad stiffness and ballast stiffness on track behavior and performance.
- Analyze the effects of track subgrade conditions on track behavior and performance.
- Conduct track granular layer thickness design.
- Conduct track modulus/stiffness analysis.

4.1.5 Track modulus analysis using GEOTRACK

As discussed earlier, GEOTRACK can be used to analyze track modulus as a function of track component properties. Because accurate values of all these properties are difficult to obtain, this analysis is useful for understanding the effects of each of these properties on track modulus.

Table 4.2 gives nominal values of material properties for the input to the GEOTRACK track modulus analysis. Table 4.3 gives the ranges of these variables considered in the parametric study. Although each variable is changed in the GEOTRACK analysis, the other variables are assigned the nominal values given in Table 4.2. Although the resilient moduli of ballast, subballast, and subgrade are stress-dependent (see Section 4.2), they are assumed constant in this analysis for the purpose of simplicity.

Table 4.2 Nominal track and subgrade parameters for GEOTRACK analysis

Variable	Value	
Rail properties		
E—GPa (ksi)	207 (30)	
I—m^4 (in.^{4})	3.95e-5 (94.9)	
Cross-sectional area—m^2 (in.^{2})	8.61e-3 (13.4)	
Gauge—m (in.)	1.50 (59.3)	
Mass—kg/m (lb/yd)	67.7 (136.2)	
Tie and fastener	Concrete	Wood
Base width—m (in.)	0.273 (10.8)	0.229 (9.0)
Base length—m (in.)	2.59 (102)	2.59 (102)
Cross-sectional area—m^2 (in.^{2})	5.59e-2 (86.6)	4.06e-2 (63)
E—GPa (ksi)	31 (4,500)	10 (1,500)
I—m^4 (in.^{4})	2.42e-4 (582)	1.07e-4 (257)
Mass—kg (lb)	363 (800)	114 (250)
Spacing—m (in.)	0.61 (24)	0.50 (19.5)
Fastener stiffness—kN/mm (kips/in.)	175 (1,000)	70 (400)
Granular layer	Ballast	Subballast
Density—Mg/m^3 (pcf)	1.76 (110)	1.92 (120)
Poisson's ratio	0.3	0.35
Modulus—MPa (ksi)	276 (40)	138 (20)
K_o	1.0	1.0
Thickness—m (in.)	0.3 (12)	0.15 (6)
Subgrade and bedrock	Subgrade	Bedrock
Density—Mg/m^3 (pcf)	1.92 (120)	2.24 (140)
Poisson's ratio	0.35	0.1
Modulus—MPa (ksi)	41 (6)	6890 (1,000)
K_o	1.0	1.0
Thickness	Infinite	

Because, in most situations, subgrade is not a uniform half-space of the same soil type, the influence of subgrade stratification is also considered in this parametric study. The subgrade is simplified as a layer of variable modulus with different thicknesses overlying bedrock.

In the GEOTRACK model, the "fastener stiffness" is the term given to the vertical compressibility between the bottom of the rail and the bottom of the tie. For wood tie track, the major contributor is the compressibility of the tie itself, whereas for concrete tie track, the major contributor is the rail seat pad.

Figure 4.4 shows the comparison of the influences of track components on track modulus from the GEOTRACK analysis. The horizontal line represents

Table 4.3 Range of variables for track modulus parametric study

Variable	Lower bound	Nominal	Upper bound
Track type	Wood	Concrete	
Tie spacing (m)	0.46	0.61	0.76
Fastener stiffness (kN/mm)	26	175	350
Ballast modulus (MPa)	138	276	551
Subballast modulus (MPa)	69	138	276
Subgrade modulus (MPa)	14	41	138
Ballast thickness (m)	0.15	0.30	0.61
Subballast thickness (m)	0.15	0.15	0.46
Subgrade thickness (m)	1.22	Infinite	

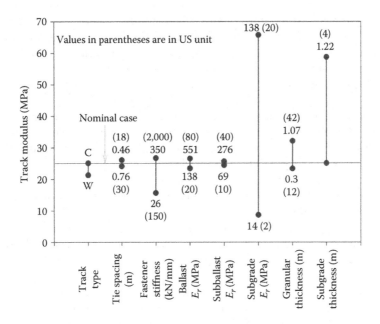

Figure 4.4 Effects of track component properties on track modulus.

the track modulus for the nominal concrete tie track, as defined by the properties given in Table 4.2. The numbers or letters in the figure represent the upper and lower bounds of the variables considered. The changes in track modulus caused by the change in each individual variable are indicated by vertical lines.

In Figure 4.4, the track moduli for the nominal concrete tie track and the nominal wood tie track are 25 MPa (3,640 psi) and 21 MPa (3,090 psi), respectively. The difference is 16% relative to the concrete tie track.

The majority of this difference can be explained by the difference in fastener stiffness between the concrete tie track and the wood tie track. If a fastener stiffness of 70 kN/mm (400 kips/in.) assumed for the nominal wood tie track is used to replace the value of 175 kN/mm (1,000 kips/in.) assumed for the nominal concrete tie track, the track modulus obtained for the concrete tie track is then 22 MPa (3,190 psi). The relative difference between these two types of track with the same fastener stiffness then becomes only 4.7%. This remaining difference is the result of the effect of the difference in tie bending stiffness and tie spacing.

In Figure 4.4, a decrease in tie spacing causes a slight increase in track modulus. However, this increase is insignificant for the change of spacing from 0.76 to 0.46 m (30 to 18 in.). On the other hand, track modulus increases significantly with increasing fastener stiffness. As shown in this figure, track modulus increases by 70% for the increase in fastener stiffness shown.

Increasing ballast modulus or subballast modulus causes a slight increase of track modulus. However, the dominant factor influencing track modulus is subgrade resilient modulus. As demonstrated in Figure 4.4, a 10-fold increase in subgrade resilient modulus from 14 to 140 MPa (2,000 to 20,000 psi) leads to an increase in track modulus of approximately by a factor of 8. Compared with the subgrade, the ballast and subballast can be considered to have much less effect on track modulus. The reasons for this are that the ballast and subballast layers are thin compared with the influence depth of the subgrade, and the ballast and, generally, subballast moduli do not vary as much as subgrade modulus.

As shown in Figure 4.4, the thicknesses of both the granular layer (ballast + subballast) and the subgrade layer overlying the bedrock affect track modulus, though the latter has a much greater effect. An increase in granular layer thickness leads to an increase in track modulus. On the other hand, an increase in the subgrade layer thickness leads to a decrease in track modulus, because the subgrade modulus is lower than that of the ballast, subballast, and bedrock.

Although Figure 4.4 shows the effect of each individual factor on track modulus independently, it does not show how the major factors interact with each other in affecting track modulus. Figures 4.5 and 4.6 are plotted to complement Figure 4.4 in examining the change of track modulus caused by the change of both material modulus and the corresponding material layer thickness. The values of the unvaried parameters are given in Table 4.2.

Figure 4.5 shows the influence of granular layer thickness and resilient modulus of the granular material on track modulus. In this figure, both the ballast and subballast are assumed to have the same modulus, and granular layer thickness is the sum of ballast layer thickness and subballast layer thickness. As this figure shows, for the change of granular layer thickness from 0.30 to 1.07 m (12 to 42 in.), track modulus increases from 24 to 34 MPa (3,460 to 5,000 psi). The effect of resilient modulus of the

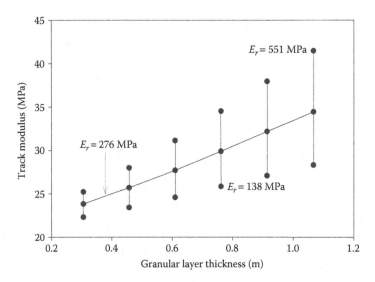

Figure 4.5 Effect of granular layer thickness and resilient modulus on track modulus.

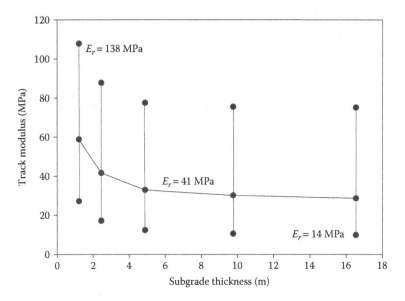

Figure 4.6 Effect of subgrade layer thickness and resilient modulus on track modulus.

granular material on track modulus with different granular layer thickness is shown by using three moduli for the granular material. The lower bound is 138 MPa (20,000 psi), the nominal value is 276 MPa (40,000 psi), and the upper bound is 551 MPa (80,000 psi). The change of track modulus caused by the change of granular material modulus is indicated by the vertical lines.

As Figure 4.5 shows, the sensitivity of track modulus to the modulus of the granular material increases with increasing thickness of the granular layer.

Figure 4.6 shows the effects of subgrade thickness and subgrade resilient modulus on track modulus. As already indicated in Figure 4.4, an increase in subgrade thickness leads to a decrease in track modulus. However, as Figure 4.6 shows, the reduction rate of track modulus decreases as the thickness of the subgrade layer increases. Also, the major contribution of a subgrade to track resiliency is from the layer down to approximately 5 m (16 ft) from the bottom of the subballast layer. However, the effect of subgrade modulus on track modulus is always significant, independent of the subgrade layer thickness.

The factor most affecting track modulus is the subgrade layer properties. The influence of subgrade condition on track modulus is further emphasized by the fact that the subgrade resilient modulus is the most variable quantity among all parameters, subject to change of soil type, environmental conditions, and stress state (see Section 4.2). Therefore, a change of track modulus in the field is primarily an indication of a change of subgrade condition.

The next most important factors affecting track modulus are the thickness of granular layer and fastener stiffness. An increase in either quantity can lead to a significant increase in track modulus. A higher track modulus is generally considered to provide better track performance. However, there is a limit above which track modulus is too high for satisfactory performance, when some track resilience is beneficial to accommodate dynamic vehicle–track interaction.

4.1.6 Track modulus as a measure of track support

Track modulus is often used as a measure of track quality. In general, a higher track modulus is considered to represent a better track foundation, and, therefore, is associated with better track performance. Hay (1982) and the American Railway Engineering and Maintenance-of-Way Association (AREMA) manual suggested that a minimum value of 14 MPa (2,000 psi) be required to ensure a satisfactory performance of railway track. Based on the field observations, Ahlf (1975) suggested that a track with a track modulus <14 MPa (2,000 psi) was poor, a track with a track modulus between 14 MPa (2,000 psi) and 28 MPa (4,000 psi) was average, and a track with a track modulus above 28 MPa (4,000 psi) was good. Raymond (1985) suggested that the optimum track modulus is 34–69 MPa (5,000–10,000 psi). None of these references suggested that the track modulus can be too high.

To understand the correlation between track modulus and track performance, GEOTRACK was used to relate track responses under loading to track modulus. For calculation of track responses, the wheel loads are chosen to represent two two-axle trucks (bogies) connected by a coupling as in a typical freight train. The axles were assumed to be 1.83 m apart. In GEOTRACK, only three axles are used with the response obtained under

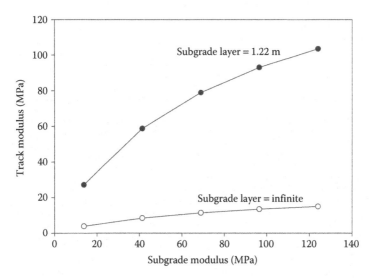

Figure 4.7 Correlations between track modulus and subgrade resilient modulus.

the center axle, because the fourth axle is far away enough to allow neglecting it. Track modulus is varied by changing the subgrade layer thickness and subgrade resilient modulus.

Figure 4.7 shows the range of track modulus obtained from the GEOTRACK analysis corresponding to the different subgrade conditions. Subgrade resilient modulus is varied from 14 to 124 MPa (2,000 to 18,000 psi) for subgrade soil thicknesses of 1.22 m underlain by bedrock and infinity. Again, the values of the parameters that are not varied are given in Table 4.2. The corresponding range of track modulus is from 8.3 to 103 MPa (1,200 to 15,000 psi). As this figure illustrates, the same track modulus may be a result of two different combinations of subgrade resilient modulus and subgrade thickness. This indicates that although subgrade resilient modulus affects track modulus significantly, it is not related to track modulus uniquely, but is also dependent on subgrade layer thickness.

GEOTRACK analysis was performed to relate track modulus to track response of concrete tie track with different subgrade conditions. The responses include rail deflection, tie deflection, subgrade surface deflection, rail-bending moment, and tie-bending moment. In addition, the maximum vertical stress at the subgrade surface and the maximum deviator stress at the subgrade surface were calculated. In general, these maximum stresses occur under the outer end of the tie for the concrete tie track. Assuming linear elastic behavior, all these responses are proportional to the magnitude of wheel load. Thus, all these responses are normalized by the magnitude of wheel load. Unless specified in the figures, the nominal values listed in Table 4.2 for the concrete tie track are used for analysis.

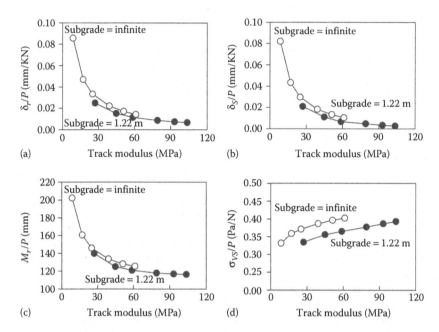

Figure 4.8 Influence of track modulus on track performance: (a) correlation with deflections of rail, (b) correlation with deflections of subgrade surface, (c) influence on rail-bending moment, and (d) influence on the maximum vertical stress.

Figure 4.8a and b show the correlations between track modulus and deflections of rail and subgrade surface. As expected, an increase in track modulus results in a decrease in all deflections. However, as these two figures show, when track modulus is approximately below 28 MPa (4,000 psi), the effect of track modulus on deflections is more dramatic. When track modulus is higher than 28 MPa, the variation of deflections with changing track modulus becomes more gradual. Another observation from these figures is that the correlation between track modulus and deflection is independent of subgrade condition or different combinations of resilient modulus and thickness, as long as track modulus is the same. This result is shown by the overlapping of two curves from two different subgrade layer thicknesses.

Figure 4.8c shows the influence of track modulus on rail-bending moment. An increase in track modulus leads to a reduction of bending moment. This means that an increase in track modulus results in lower bending stress in the rail. Again, track modulus of approximately 28 MPa can be considered as a dividing point below which the increase in the moments for both the rail and tie with decreasing track modulus is more rapid.

Figure 4.8d shows the influence of track modulus on the maximum vertical stress at the subgrade surface. An increase in track modulus results in an increase in stresses in the subgrade.

The trend in Figure 4.8d is potentially of concern. A higher track modulus leads to a higher stress in the subgrade, which would seemingly lead to a more detrimental situation in the subgrade. However, this is not actually true because track performance depends not only on stresses but also on soil strength. Track modulus is also an indication of subgrade soil strength. A higher track modulus represents a stiffer subgrade, and a stiffer subgrade generally has higher soil strength. An approximate correlation between subgrade resilient modulus and soil compressive strength is given next, which demonstrates that the strength of soil increases rapidly with increase in soil resilient modulus.

Soil condition	Resilient modulus (MPa)	Compressive strength (kPa)
Soft	7–28	34–103
Medium	28–69	103–207
Stiff	69–138	207–345

Therefore, although higher soil stresses will be generated for a track with higher track modulus, subgrade soil will have a higher strength to resist higher stresses. In general, the increase in strength is more rapid than the increase of stresses in the subgrade for a stiffer subgrade.

The selection of a minimum value of track modulus to prevent excessive stresses and deformation depends on many factors. However, to avoid a rapid change of track responses caused by the change of track modulus under any magnitude of axle loads, based on this modeling parametric study, track modulus above approximately 28 MPa (4,000 psi) may be considered favorable.

4.1.7 KENTRACK

KENTRACK is a multilayer elastic finite element computer program that was developed by Prof. Jerry Rose and his colleagues and students in the University of Kentucky (Rose et al. 2014). It is similar to GEOTRACK in that track substructure layers are modeled as Burmister's multilayer system, but it uses FEM to represent rails and ties.

Figure 4.9 shows a sketch of the track components included in KENTRACK. When calculating stress and strain of rails and ties, the finite element method is used. Rail and tie are classified as beam elements and a spring element is used to simulate the tie plate and fastener between rails and ties. For the track substructure, multiple layers are used: ballast, subballast or hot mix asphalt (HMA) layer, subgrade soil, and bedrock.

Similar to GEOTRACK, this computer program can be utilized for track response analysis and for the structural design of railway track foundation. Properties that are used to characterize track substructure layers include resilient modulus and Poisson's ratio.

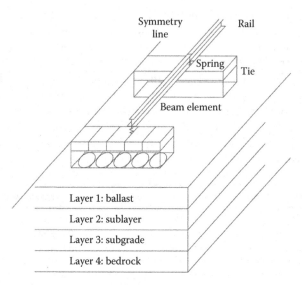

Figure 4.9 KENTRACK. (Courtesy of Professor Jerry Rose.)

This model is especially useful when the track structure includes a layer of asphalt. HMA is a temperature-dependent material, and KENTRACK allows the calculation of its modulus based on temperature and several properties specific to HMA, such as volume of voids and volume of bitumen for HMA, and HMA viscosity. Table 4.4 lists the major input and output parameters for KENTRACK.

KENTRACK also includes an element of track foundation design (granular layer and/or HMA layer thickness design) based on two criteria: limiting

Table 4.4 Input and output parameters in KENTRACK

Input	Output
• Rail—weight, Young's modulus, moment of inertia, and cross-sectional area	• Subgrade vertical compressive stress
• Fastener/pad—vertical stiffness	• Asphalt strain
• Ties—tie spacing, length, height, width, Young's modulus, and moment of inertia	• Service life of subgrade
• Substructure layers: modulus, Poisson's ratio, and layer thickness	• Service life of asphalt
• HMA material: Poisson's ratio, thickness, volume of voids, temperature and volume of bitumen for HMA, and HMA viscosity	
• Wheel loads: magnitudes and spacing	
• Traffic volume	

vertical compressive stress on the subgrade surface, and limiting tensile strain on the bottom of HMA layer. Use of KENTRACK for the track foundation design will be described in more detail in Chapter 5.

4.1.8 Finite element model

With commercial FEM software packages available, such as ANSYS, ABAQUS, and LS-DYNA, the use of FEM for simulation and analysis of track and subgrade has become a common approach. FEM is especially useful for simulation of dynamic responses of track and subgrade under dynamic wheel loads.

In general, a FEM would include a segment of track and may only include half of the track cross section assuming symmetry along its longitudinal centerline. Depending on the degree of accuracy and the computer running time, the size and shape of the FEM mesh (element) can vary, and can include various sized elements of triangular, rectangular, brick, or other shapes. In FEM, special attention needs to be given to the boundary conditions and linear or nonlinear interactions between components of track (such as from tie to ballast, subballast to subgrade).

With a commercial software package, a two-dimensional (2D) or 3D FEM can be developed for the track and track substructure. Figures 4.10 and 4.11 show two examples: a 2D FEM assuming plain strain in the longitudinal direction, and a 3D FEM (in general, a 2D FEM model is not as representative as a 3D FEM for railway track simulations because of discrete wheel load and tie conditions along the track). Both examples can be further simplified using half of the track, assuming symmetrical geometry of the track. In general, both sides can be fixed in the horizontal direction, and the base of the model can be fixed in the horizontal as well as vertical directions.

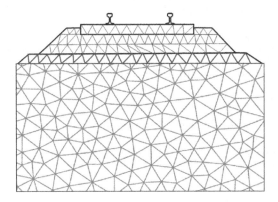

Figure 4.10 2D FEM for track and subgrade.

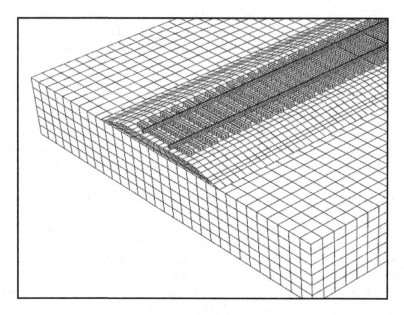

Figure 4.11 3D EFM for track and subgrade.

4.2 RESILIENT MODULUS

Resilient modulus of granular ballast and cohesive subgrade soil is the most important material property for the track substructure. Track performance depends to a great degree on the resilient behavior of ballast, subballast, and subgrade layers. Characterization of resilient modulus for ballast and subgrade materials is essential for the analysis and design of track and track substructure under repeated dynamic wheel loads. Knowledge of resilient modulus is necessary for calculating stress, strain, and deflection in the track foundation layers as well as for analyzing the system performance.

The granular and soil layers forming the foundation of railway tracks are subjected to repeated traffic loading. Under each individual cycle of loading, the layers behave essentially elastic as plastic deformation accumulates with repeated cycles. Resilient modulus is the property to characterize the elastic stiffness of granular and soil materials.

Resilient modulus is usually determined by repeated load triaxial tests with constant confining pressure, σ_3, and with deviator stress $(\sigma_1 - \sigma_3)$ cycled (see Figure 4.12). From Figure 4.12, resilient modulus is defined as follows:

$$M_r = \frac{\sigma_d}{\varepsilon_r} \qquad (4.25)$$

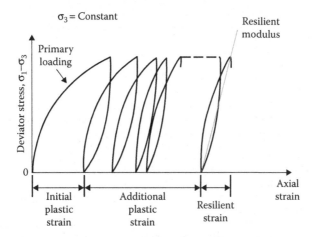

Figure 4.12 Resilient modulus definition from repeated loading test.

where:

M_r is the resilient modulus

σ_d is the repeated deviator stress ($\sigma_1 - \sigma_3$)

ε_r is the recoverable (i.e., resilient) strain in the direction of axial stress σ_1 (major principal stress), with the confining stress σ_3 (minor principal stress) being constant

In the past, studies of resilient modulus of various types of aggregates and soils were mainly motivated by the fact that the failure of highway pavement could result not only from the excessive accumulated permanent deformation of subgrade under repeated traffic loading, but also from the fatigue cracking of the asphalt concrete surface caused by repeated resilient deformation. The characterization of resilient deformation of subgrade soils under repeated stress application requires an understanding of their resilient moduli. Since 1986, when the American Association of State Highway and Transportation Officials (AASHTO) published a design guide (AASHTO 1986) for pavement structures that uses resilient moduli to characterize material properties, it has become essential for pavement design to understand and quantify the characteristics of resilient modulus of each layer.

4.2.1 Resilient modulus of granular materials

For granular materials such as ballast, the magnitude of resilient modulus is stress dependent. Past studies (Selig and Waters 1994) have found that the resilient modulus of unbound granular materials greatly increased as the confining pressure increased and was affected to a much smaller extent by the magnitude of repeated deviator stress. This has led to the following commonly used relationship between resilient modulus and stress state:

$$M_r = k_1 \left(\theta'\right)^{k_2} \tag{4.26}$$

where:

k_1 and k_2 are the soil constants determined from laboratory test results
θ' is the bulk effective stress in the loaded state

Bulk effective stress is equal to the sum of the three principal effective stresses as follows:

$$\theta' = \sigma'_1 + \sigma'_2 + \sigma'_3 \tag{4.27}$$

Equation (4.26) is essentially a power or log–log model between resilient modulus and bulk stress. In addition, a semilog model and arithmetic models have been found to be valid for various ballast materials (Selig and Waters 1993):

$$M_r = k_3 + k_4 \theta' \tag{4.28}$$

$$M_r = k_5 + k_6 \log \theta \tag{4.29}$$

Where k_3, k_4, k_5, and k_6 are soil constants determined from laboratory test results.

4.2.2 Resilient modulus of fine-grained subgrade soils

From many past tests, resilient modulus of fine-grained soils is not a constant stiffness property, but varies, dependent on many different factors. A change of resilient modulus from 14,000 to 140,000 kPa for a fine-grained subgrade soil can result for the same soil from a change of factors such as stress state or moisture content and can lead to a dramatic difference of subgrade response under traffic loads.

Factors that have significant influence on the magnitude of resilient modulus can be grouped into three categories: (1) loading condition or stress state, which includes the magnitude of deviator stress and confining stress, and the number of repetitive loadings and their sequence; (2) soil type and its structure, which primarily depends on compaction method and compaction effort for a new subgrade; and (3) soil physical state, which is defined by moisture content and dry density and is subject to the change of environment.

Under the first category, the most important factor affecting resilient modulus is the repeated deviator stress. Although resilient modulus increases with increasing confining stress, it was found (Tanimoto and Nishi 1970; Barksdale 1975; Fredlund et al. 1975; Townsend and Chisolm 1976) that confining pressure has much less of an effect than deviator stress for fine-grained subgrade soils, especially on clay soils. There is also some influence from the number of stress applications, but resilient modulus tends to become constant with increasing number of stress applications if deviator stress is below a certain level with regard to failure. Therefore, models are primarily established between resilient modulus and deviator stress for fine-grained subgrade soils.

Many models have been proposed to represent the characteristic trend of resilient modulus with deviator stress for fine-grained soils. The main models are the following:

1. Bilinear model:

$$M_r = K_1 + K_2\sigma_d \quad \text{when} \quad \sigma_d < \sigma_{di}$$

$$M_r = K_3 + K_4\sigma_d \quad \text{when} \quad \sigma_d > \sigma_{di}$$

(4.30)

where:
σ_{di} is the deviator stress at which the slope of M_r versus σ_d changes
K_1, K_2, K_3, and K_4 are the model parameters dependent upon soil type and its physical state (K_2 and K_4 are usually negative)

This model was proposed by Thompson and Robnett (1976).

2. Power models:

$$M_r = k\sigma_d^n$$

(4.31)

where:
k and n are the parameters dependent on soil type and its physical state (n is usually negative)

Moossazadeh and Witczak (1981) used this model and obtained good agreement with test results on three fine-grained soils, with the determination of $k = 0$–200 and $n = -1.0$ to 0 for resilient modulus (ksi) (1 ksi = 6,895 kPa) and deviator stress (psi) (1 psi = 6.895 kPa). Pezo et al. (1991) also used this model and obtained a range of $k = 6$, 000–$55,000$ and $n = -0.34$ to -0.04 for an Austin soil (A-7-6), with resilient modulus and deviator stress in units of psi. Furthermore, Brown et al. (1975) and Brown (1979) proposed a similar model, but with the consideration of effective confining stress σ_3' for saturated overconsolidated soils as follows:

$$M_r = k\left(\frac{\sigma_d}{\sigma_3'}\right)^n$$

(4.32)

3. Semilog model:

$$M_r = 10^{(k-n\sigma_d)} \tag{4.33}$$

or

$$\log(M_r) = k - n\sigma_d \tag{4.34}$$

Fredlund et al. (1977) proposed this model for a moraine glacial till and obtained the range of parameters $k = 3.6–4.3$ and $n = 0.005–0.09$ for resilient modulus and deviator stress in units of kPa. Raymond et al. (1979) proposed a similar model with replacement of $\sigma_d/\sigma_{d(\text{failure})}$ for σ_d and also got consistent results with tests on Leda clay.

4. Hyperbolic model:

$$M_r = \frac{k + n\sigma_d}{\sigma_d} \tag{4.35}$$

This model was proposed by Drumm et al. (1991) for fine-grained soils. For the Tennessee soils tested, the parameters ranged from $k = 2–70$ and from $n = 2–12$, with resilient modulus in ksi and deviator stress in psi.

5. Octahedral model:

$$M_r = k \frac{\sigma_{\text{oct}}^n}{\tau_{\text{oct}}^m} \tag{4.36}$$

where:

σ_{oct} and τ_{oct} are the octahedral normal and shear stresses

This model was derived by Shackel (1973) and is more difficult to apply.

As indicated in the literature, each model was able to fit the relationship between resilient modulus and stress state for the soils tested by the respective researchers. Comparisons have been made between these models and test data, as shown in Figure 4.13. The data used in Figure 4.13 are typical of resilient modulus values for fine-grained subgrade soils from the test results by Seed et al. (1962) and Thompson and Robnett (1979), respectively. All four models are fit to the actual data points shown in the figure. The results are summarized in Table 4.5. As shown, all models are adequate to represent the test results by the fitting of model parameters, although there exists a ranking of models in terms of their coefficients of correlation, as listed in Table 4.5. The best representation is the bilinear model, followed by the power model, then the semilog model, and finally the hyperbolic model.

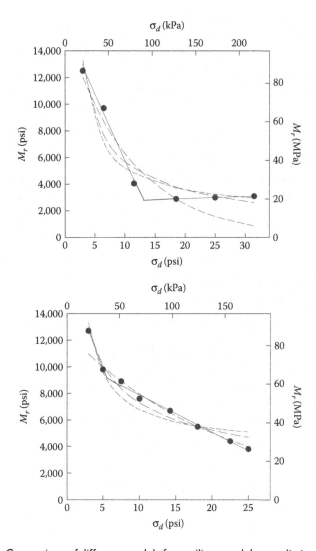

Figure 4.13 Comparison of different models for resilient modulus prediction.

The bilinear model gave the best fit but requires five parameters. The power model was second best and only requires two parameters. The remaining models showed no apparent advantages over the other two.

Also, it is apparent that the correct prediction of resilient modulus using these models is strongly dependent on the values of the model parameters. As will be discussed in the following section, soil physical state has an influence on resilient modulus that is similar to or even greater than that of stress state. Therefore, values of the parameters can be significantly dependent on soil moisture content and dry density in addition to soil type.

Table 4.5 Comparisons of different models for resilient modulus prediction

Model	Coefficient of correlation
Data (both M_r and σ_d are in kPa)	
$M_r = 109{,}000 - 1{,}000\sigma_d$ and $M_r = 18{,}100 + 15.4\sigma_d$	$r^2 = 1.0$
$M_r = 705{,}000\, s_d^{-0.68}$	$r^2 = 0.92$
$M_r = 10^{(5.04 - 0.0058\sigma_d)}$	$r^2 = 0.89$
$M_r = (1{,}630{,}000 + 12{,}900\sigma_d)/\sigma_d$	$r^2 = 0.89$
Data (both M_r and σ_d are in kPa)	
$M_r = 118{,}000 - 1{,}450\sigma_d$ and $M_r = 73{,}800 - 281\sigma_d$	$r^2 = 1.0$
$M_r = 383{,}000\, s_d^{-0.48}$	$r^2 = 0.97$
$M_r = 10^{(4.94 - 0.0029\sigma_d)}$	$r^2 = 0.95$
$M_r = (1{,}330{,}000 + 27{,}400\sigma_d)/\sigma_d$	$r^2 = 0.91$

4.2.3 Influence of soil physical state on resilient modulus

In this discussion, soil "physical state" is defined by two variables: moisture content and dry density. In turn, these two variables can be quantified using the soil-compaction curve. As such, the influence of soil physical state on resilient modulus is discussed by means of compaction curve.

Many test results showing the influence of moisture content and dry density on resilient modulus of fine-grained subgrade soils are available in the literature. A systematic approach to considering their influences on resilient modulus follows. Moisture content and dry density can change in many different ways, but can be represented using a combination of two basic paths as shown in Figure 4.14: the path of moisture content variation with constant dry density, and the path of moisture content variation with constant compactive effort. The difference between these two paths is the variation of dry density that may also lead to a significant change of resilient modulus. Consideration of the effect of moisture content variation on resilient modulus must be accompanied by information on the variation of dry density. Influence of moisture variation with and without the variation of dry density, as will be discussed later, may be significantly different.

To reduce the effect of other factors on the relationships between resilient modulus and the change of moisture content in terms of these two paths, or in other words to make these relationships relatively unique, the change of resilient modulus for any individual soil from the literature at any physical state is normalized by a reference resilient modulus obtained at optimum moisture content and maximum dry density. It is felt that the choice of resilient modulus at optimum moisture constant and maximum dry density as a reference value is logical because of its importance in geotechnical engineering. Furthermore, all other parameters, including stress conditions and

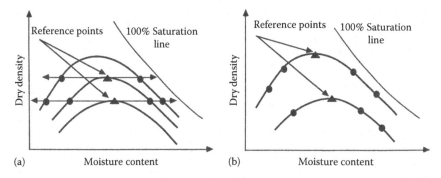

Figure 4.14 Paths of moisture content variation: (a) constant dry density; (b) constant compactive effort.

soil type, are kept constant for resilient moduli along each path when either dry density or compactive effort is constant. By this way, the change of resilient modulus primarily caused by soil physical state along each path is considered.

Following this approach, available repeated load triaxial test results from the literature on various fine-grained subgrade soils were used to develop the equations that describe the relations between resilient modulus and the change of soil physical state. Figure 4.15 shows the relationship between resilient modulus and moisture content with values of dry density the same as for the reference resilient modulus, as illustrated previously in Figure 4.14a. The correlation includes 27 repeated load triaxial test results

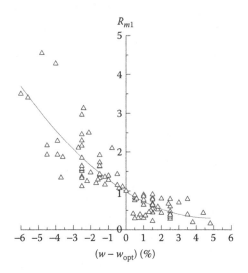

Figure 4.15 Relationship between M_r and w with constant dry density.

on 11 fine-grained soils from the literature (Seed et al. 1962; Sauer and Monismith 1968; Culley 1971; Robnett and Thompson 1976; Fredlund et al. 1977; Edil and Motan 1978; Kirwan et al. 1979; Elfino and Davidson 1989). The best-fit polynomial equation for these data is as follow:

$$R_{m1} = 0.98 - 0.28(w - w_{\text{opt}}) + 0.029(w - w_{\text{opt}})^2 \tag{4.37}$$

where:

R_{m1} is the $M_r/M_{r(\text{opt})}$ for the case of constant dry density

M_r is the resilient modulus at moisture content w (%) and the same dry density as $M_{r(\text{opt})}$

$M_{r(\text{opt})}$ is the resilient modulus at maximum dry density and optimum moisture content w_{opt} (%) for any compactive effort

The correlation coefficient, r^2, is equal to 0.76.

Figure 4.16 shows that the relationship between resilient modulus and moisture content with compactive efforts is the same as that for the reference resilient modulus, as illustrated previously in Figure 4.15b. The correlation included 26 repeated load triaxial test results of 10 fine-grained subgrade soils from the literature (Seed et al. 1962; Kallas and Riley 1967; Shifley and Monismith 1968; Tanimoto and Nishi 1970; Edris and Lytton 1976; Fredlund et al. 1977; Kirwan et al. 1979; Pezo et al. 1991). The best-fit polynomial equation for these data is as follows:

$$R_{m2} = 0.96 - 0.18(w - w_{\text{opt}}) + 0.0067(w - w_{\text{opt}})^2 \tag{4.38}$$

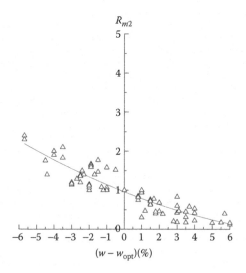

Figure 4.16 Relationship between M_r and w with constant compactive effort.

where:

$R_{m2} = M_r/M_{r(opt)}$ for the case of constant compaction effort

M_r is the resilient modulus at moisture content w (%) and the same compactive effort as $M_{r(opt)}$

The correlation coefficient, r^2, is equal to 0.83.

Correlations were also performed between $M_r/M_{r(opt)}$ and the ratio w/w_{opt} for both cases. The trends shown by these correlations are similar to the trends as indicated by the aforementioned correlations (Li and Selig 1991). Because no better coefficients of correlation than those of Equations 4.37 and 4.38 were obtained, those correlations are not shown in this book.

As illustrated in Figures 4.15 and 4.16, both cases indicate that an increase of moisture content leads to a significant decrease of soil stiffness. If a value of 69,000 kPa for resilient modulus at optimum moisture content is assumed, an increase of moisture content by 8% can lead to a decrease of resilient modulus from a value of 280,000 kPa to a value of 28,000 kPa.

A comparison between these two cases shows that the case with constant dry density causes a greater change of resilient modulus with change in moisture content below optimum than the case with constant compactive effort. The difference between the two cases becomes smaller when moisture content is above the optimum. This trend can be explained by the effect of dry density on resilient modulus.

Figure 4.17 illustrates the trend of the influence of dry density on resilient modulus based on tests by Seed et al. (1962). Whether resilient modulus increases, decreases, or does both with increase in dry density depends on the moisture content. In general, at lower moisture content, resilient modulus tends to increase with increasing dry density, whereas, at higher moisture content, resilient modulus tends to decrease with increasing dry density.

The trends in Figure 4.17 can explain the differences between the two cases shown in Figures 4.15 and 4.16. If moisture content is low, that

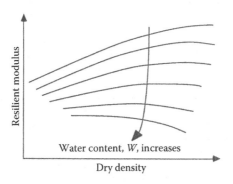

Figure 4.17 Influence of dry density on resilient modulus.

is, below optimum, a decrease of moisture content is accompanied by a decrease of dry density along the path of constant compactive effort. According to the trend shown in Figure 4.17 for lower moisture content, a decrease of dry density generally leads to a decrease of resilient modulus. Therefore, a decrease of moisture content accompanied by a decrease of dry density along the path of constant compactive effort should result in a smaller increase of resilient modulus due to the opposite effects of dry density and moisture content on resilient modulus than a decrease of moisture content along the path of constant dry density. Similarly, if moisture content is high, that is, above optimum, an increase of moisture content along the path of constant compactive effort is accompanied by a decrease of dry density. According to the trend shown in Figure 4.17 for higher moisture content, a decrease of dry density generally leads to an increase of resilient modulus. Therefore, an increase of moisture content above optimum along the path of constant compactive effort should lead to a smaller decrease of resilient modulus due to the offsetting influence of dry density on resilient modulus relative to an increase of moisture content along the path of constant dry density.

The two relatively unique correlations of Equations 4.37 and 4.38 between resilient modulus and soil physical state for the two cases establish the basis to predict the change of resilient modulus with a change in soil physical state in the Li–Selig method described next.

4.2.4 Li–Selig method for prediction of resilient modulus

As mentioned before, resilient modulus is primarily dependent on three factors in most situations: (1) stress state, (2) soil type and its structure, and (3) soil physical state. As a result, prediction of resilient modulus should take into account these three factors. To do so, the following two basic assumptions are made.

First, the relationships between resilient modulus and soil physical state from Equations 4.37 and 4.38 are of the form:

$$R_{m1} = f_1\left(w - w_{opt}\right) \tag{4.39}$$

$$R_{m2} = f_2\left(w - w_{opt}\right) \tag{4.40}$$

One is for water content variation with constant dry density, and the other is for water content variation with constant compactive effort.

Second, the bilinear model or the power model is valid at optimum moisture content and maximum dry density for any compactive effort. The model parameters will only depend on soil type and its initial structure. For the same soil type and initial structure, they are constants, not dependent on soil physical state. Because the power model, Equation 4.31, involves fewer parameters than the bilinear model, Equation 4.30, the power model is selected in the Li–Selig method, although the bilinear model will be mentioned when the concept of the breakpoint is used.

To predict resilient modulus, the following are the steps in the Li–Selig method:

1. Obtain resilient modulus at maximum dry density and optimum moisture content, $M_{r(opt)}$, either directly from a test, or using Equations 4.30 or 4.31 if model parameters are available from previous tests on a similar soil.
2. If a soil for which resilient modulus is needed has another optimum moisture content corresponding to another compactive effort, or exists in a physical state as defined by moisture content and dry density other than that for step 1, then Equation 4.37 or 4.38 or their combinations need to be used, as follows:

    a. If the physical state of the soil for which resilient modulus is needed differs from the known state in a manner shown in Figure 4.14a or b, then get resilient modulus from either

$$M_r = R_{m1} M_{r(opt)} \tag{4.41}$$

or

$$M_r = R_{m2} M_{r(opt)} \tag{4.42}$$

    b. If the physical state of the soil for which resilient modulus is needed is neither on the same compaction curve nor at the same dry density as that for which resilient modulus is known from step 1, then calculation of resilient modulus requires several steps.

    Let us assume that M_r at point Q in Figure 4.18 is needed and that M_r at point O is known, that is, conversion from one optimum to another optimum. Path OQ is the sum of paths OA and AQ. First, get M_r at point A by

$$M_{rA} = f_1 \left(w_1 - w_{opt} \right) M_{r(opt1)} \tag{4.43}$$

Then, get M_r at point Q by

$$M_{rQ} = \frac{M_{rA}}{f_2 \left(w_1 - w_{opt2} \right)} \tag{4.44}$$

    c. Repeat example (b) but estimate M_r at point P from M_r at point O. Path OP is the sum of paths OA, AQ, and QP. First, get M_r at point Q as in example (b). Then, get M_r at point P by

$$M_{rP} = f_2 \left(w_2 - w_{opt2} \right) M_{rQ} \tag{4.45}$$

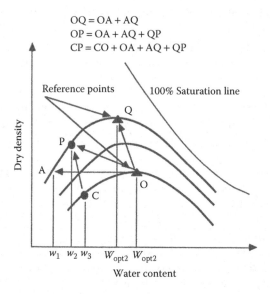

$$OQ = OA + AQ$$
$$OP = OA + AQ + QP$$
$$CP = CO + OA + AQ + QP$$

Figure 4.18 Paths for determining resilient modulus at any soli physical state.

d. Finally, consider the most general case in which M_r at point C is known and M_r at point P is desired. Path CP is the sum of paths CO, OA, AQ, and QP. First, get M_r at point 0 from M_r at point C by

$$M_{r(\text{opt})} = \frac{M_{rc}}{f_2(w_3 - w_{\text{opt}})} \qquad (4.46)$$

Then, find M_r at point P using the steps in example (c).

It can be seen that the Li–Selig method can give an estimation of resilient modulus at any stress state and any soil physical state by a straight forward procedure.

To predict resilient modulus for fine-grained soil at any soil physical state under any stress condition by the Li–Selig method, a reference value of resilient modulus is required either directly from tests or from Equations 4.30 to 4.31. When an equation is used, the model parameters are needed to carry out the calculation, where these parameters are defined at optimum moisture content and are dependent on soil type and its structure but not on soil physical state.

It is logical to first find resilient modulus at optimum moisture content and maximum dry density as a reference value because of the common use of this reference state in practice. Therefore, values of these model parameters of Equation 4.31 for various soil types were compiled from the literature (Li and Selig 1991). The available data were mostly for compacted soil specimens, and, thus, the values of the model parameters compiled are primarily for these compacted soils. If a resilient modulus value from a test was not at optimum moisture, then its value at optimum moisture was calculated using the approach discussed earlier, as long as moisture content and compaction curve for the tested soil were available in the literature. In general, for a total of 48 fine-grained soil types compiled from the literature, k value ranges from 50,000 to 6,000,000 while n value ranges from -0.1 to -0.9, with the unit of kPa for resilient modulus and deviator stress.

An equation developed by Thompson and LaGrow (1988) significantly contributes to the Li–Selig method, although it cannot replace the compilation of model constants, because it is only applicable to a specific stress state: the breakpoint of the bilinear model. The following is the relationship found by Thompson and LaGrow between the breakpoint resilient modulus at optimum moisture content and 95% of maximum dry density:

$$M_{r(\text{opt})} = 30,800 + 677\left(\% \text{ clay}\right) + 821\left(PI\right) \tag{4.47}$$

where:

$M_{r(\text{opt})}$ is the resilient modulus (kPa) at optimum water content and 95% maximum dry density

% clay is the % particles finer than 2 μm

PI is the plasticity index

By using Equation 4.47 together with Equations 4.37 and 4.38, resilient modulus at any physical state for any soil type can be predicted for a specific stress state, that is, the breakpoint. In other words, resilient modulus at this specific stress state for any soil type under any physical state can be determined by the following steps: (1) perform index tests and compaction tests to obtain the percentage of clay, the plasticity index, and the compaction curve; (2) determine resilient modulus at optimum moisture content using Equation 4.47; (3) compare the soil moisture content and dry density at which resilient modulus is desired with known w_{opt} and $\gamma_{d(\text{max})}$; and (4) use Equation 4.37 or 4.38 or their combinations in terms of the paths of variation of soil physical state to obtain desired resilient modulus.

4.2.5 Examples of prediction using Li–Selig method

Figure 4.19 shows a compaction curve for a clay soil ("Vicksburg clay" or "Buckshot clay"). Suppose that for the three different physical states designated as A, B, and C, resilient modulus values are to be desired. To use the Li–Selig method to predict resilient modulus for these three different physical states, the prediction paths are illustrated in this figure by arrows. The point O is the reference point, corresponding to the resilient

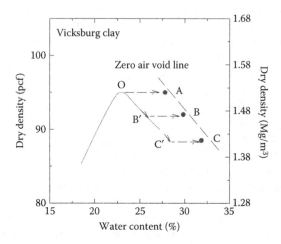

Figure 4.19 Compaction curve and prediction paths for resilient modulus.

modulus at optimum moisture content and maximum dry density, which is determined by Equation 4.31, with model parameters $k = 621,000$ and $n = -0.42$ for resilient modulus and deviator stress in units of kPa (Li and Selig 1991). Thus, $M_{r(opt)} = 621,000\sigma_d^{-0.42}$.

For the prediction of resilient modulus at point A from the reference resilient modulus at O, the path OA is used. Equation 4.37 is then used to obtain M_r at the state A. For the prediction of M_r at the state B or C, the path OB or OC consists of two parts, OB′ or OC′; and B′B or C′C. Both Equations 4.37 and 4.38 are then used to obtain M_r at the state B or C.

Figure 4.20a shows the comparisons of predicted resilient modulus from the Li–Selig method with resilient modulus test results measured by Thompson (1990) for the Vicksburg clay using repeated load triaxial tests for three different soil physical states under different stress states. As can be seen, the agreement between predicted and measured resilient modulus is reasonable.

Figure 4.20b shows further comparisons of predicted resilient modulus and tested resilient modulus on the same Vicksburg clay by Townsend and Chisolm (1976). Repeated load triaxial tests were used for the soil specimens compacted with two different moisture contents (24% and 27%, respectively), but on the same compaction curve. The data points shown in the figure are the results from the tests for these two different soil states. For the prediction of resilient modulus, the reference values of resilient modulus at optimum moisture content (22%) and maximum dry density were determined through Equation 4.31, with model parameters of $k = 5,980,000$ and $n = -0.76$ for resilient modulus and deviator stress in units of kPa (Li and Selig 1991). Again, these model parameters at the

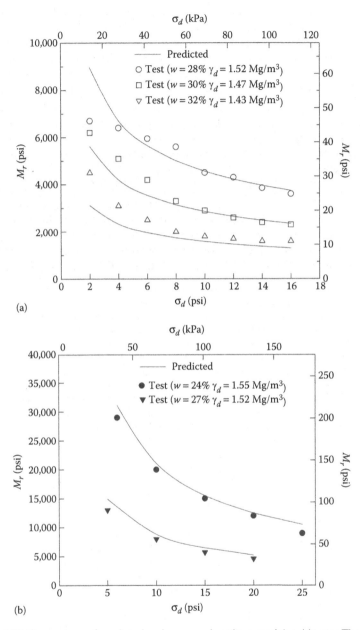

Figure 4.20 Comparison of predicted and measured resilient modulus: (a) using Thompson model (Data from Thompson, M. R. (December 1990). *Results of Resilient Modulus of Vicksburg Clay for FAST Test center.* Internal Report for American Association of Railroads, University of Illinois, Urbana, IL.); (b) using Townsend et al. model (Data from Townsend, F. C. and Chisolm, Ed. E. November 1976. *Plastic and Resilient Properties of Heavy Clay Under Repetitive Loadings.* Technical Report S 76 16, Soils and Pavement Laboratory, U.S. Army Engineer Waterways Experiment Station.).

reference point were compiled from tests by Townsend and Chisholm on buckshot clay. Thus, $M_{r(opt)} = 5,980,000\sigma_d^{-0.76}$. Then, the prediction path as illustrated in Figure 4.14b was used with Equation 4.38 for the prediction of resilient modulus at these two states. The predicted results are shown in the figure. The predicted and the tested values agree well with each other. Moreover, a comparison between Figure 4.20a and b indicates that test results by two different agencies showed similar trends of resilient modulus change caused by both soil physical state and stress state for this same type of subgrade soil.

Figure 4.21 shows the repeated load triaxial test results and compaction test results on a subgrade soil by Seed et al. (1962). The upper part of this figure shows the variation of resilient strain with different water content under different compactive efforts by kneading compaction at a deviator stress of 69 kPa. From the upper figure, the test result of resilient modulus can be determined in terms of $M_r = 10/\varepsilon_r$ at any point on this upper part. The lower part of this figure shows the corresponding soil physical state, that is, water content and dry density, for different compactive efforts.

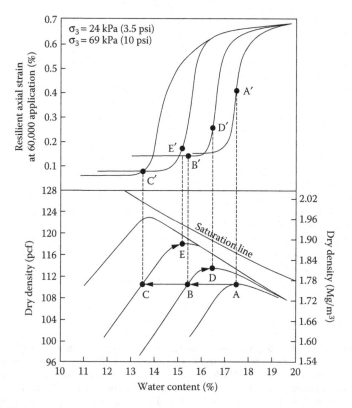

Figure 4.21 Relationship between dry density, moisture content, and resilient strain with superimposed prediction paths.

Table 4.6 Comparison of predicted and test resilient modulus
results

Soil physical state	Test results (kPa)	Predicted results (kPa)
A (at optimum water content)	21,000	21,000
B	49,000	38,000
C	83,000	57,000
D	28,000	33,000
E	48,000	45,000

In addition to the test results shown in this figure, the prediction paths of resilient modulus are also shown in the figure. Point A is taken as a reference point for the prediction of resilient modulus of other points such as B, C, D, and E. As indicated by these paths, only Equation 4.37 for constant dry density path is required to predict the resilient modulus at points B or C, whereas both Equations 4.37 and 4.38, the second of which is for constant compactive effort path, are required for points D or E, that is, the prediction path for D or E consists of AB and BD, or AC and CE.

The test results of resilient modulus based on the resilient strains at these points A', B', C', D', and E' are listed in Table 4.6. The prediction results of resilient modulus at these points B, C, D, and E based on the reference value of resilient modulus are also listed in this table. The resilient modulus at the reference point A is determined directly from the resilient strain at point A' through Equation 4.25. Again, as can be seen, the predicted results for different physical states agree well with the measured results.

4.3 CUMULATIVE PLASTIC DEFORMATION

Cumulative plastic deformation is the total permanent deformation accumulated under repeated loading (see Figure 4.12). One of the major criteria of track foundation design is to prevent excessive plastic deformation due to repeated wheel loads (Chapter 5). Much track maintenance effort and expenditure is directed at providing a smooth track surface, as track settlement arises from the plastic deformation of the ballast, subballast, and subgrade: the geotechnical components of track. In most cases, settlement is not uniform, and the associated differential or nonuniform settlement leads to track roughness. Therefore, understanding of how plastic deformation accumulates under traffic and how it is related to the growth of track roughness is essential for the proper design of track foundations as well as for developing effective procedures for track surfacing operations.

4.3.1 Cumulative plastic deformation of ballast

An understanding of the mechanics of ballast settlement, compaction, and breakdown under loading provides the basis for estimating the improved track geometry resulting from using a better ballast material with longer life, or of an improved and more durable ballast maintenance method. The following discusses the basis for such a mechanistic model, which predicts the timing for track geometry correcting maintenance and/or ballast renewal, given initial track conditions. The model calculates the cumulative strain in the ballast layers due to traffic loading and converts this to settlement and track roughness.

Ballast permanent strain under repeated loading often accounts for the largest portion of track settlement of all the geotechnical layers. Although a soft and deforming subgrade soil can be a source of a large amount of track settlement over time, ballast settlement typically is larger than subgrade settlement: particularly when the time between required track maintenance cycles (tamping) is short. Therefore, ballast settlement rate usually controls the need for repeated geometry correcting maintenance. The ability of ballast to maintain stable track geometry depends primarily on: (1) its material quality, (2) physical state, and (3) load magnitudes.

As described in Chapter 3, ballast life can be defined in terms of the amount of traffic that produces ballast mechanical wear and generation of fine material in sufficient quantity to fill the ballast voids. This ballast breakdown rate, as calculated by the model, is dependent primarily on the ballast material quality, gradation, layer thickness, and the imposed loading.

The fundamental equation used for vertical ballast plastic strain is

$$\varepsilon_N = \varepsilon_1 N^b \tag{4.48}$$

where:

ε_N is the plastic strain after N load cycles
ε_1 is the first cycle plastic strain
b is a constant exponent

This "power equation" was chosen because it provided the best fit to ballast strain data from field and laboratory tests (Chrismer 1993). Because typical ballast life is represented by N values in the order of 10 million cycles, it was necessary to have a predictive equation that followed the observed trend of actual ballast strain over this range. With proper selection of coefficients, the power equation provides a best fit to the data for this wide range of N.

Both laboratory and field measurements are available to select representative values of ε_1 and b under the influence of ballast materials and loading conditions. Laboratory box test results are available to model the effects of less than ideal conditions, such as fouled ballast, the presence of water, and impact wheel loads.

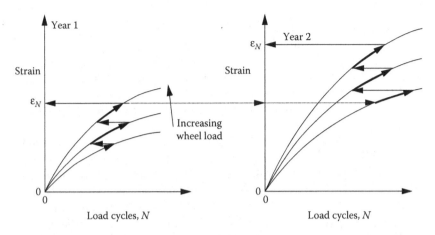

Figure 4.22 Ballast strain accumulation with load cycles for increasing wheel loads.

Because the load–settlement relationship is nonlinear, Figure 4.22 shows the summation of cumulative plastic strains from different load levels. The strain calculation sums the individual strains resulting from the number of wheel loads for each load increment by the process shown graphically in Figure 4.23 for a few wheel load magnitudes.

A separate strain equation can be written for each load increment as in the indicial form of Equation 4.48:

$$\varepsilon_{Ni} = \varepsilon_{1i} N_i^{bi} \tag{4.49}$$

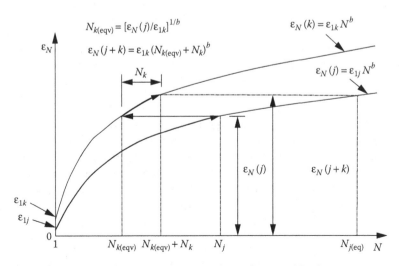

Figure 4.23 Method used for ballast strain calculation for mixed loading.

where:

ε_{Ni} is the strain resulting from N load cycles of magnitude i

ε_{1i} is the corresponding first cycle strain

The exponential term b_i was found to be reasonably represented as a constant and, therefore, the subscript i is not needed

The strain equation, starting with the smallest load, is summed as follows:

$$\varepsilon_N = \sum \varepsilon_{Ni} = \sum_{i=1}^{10} \varepsilon_{1i}\left(N_i + N_{eqv}\right)^b \tag{4.50}$$

Where N_{eqv} pertains to the equivalent number of wheel loads of magnitude i, yielding a strain equivalent to all of the applications of lesser loads. This can be visualized as shown in Figure 4.23, where the strain equations for two load levels, j and k (load k > load j), are shown. After N_j cycles of load level j, an amount of plastic strain $\varepsilon_{N(j)} = \varepsilon_{1j}N^b$ will develop. Additionally, a strain is developed from N_k cycles of wheel load k. However, this strain cannot be directly obtained from the corresponding equation $\varepsilon_{N(k)} = \varepsilon_{1k}N^b$ due to the nonlinear relationship of strain with N for the different wheel loads. To obtain the resulting plastic strain from the combination of $N(j)$ and $N(k)$, the equivalent number of cycles (N_{eqv}) at load level k that would have been necessary to cause strain equal to $\varepsilon_{N(j)}$ must first be found by equating $\varepsilon_{N(j)}$ to $\varepsilon_{N(k)}$. The result is

$$\varepsilon_{N(j)} = \varepsilon_{N(k)}$$

$$\varepsilon_{N(j)} = \varepsilon_{1k}\left[N_{k(eqv)}\right]^b$$

$$N_{k(eqv)} = \left|\frac{\varepsilon_{N(j)}}{\varepsilon_{1k}}\right| 1/b \tag{4.51}$$

The total cumulative strain after N_j cycles of load level j plus additional N_k applications of load level k is then given by

$$\varepsilon_N = \varepsilon_N(j + k) = \varepsilon_{1k}\left(N_k + N_{k(eqv)}\right)^b \tag{4.52}$$

This is the same as Equation 4.50 with two load levels.

However, the ballast strain is not assumed to continue along the line described by the power curves in Figure 4.23 for the entire ballast life. As fouling progresses the amount of fines inhibiting ballast drainage and begin to control the mechanical stress–strain behavior. Therefore, the rate of strain is expected to increase toward the end of ballast life, as illustrated in Figure 4.24a. Tamping would be required with increasing frequency (as shown in Figure 4.24b) to maintain acceptable geometry as the end of ballast life approaches. The manner in which the ballast-fouling condition affects the strain rate is discussed later in this chapter.

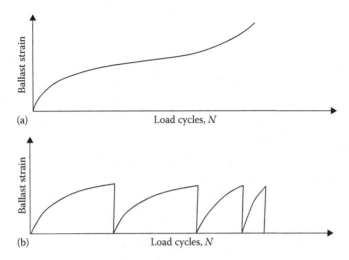

Figure 4.24 Strain of ballast with load cycles: (a) without tamping; (b) with tamping.

The ballast strain rate increases as ballast degrades with mechanical wear due to the influence of the generated fine material on the mechanical properties of the ballast. In dry conditions, the experimental results have indicated no clear trend of either increasing or decreasing strain accumulation with increasing amount of fines. But, under saturated conditions, there is no ambiguity; the strain rate is markedly increased with added fines. To account for this effect in the model, the ballast strain rate equation is increased according to the amount of fouling fine material present and the rainfall rate. This correction to the strain rate due to fouling and moisture is based on some relatively simple tests (Chiang 1988), and is somewhat limited and speculative as a result. Since the model was developed, Ebrahimi (2011) has performed much more advanced research into the effect of fouling material characteristics and moisture on the deformational behavior of ballast, and his results should be incorporated into ballast strain models.

Ballast material quality affects its deformation rate under loading. The ballast strain data shown in Figure 4.25 was obtained in a triaxial cell by Raymond (1977), and clearly shows how the material quality, as expressed by Abrasion Number (An), is a reasonably good predictor of first cycle strains and long-term ballast strain under loading. Raymond used a relatively small deviator stress and a large compaction effort applied to the ballast in the triaxial cell, which resulted in small plastic strains compared with that of ballast in track. Although Raymond's strain data may not be of the same magnitude as that of ballast in track, the data was used in the mechanistic model (Chrismer 1993) to indicate the trend of strain with material quality.

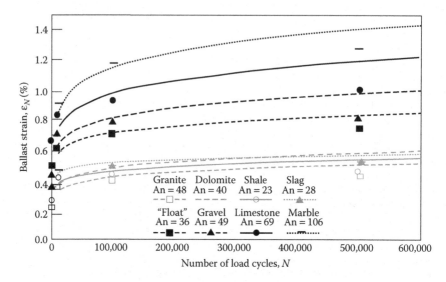

Figure 4.25 Ballast strain in triaxial cell for ballast of varying material quality.

When a full depth of ballast is placed during construction or when the ballast layer is being completely renewed, the entire ballast layer is assumed to accumulate strains under loading according to Equation 4.48. However, the model makes a distinction between the strain rate of the upper ballast layer, which is subjected to tamping, and the layer underneath that was not disturbed by tamping. Basically, tamping was assumed to create two ballast layers that strain at different rates: an upper layer of approximately 6 in. deep under the tie (the typical penetration depth of tamping tines), and the bottom ballast layer that continues to strain at the rate from when it was first placed. The settlement of the upper tamped layer was assumed to start from the loose, disturbed condition, and so subsequent traffic loading produces most of the ballast strain from within this upper layer.

In addition to these considerations of wheel loading magnitudes, ballast material quality, and fouling, the ballast model also considers the effect on ballast mechanics and deformation resulting from tamping degradation, the internal ballast stresses resulting from the type of track construction, and the track modulus.

4.3.2 Cumulative plastic deformation of subballast

The rate of subballast layer strain is assumed to grow at a rate measured during an Association of American Railroads (AAR) investigation (Selig 1980) for newly constructed track. The growth rate of newly placed subballast was found to accumulate according to a "power" form of equation just as with ballast.

However, if the track was constructed many years in the past, it may be assumed that the subballast strain rate is basically linear. For example, the average rate of subballast strain for several subballast types from different test sections was found to be as simplified as

$$\varepsilon_N(\%) = 1.7 \times 10^{-7} N \tag{4.53}$$

4.3.3 Cumulative plastic deformation for fine-grained soils

Study of cumulative plastic deformation and failure mechanisms for fine-grained soils under repeated loading is essential to design and maintenance of railway track. Excessive subgrade plastic deformation can be responsible for a major portion of maintenance and rehabilitation costs.

Three mechanisms are mainly responsible for the cumulative plastic deformation of fine-grained soils under repeated loading. They are cumulative plastic shear deformation, cumulative consolidation, and cumulative compaction deformation.

4.3.3.1 Soil critical state

Prior to the discussion of predicting cumulative plastic deformation, it is useful to understand that there exists a critical level of repeated deviator stress for fine-grained soils, above which the soil plastic deformation increases rapidly with repeated loading (first assumed by Larew and Leonards in 1962, and then confirmed by others). This critical level of deviator stress has often been called the dynamic strength of soil and usually is smaller than the soil static strength determined under the condition of monotonic loading.

Although it is difficult to determine the drainage condition and to quantify the change of pore water pressure for a soil under repeated loading, some studies were conducted using effective stress analysis. Wilson and Greenwood (1974) as well as O'Reilly et al. (1991) found that there existed a correlation between cumulative plastic strain and cumulative residual excess pore water pressure. The residual excess pore water pressure increased with repeated loading as the cumulative plastic strain. Experimental results (Andersen et al. 1976; Mitchell and King 1977; Sangrey et al. 1978; Matsui et al. 1980; Takahashi et al. 1980) showed that under repeated loading, not only did normally consolidated soils exhibit the behavior of cumulative pore water pressure, but overconsolidated soils tended to accumulate pore water pressure as well. For overconsolidated soils, the pore water pressure tended to be negative at first due to its dilative behavior. However, after a certain number of loading cycles, the pore water pressure changed direction. The pore pressure thus became positive and continued to accumulate due to the gradual soil plastic compressive deformation under repeated loading.

The accumulation of excess pore water pressure, however, decreased with increasing overconsolidation ratio. In other words, an overconsolidated soil is more resistant to failure in undrained repeated loading.

However, repeated loading does not always impose a destructive effect on soil strength when drainage is introduced. Depending on the stress history of soil, repeated loading may improve soil strength (Andersen et al. 1976; Sangrey et al. 1978; Sangrey and France 1980; O'Reilly et al. 1991). For normally consolidated clay, a dissipation of excess pore water pressure caused by repeated loading will lead to reconsolidation. This reconsolidation will have two effects. The first one is the soil volumetric compression that leads to a larger soil settlement. On the other hand, it has a stabilizing effect on the normally consolidated soil and makes it more resistant to further undrained cyclic loading and cyclic shear deformation.

4.3.3.2 Prediction model

Various models have been developed for predicting cumulative plastic strain under repeated loading. The following power model is the most commonly used (e.g., Monismith et al. 1975; Knutson et al. 1977):

$$\varepsilon_p = AN^b \qquad (4.54)$$

where:
   ε_p is the cumulative plastic strain (%)
   N is the number of repeated load applications
   A and b are two parameters representing the influences of other factors on the plastic strain

Other models have also been developed (e.g., Hyde and Brown 1976; Majidzadeh et al. 1976; Baladi et al. 1983). In general, the relationship between cumulative plastic strain and the number of repeated stresses is represented in a similar approach to this equation, although some models have also included stresses and soil properties.

A prediction model for cumulative plastic strain should take into account major influential factors, such as soil stress state, number of repeated load applications, soil type, and soil physical state. In this model, Equation 4.54 quantifies the relationship between the cumulative plastic strain and the number of repeated stress applications. Influences of other major factors are represented by parameters A and b. As other factors, in addition to N, may impose significant influences on the development of plastic deformation, a constitutive model without a reasonable quantification of those factors may have limited practical use.

For the relationship between the cumulative plastic strain and the number of repeated stress applications, Equation 4.54 can be justified to represent the general trend for most fine-grained soil, because most constitutive equations in the literature can be considered as a power equation or represent a similar relationship. This power model for the relationship between ε_p and N has also been verified on the statistical basis of many test results (Li and Selig 1996).

In addition to the number of stress applications, three other major factors need to be considered. These are soil stress state, soil physical state, and soil type, all of which are subject to significant changes due to variations of load level, season and weather, and subgrade location. For the stress state, many test results (such as Seed et al. 1955; Monismith et al. 1975; Brown et al. 1977) showed that the deviator stress ($\sigma_d = \sigma_1 - \sigma_3$) is the main stress factor influencing the cumulative plastic strain for fine-grained soils under repeated loading. An increase in deviator stress will lead to an increase in plastic strain. Soil physical state is defined here as soil moisture content and dry density. Many test results (such as Saucer and Monismith 1968; Edil and Motan 1978) have indicated a significant effect of soil physical state on cumulative plastic strain for fine-grained soils. For example, an increase in soil moisture content will increase plastic strain accumulation. Influences of these factors are quantified in terms of the parameters A and b in power model, Equation 4.54. Following is the analysis showing how these three factors are related to these two parameters.

4.3.3.2.1 Exponent b

Studies by Monismith et al. (1975) and Knutson et al. (1977) showed that exponent b was independent of, or not correlated with, the deviator stress for the soils they tested. To verify if this is a general fact and to determine the influence of other factors on exponent b, many available test results in the literature were compared using the power model (Li and Selig 1996). The influences of the major factors on exponent b for various soils were calculated by regression using Equation 4.54 with the available test results in the literature. The average and standard deviation (SD) of exponent b were determined to evaluate the degree of influence of each major factor. From that study, a change in deviator stress level did not cause a significant change of exponent b, and the variation of soil physical state did not impose a significant influence on exponent b either. In addition, the change in this exponent caused by changes in both deviator stress and soil physical state was insignificant.

Figure 4.26 gives a graphic example of influences of both deviator stress and soil physical state on exponent b. The test results of cumulative plastic strain on Vicksburg Buckshot clay (Townsend and Chisolm 1976) are replotted in a log–log coordinate system. As shown in this figure, for changes of water content from 27.6% to 33.1%, dry density from 1.38 to 1.47 Mg/m^3, and deviator stress from 9.7 to 111 kPa, the change of slope b can be considered to be insignificant.

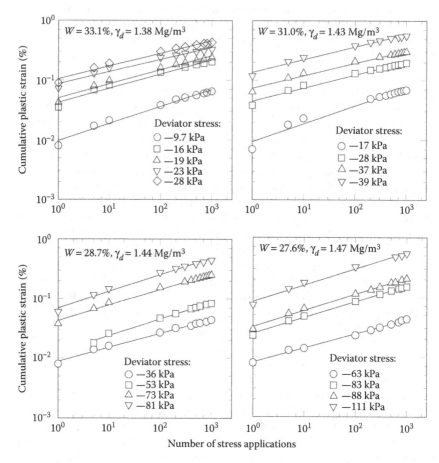

Figure 4.26 Influence of deviator stress and soil physical state on slope exponent b and intercept A.

Therefore, for the same soil type, exponent b in Equation 4.54 can be considered to be independent of soil deviator stress as well as soil physical state. A study by Li and Selig (1996) also proved that this power model, that is, Equation 4.54, gives a good correlation between the cumulative plastic strain and the number of repeated stress applications, as mentioned previously.

4.3.3.2.2 Coefficient A

To use Equation 4.54 for predicting cumulative plastic strain, coefficient A also needs to be quantified. Coefficient A can be written as

$$A = \varepsilon_p(N = 1) \tag{4.55}$$

In other words, coefficient A is the soil plastic strain after the first cycle of repeated loading. Therefore, logically this coefficient should depend on not only soil type, but soil physical state and deviator stress as well. The influences of deviator stress, soil physical state, and soil type on coefficient A were studied by Li (1994). From that study, this coefficient varied greatly from 0.0005% to 6.3% with variation in these factors.

Figure 4.27 shows an example of the influences of the deviator stress and soil physical state on coefficient A. The plotting is based on the test results of cumulative plastic strain for the Vicksburg Buckshot clay by Townsend and Chisolm (1976). The influence of the deviator stress on coefficient A at four different soil physical states can be seen to be very significant. The trend shown between A and σ_d is similar to the trend between ε_p and σ_d (Monismith et al. 1975; Knutson et al. 1977), and is representative of most test results in the literature. Figure 4.27 also shows the important influence of the soil physical state on this coefficient, as coefficient A changes significantly as a result of four different soil physical states.

Based on the trend shown in Figure 4.27, a relationship between coefficient A and deviator stress can be assumed to be as follows:

$$A = a' \sigma_d^m \qquad\qquad (4.56)$$

Where a and m are correlation parameters.

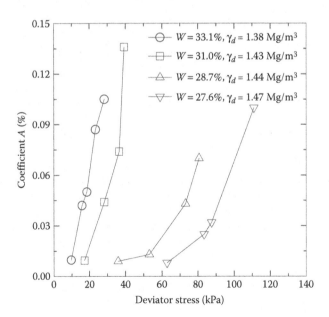

Figure 4.27 Influence of deviator stress and soil physical state on coefficient A.

For the influence of soil physical state, it is neither convenient nor common to introduce moisture content and dry density directly into the equation. However, a soil strength parameter under monotonic static loading can indirectly represent the influence of soil physical state on coefficient A. As soil compressive strength, which can be determined by conventional test techniques, varies with soil moisture and dry density, it is introduced to Equation 4.56 to represent the influence of soil physical state in the following format:

$$A = a\left(\frac{\sigma_d}{\sigma_s}\right)^m \tag{4.57}$$

where:

  a and m are the material parameters
  σ_s is the soil static strength

Use of soil static strength may also represent the influences of other factors such as soil structure. Change of soil structure and properties will lead to a change of soil static strength; thus, their influences on cumulative plastic strain may also be represented indirectly in Equation 4.57.

The change of soil plastic strain after the first cycle of repeated loading (i.e., the coefficient A) caused by the change of soil physical state is expressed as a function of soil static strength, which, in turn, depends on the soil physical state. However, the relationship between A and the two variables σ_d and σ_s may be different for different soil types, and is characterized by the two parameters a and m.

Available test results in the literature have been back calculated for both parameters a and m using Equation 4.57. To do this, test results of the plastic strain at $N = 1$ (i.e., coefficient A), soil static strengths (σ_s) at different physical states, and deviator stresses (σ_d) are used. If test results lack information on either σ_s or σ_d, only the coefficient m can be back calculated from Equation 4.57, assuming that either σ_s or σ_d is a constant. Back calculated values of parameters a and m for various soil types were given in a reference by Li (1994).

After the influences of deviator stress and soil physical state are introduced into Equation 4.57, the ranges of parameters a and m become much smaller than the range of coefficient A. For a total number of 22 soils studied, exponent m has a range of 1.0 to 4.2, while coefficient a has a range of 0.3 to 3.5, as compared with the range of coefficient A from 0.0005 to 6.3.

4.3.3.3 Li–Selig model for cumulative plastic deformation

Substituting Equation 4.57 into Equation 4.54 gives the following Li–Selig constitutive model for cumulative plastic strain:

$$\varepsilon_p = a \left(\frac{\sigma_d}{\sigma_s} \right)^m N^b$$

or (4.58)

$$\varepsilon_p = a\beta^m N^b$$

where:

$$\beta = \frac{\sigma_d}{\sigma_s}$$

To predict cumulative plastic strain by Equation 4.58, values of these three material constants a, m, and b for various soil types will need to be determined. Although influences of stress state and soil physical state are represented quantitatively in Equation 4.58, these material constants vary for different soils.

Unfortunately, there is no simple test technique to determine parameters a, m, and b. When a more sophisticated repeated loading test is not available, an estimation of these parameters based on soil type is essential. For this purpose, the values of these parameters for different soil types were determined by Li (1994) and are given in Table 4.7. As shown, for each soil classification, an average value and a possible range are listed.

As shown in Table 4.7, all three parameters generally increase with increasing clay content and soil plasticity, that is, with the trend of ML to MH to CL to CH. This is considered reasonable because, for all three parameters, a higher value is related to a higher plastic strain accumulation, and a more cohesive soil generally is more susceptible to plastic strain accumulation.

Among these three parameters, coefficient a may be subject to the greatest uncertainty and variability. However, this coefficient may be considered conceptually as the soil plastic strain when the soil is loaded to its static strength prior to the repeated loading, that is, $\beta = N = 1$, thus $\varepsilon_p = a$.

Table 4.7 Values of soil parameters a, b, and m for various soil types

Soil type	a	b	m
CH (fat clay)	1.2	0.18	2.4
	(0.82–1.5)	(0.12–0.27)	(1.3–3.9)
CL (lean clay)	1.1	0.16	2.0
	(0.3–3.5)	(0.08–0.34)	(1.0–2.6)
MH (elastic silt)	0.84	0.13	2.0
		(0.08–0.19)	(1.3–4.2)
ML (silt)	0.64	0.10	1.7
		(0.06–0.17)	(1.4–2.0)

Source: Li, D. and Selig, E.T., J. Geotech. Eng., ASCE, 122(12), 1006–1013, 1996.

However, "*a*" should not be obtained using the static strain at failure because this strain often is not well defined.

Exponent *b* can be considered as a constant to reflect the accumulation rate of plastic strain under repeated loading. Soil with a higher value of *b* has a higher cumulative rate of plastic strain. For a soil of the same physical state under a constant repeated deviator stress, the ratio of cumulative plastic strains ε_{p2} and ε_{p1}, $\varepsilon_{p2}/\varepsilon_{p1}$, at two traffic levels (N_2 and N_1) is equal to $(N_2/N_1)^b$, depending on the magnitude of exponent *b*.

The fact that exponent *m* for fine-grained soils is larger than 1 indicates a softening phenomenon of deviator stress on soils. As indicated in Table 4.7, the degree of softening increases (i.e., higher value of exponent *m*) with increasing clay content or plasticity. In fact, the relationship between deviator stress and plastic strain for some granular materials appears to be linear (i.e., *m* = 1), according to some test results (e.g., Brown 1974; Loo 1976).

The constitutive model or Equation 4.58 is for a given soil physical state with repeated cycles of a given deviator stress state. For a railway track, repeated traffic load levels are not constant, but consist of a spectrum of different numbers of cycles of different load levels. In addition, the subgrade physical state at a given location varies over time due to seasonal changes and traffic action.

The spectra for deviator stresses and soil strengths can be written as follows:

$$\left(\sigma_d\right) = \left(\sigma_{d1}, \sigma_{d2}, \ldots, \sigma_{di}\right) \tag{4.59}$$

$$\left(\sigma_s\right) = \left(\sigma_{s1}, \sigma_{s2}, \ldots, \sigma_{sj}\right) \tag{4.60}$$

where:
  σ_{di} is one of deviator stress levels with N_i cycles of stress applications
  σ_{sj} is one of soil static strengths corresponding to N_j cycles of stress applications

In Equation 4.60, soil physical states are represented indirectly by soil static strengths.

For each level of deviator stresses, σ_{di}, or soil static strengths, σ_{sj}, the prediction of cumulative plastic strain for a corresponding number N_i or Nj of stress applications can be determined by Equation 4.58.

To consider both multistress levels and soil physical states, a procedure, as illustrated in Figure 4.28, can be used. This procedure is based on a procedure often used for calculating the cumulative plastic strain only for a spectrum of multistress levels (Monismith et al. 1975; Poulsen and Stubstad 1977). Thus, the final cumulative plastic strain, ε_p, can be determined as follows:

$$\varepsilon_p = \sum \left(\varepsilon_{p1} + \varepsilon_{p2} + \ldots\right) \tag{4.61}$$

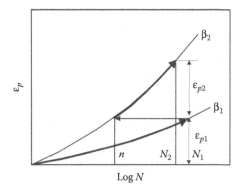

Figure 4.28 Model for total cumulative plastic strain.

where:

$$\varepsilon_{p1} = a\beta_1^m N_1^b \tag{4.62}$$

$$\varepsilon_{p2} = a\beta_2^m \left(N_2^b - n^b \right) \tag{4.63}$$

The meanings of N_1, N_2, and n for the procedure are illustrated in Figure 4.28.

Equation 4.61 can also be expressed in a simple form as Equation 4.58. However, various numbers of β_i must be converted to a reference value β_0 with equivalent numbers of cycles. For example, N_i cycles of β_i can be converted to N_{i0} cycles of β_0. N_{i0} is determined as follows (Li 1994):

$$N_0^i = N_i \left(\frac{\beta_i}{\beta_0} \right)^{m/b} \tag{4.64}$$

Thus, the total equivalent number of repeated load applications is given by

$$N = N_0 + N_0^1 + \ldots + N_0^i \tag{4.65}$$

And, Equation 4.61 then becomes the format as follows:

$$\varepsilon_p = a \left(\beta_0 \right)^m N^b \tag{4.66}$$

4.3.3.3.1 Applicability of Li–Selig model

To verify the applicability of the Li–Selig model described earlier, the following presents test results of cumulative plastic strain compared with predicted results.

Figure 4.29a and b shows the comparisons of predicted and experimental cumulative plastic strains for a Vicksburg Buckshot clay. The test results for two soil physical states and seven stress states were obtained by Townsend

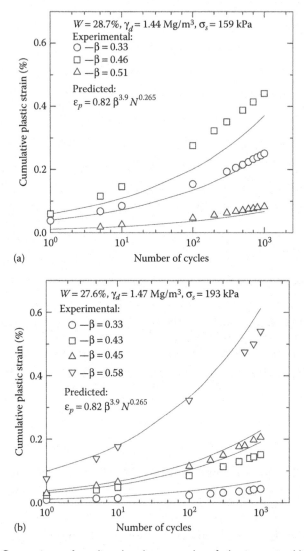

Figure 4.29 Comparison of predicted and test results of plastic strain: (a) w = 28.7%, γ_d = 1.44 Mg/m³, σ_s = 159 kPa; (b) W = 27.6%, γ_d = 1.47 Mg/m³, σ_s = 193 kPa.

and Chisolm (1976), and are shown by various symbols in these two figures. Both the repeated load tests and static soil strength tests were conducted under zero confining pressure. In the prediction for seven deviator stress levels ranging from 36 to 111 kPa, two soil physical states (W = 28.7%, γ_d = 1.44 Mg/m³; and W = 27.6%, γ_d = 1.47 Mg/m³) are taken into account by means of unconfined compressive strength (σ_s = 159 and 193 kPa, respectively). The material constants *a*, *m*, *b* are 0.82, 3.9, and 0.26, respectively.

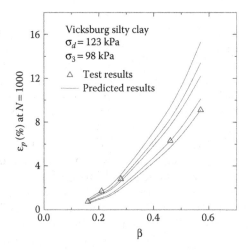

Figure 4.30 Comparison of predicted and experimental plastic strain at $N = 1,000$.

Predictions of cumulative plastic strains under various stress states and for the two soil physical states are carried out by Equation 4.58. The predicted results are shown by the solid lines in these two figures. As can be seen, the influences of deviator stress and soil physical state on cumulative plastic strain are reasonably reflected in the prediction.

Figure 4.30 shows a comparison of predicted results with the test results of the plastic strain at $N = 1,000$ for a Vicksburg silty clay (Seed and McNeill 1956). Only one stress state ($\sigma_d = 122$ kPa and $\sigma_3 = 98$ kPa) was employed for the tests. However, the soil was tested with five different physical states (moisture content ranging from 14.6% to 19.2%). The compressive strengths corresponding to the five different soil physical states are 785, 588, 432, 265, and 216 kPa, respectively. The plastic strains are plotted as a function of β, the ratio of deviator stress to the five compressive strengths. The test results are shown by the triangles.

In this example, the prediction of plastic strains at any four different soil physical states is based on the test result of the plastic strain at a reference soil physical state. The equation used for prediction, $\varepsilon_p = \varepsilon_{pr}(\beta / \beta_r)^m$, is derived from Equation 4.58, in which ε_{pr} is the plastic strain from the test at a reference soil physical state, and β_r is the corresponding ratio of deviator stress to soil static strength. For this silty clay soil, a value of 2.2 for exponent m is used.

By using each of the five test results of plastic strain at the five soil physical states as the reference, the plastic strains at the other soil physical states were predicted. These results are shown as solid lines in Figure 4.30. As can be seen, the predicted trends of plastic strain with β agree reasonably well with the test trends.

4.3.3.3.2 Results of test subgrade

At the TTC located near Pueblo, Colorado, a test track was constructed on a soft soil subgrade in October 1991. One of the main objectives of this test track subgrade was to study the effect of the soft soil subgrade on track performance under repeated heavy axle loading. As the natural subgrade soil at TTC is a silty sand and does not represent a low-strength subgrade soil, a trench of 3.66 m in width and 183 m in length was excavated in the natural subgrade to a depth of 1.52 m from the ground surface, and was filled with the Vicksburg Buckshot clay ($PI = 40$–45, $LL = 60$–70). The clay was compacted with a specified moisture content and dry density to achieve a subgrade of low stiffness. The specified soil moisture content was 30%, and the specified dry density was 90% of its maximum dry density (1.52 Mg/m^3), according to the ASTM D698. To isolate the clay moisture content from the existing silty sand, the trench was lined on the sides and bottom with an impermeable plastic membrane. The clay subgrade was then covered by the subballast, ballast, and track superstructure. Figure 4.31 shows the cross section of the test track and subgrade.

Extensive experimental and analytical work was conducted, and the details can be found in Ebersohn et al. (1995) and Li (1994). Figure 4.32a shows the subgrade settlements under both ends of ties along the track after 62 million gross tons (MGT) of traffic (static wheel load = 173 kN and train speed = 64 km/h). The initial and final subgrade surface elevations

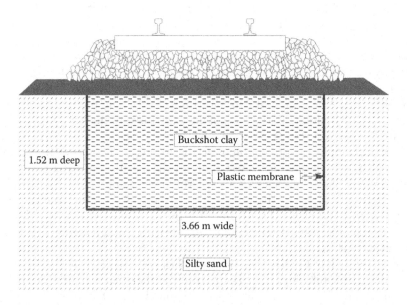

Figure 4.31 Cross section of test track and subgrade. (Courtesy of Transportation Technology Center Inc., Pueblo, Colorado.)

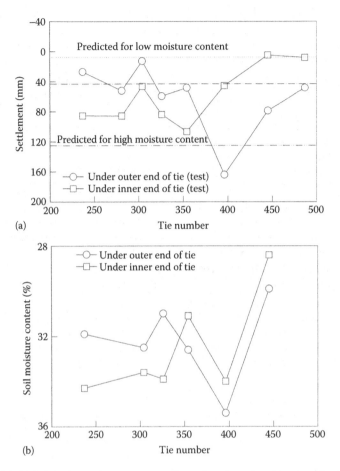

Figure 4.32 (a) Subgrade settlement: test and predicted results; (b) subgrade soil moisture content distribution.

along the track were surveyed at a number of locations, and their differences gave subgrade settlements due to the traffic. The maximum subgrade settlements usually occurred under tie ends, as the maximum deviator stress usually occurs there (Li 1994). The subgrade settlements under the two ends of the same tie are generally not equal, indicating a differential settlement across the track. The subgrade settlement along the track was also nonuniform. The largest subgrade settlement for the subgrade was approximately 165 mm and occurred under the outer end of tie #390. The average of the subgrade settlements under both ends of ties was roughly 58 mm. As shown in Figure 4.32a, the accumulation of subgrade plastic deformation was excessive due to low stiffness of the subgrade and heavy traffic loads applied.

The differential subgrade settlement across and along the track was mainly caused by the nonuniform distributions of the subgrade soil strength due to the soil moisture content variation and the soil deviator stress. Figure 4.32b shows the soil moisture content distributions under both ends of ties along the track. A comparison between Figure 4.32a and b indicates that the distributions of the differential subgrade settlements are approximately consistent with the distributions of soil moisture content. Note that the largest subgrade settlement occurred under the outer end of tie #390, where the clay also had the highest soil moisture content.

Predicted results of the subgrade settlements for three soil strengths corresponding to three moisture contents are given in Table 4.8. Three moisture contents, that is, lowest ($W = 28.4\%$), average ($W = 32.4\%$), and highest ($W = 35.4\%$) are considered for predictions. For prediction, the soft subgrade is divided into five layers, each 0.3 m thick. The contribution of the bottom silty sand subgrade to the total settlement is considered to be negligible because of its high stiffness. The deviator stress under the tie end at the center of each layer in the soft subgrade was determined using the GEOTRACK model under a dynamic wheel load of 236 kN, and the total equivalent number of stress applications for 62 MGT of traffic is 770,000 (Li 1994). The material parameters (a, m, and b) for the clay are 1.2, 2.4, and 0.18, respectively, from Table 4.7. The plastic strain at the center of each layer was then calculated by Equation 4.58. The final subgrade settlement

Table 4.8 Prediction of cumulative plastic deformation for test subgrade

Soil condition (σ_s—w)	Subdivided layer number	σ_d (kPa)	β	ε_p (%)	P (mm)
$\sigma_s = 193$ kPa, W = 28.4%	1	66	0.34	1.03	
	2	59	0.30	0.76	
	3	51	0.26	0.54	
	4	44	0.23	0.40	
	5	40	0.21	0.34	9.4
$\sigma_s = 90$ kPa, W = 32.4%	1	59	0.65	4.9	
	2	55	0.62	4.4	
	3	48	0.53	3.0	
	4	44	0.49	2.5	
	5	40	0.45	2.0	51
$\sigma_s = 48$ kPa, W = 35.4%	1	47	0.97	12.8	
	2	44	0.91	11.0	
	3	40	0.83	8.9	
	4	37	0.77	7.4	
	5	34	0.71	6.1	142

Note: Each subdivided layer is 0.3 m thick.

is determined by summing up the deformations of all five layers using the following formula:

$$\rho = \sum_{i=1}^{5} \varepsilon_p^i h_i \qquad (4.67)$$

Where h_i is the thickness of each layer (0.3 m).

Note that the deviator stress distributions with depth for the three different soil strengths are different. A softer subgrade will result in lower deviator stress in the subgrade, as indicated in this table.

The predicted results showed significant influence of subgrade soil strength on the total subgrade settlement. An increase in soil moisture content from 28.4% to 35.4% led to large increase of subgrade settlement, as shown in Table 4.8. The predicted settlements for the three different soil strengths (i.e., three different moisture contents) are also plotted in Figure 4.32a with the test results. The range of predicted results due to three different soil strengths can be seen to be consistent with the range of test results.

4.3.4 Effects of traffic, ballast, and subgrade conditions on total settlement

By using the plastic deformation prediction models presented in this chapter, a series of example cases are presented to illustrate the effects of heavy axle load (HAL) traffic and track substructure properties on track settlement. In these examples, the track substructure conditions considered are: poor, fair, and good subgrade; various degrees of fouling condition of the ballast layer; and total thickness for the granular layer (ballast + subballast).

Table 4.9 lists the combinations of traffic that are used in the examples. Traffic is represented by car loading and annual MGT. Cars with 263-kip

Table 4.9 Traffic mix characteristics

Mix	Annual traffic mix (MGT)			Traffic (MGT)
	263k cars	315k cars	133k cars	
1	50	0	0	50
2	40	10	0	50
3	0	50	0	50
4	50	0	10	60
5	0	10	50	60
6	0	0	50	50

capacity are used to represent standard freight traffic, 315-kip cars were used for HAL traffic, and 133-kip cars are used to represent passenger train traffic.

Mix 1 represents the nominal case of only standard freight traffic. Mix 2 represents 80% standard freight traffic and 20% HAL traffic, and Mix 3 is only HAL traffic. Mixes 4, 5, and 6 represent various combinations of passenger traffic with either the standard or HAL freight traffic. Mixes 4, 5, and 6 are used in the first example to illustrate the effects of combining light loads with heavy loads on settlement for various subgrade conditions.

Example 4.1

The settlements are compared in Example 1 for poor, fair, and good subgrade conditions under various mixes of traffic. The deformation and strength characteristics for the three subgrade conditions are given in Table 4.10. The results of the settlement analysis are presented in Table 4.11.

The nominal case (Mix 1) shown in Table 4.11 is for 50 MGT of standard freight traffic (263-kip cars) with 33-kip wheel loads. For this mix, the deformation increases from 0.6 to 5.3 in. as the subgrade changes from good to poor. With the addition of 20% HAL traffic to the standard freight traffic (Mix 2), settlement increases by 15%–20% for all subgrade conditions. This occurs even though the total annual traffic remains at 50 MGT. Going from the standard freight traffic (Mix 1) to only the HAL traffic (Mix 3) results in at least a 40% increase in track substructure settlement.

Table 4.10 Properties of subgrade soils for Example 1

Subgrade condition	Resilient modulus, E_R (psi)	Compressive strength (psi)
Poor	1,000	4
Fair	2,000	8
Good	8,000	16

Table 4.11 Example 1 results

| Mix | Total settlement (in.) for subgrade condition of | | |
	Poor	Fair	Good
1	5.3	1.7	0.6
2	6.1	2.0	0.7
3	7.4	2.4	0.8
4	5.3	1.7	0.6
5	6.4	2.1	0.8
6	1.8	0.6	0.2

The addition of 10 MGT of the light wheel load passenger traffic to the 50 MGT standard freight traffic (Mix 4) caused negligible increase in settlement (compared to Mix 1). This result illustrates that the substructure deformation is driven by the heaviest load it experiences. The results in Table 4.11 for Mixes 5 and 6 illustrate this even more dramatically. Mix 6 represents all relatively light passenger traffic resulting in a small amount of substructure deformation. Adding only 10 MGT of HAL traffic (Mix 5) results in a 2.5 to 4.0 times greater substructure-related track settlement.

Example 4.2

The track settlement is compared for Mixes 1 through 3 for clean ballast and fouled ballast conditions. The clean ballast was modeled with a resilient modulus (stiffness) of 45,000 psi, and the fouled ballast was modeled with a modulus of 11,000 psi. These results are based on "fair" subgrade conditions. Table 4.12 gives the results.

Table 4.12 shows that the fouled ballast condition results in 15%–20% increase in settlement for all three traffic mixes. The impact of HAL on substructure settlement is similar for fouled and clean ballast condition. Changing from Mix 1 to Mix 2 results in approximately 15% increase in settlement for both clean and fouled ballast condition. Changing from Mix 1 to Mix 3 traffic results in a 40% increase in settlement for both clean and fouled ballast.

Example 4.3

The effect of the thickness of the granular layer on the overall settlement of the track substructure (granular layer and subgrade settlement) is shown in Example 3. "Fair" subgrade conditions are assumed. The results are given in Table 4.13. Adding a relatively small amount of HAL traffic (10 MGT) to standard freight traffic (Mix 2) results in approximately 15% increase in track substructure settlement over Mix 1 for all three granular layer thicknesses. Increasing from 50 MGT of all standard freight traffic (Mix 1) to 50 MGT of all HAL traffic (Mix 3) results in 35%–40% increase in settlement for all thickness of granular layer.

Table 4.12 Example 2 results

Mix	Total settlement (in.) for upper ballast condition of:	
	Clean	Fouled
1	1.7	2.0
2	2.0	2.4
3	2.4	2.8

Table 4.13 Example 3 results

Mix	Total settlement (in.) for G.L.[a] thickness of		
	8 in.	18 in.	28 in.
1	1.9	1.7	1.4
2	2.1	2.0	1.6
3	2.6	2.4	1.5

[a] G.L. = granular layer.

4.3.5 Track settlement and roughness

As stated earlier, the goal of determining track settlement in the track substructure layers is to relate this to a rate of differential settlement or track roughness accumulation. This is to determine the timing for needed intervention to correct track geometry deviations. A linear relationship was found (Chrismer 1993) to best describe the relationship between the differential and average track settlement. Track roughness is often expressed in terms of SD of a measured track geometry parameter, and the initial track roughness just after tamping is at the minimum roughness value, SD_{min}. As track settles following tamping, it was found that the rate of increase in SD profile roughness would increase by 0.15 for every unit of average track settlement. Thus, the rate of SD roughness increase with traffic is then expressed as

$$SD = SD_{min} + \alpha S_L \qquad (4.68)$$

where:
    S_L is the total average track settlement
    α is 0.15 as mentioned
    a typical value for SD_{min} was found to be approximately 2.5 mm

The intervention level SD_{max} for geometry-correcting maintenance depends on the class of track. For high-speed train operations, only a small amount of track settlement/roughness is allowable.

A ballast maintenance model was developed by Chrismer (1993), which used the aforementioned mechanistic approach to calculate the required timing of needed tamping cycles and the end of ballast life. With this information and the user-supplied cost data on ballast and maintenance activities, this ballast maintenance model provides the life cycle cost of the ballast. The user may then try different ballast materials to determine if a lower life cycle cost is provided. An example output of the ballast maintenance model is shown in Figure 4.33, where the ballast is initially renewed after an undercutting/cleaning operation (U/C), followed by the model-determined need for three tamping cycles before the end of ballast life is reached.

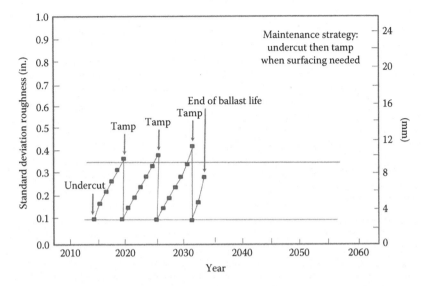

Figure 4.33 Top-of-rail longitudinal roughness with traffic and maintenance.

Figure 4.34 shows one of the model input pages together with the output that is provided after the user enters all needed information and pushes the "Run" button. To obtain total track settlement the model includes the subgrade deformation model, as described in this chapter. Lastly, the model multiplies the subballast strain rate by the user-specified layer thickness as shown, and, thus, obtains the total track settlement due to permanent deformation in all track substructure layers.

4.3.6 Track settlement at bridge approaches

A study of track settlement behavior at track transitions such as road crossings and bridge approaches has led to the development of a method to predict the amount of differential track settlement at track transitions caused by ballast and subgrade deformation. This method uses the ballast and subgrade models described in this chapter; and its applications have been validated with field settlement measurements at bridge approaches, where the ballast and subgrade properties were well documented. As an analytical tool, the method can determine the rate of track differential settlement at transitions with traffic loading, and can predict when tamping is required to correct track geometry. Also, the method can determine the relative amount of deformation from individual ballast layers and subgrade layers. These capabilities provide a means to assess whether the main source of deformation is the subgrade or ballast, and the best means to reduce differential settlement and decrease the frequency of tamping at these track transitions.

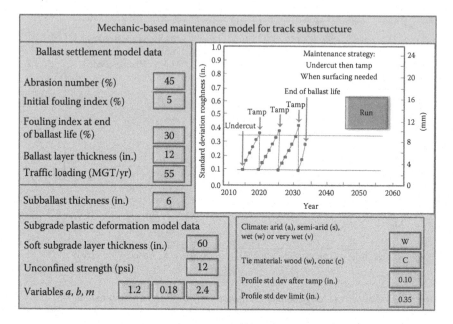

Figure 4.34 Data input page for model to calculate track roughness and maintenance timing.

4.3.6.1 Ballast layer deformation

The ballast deformation model, Equation 4.48, is used to predict the timing of required ballast tamping maintenance, and when ballast renewal is needed at the end of its life. The basic form of the ballast settlement equation is

$$\varepsilon_p = \varepsilon_1 * \frac{(N^b)}{100}\%$$

Typical nominal values for ε_1 and b parameters are 0.22% and 0.21%, respectively. The N value is derived from the design wheel load, P, and the number of axle loads that cause a load cycle on the ballast (see Section 2.2), as follows:

$$N = [\text{MGT} * 1{,}000{,}000 \text{ tonnes} * (2000 \text{ lb/tonne})]/(2 \text{ Wheels/cycle} * P)$$

Where the term "2 wheels/cycle" relates to the two wheels per axle. If P is 33,000 pounds, and the annual tonnage is 75 MGT/year, then

$$N = [75 * 1{,}000{,}000 \text{ tonnes} * (2000 \text{ lb/tonne})]/(2 * 33{,}000 \text{ lb})$$

$$= 2{,}272{,}700 \text{ cycles}$$

Using the ballast strain equation, the ballast settlement δ_P in one year following construction is obtained by multiplying this strain by the ballast layer thickness which, assuming a 12-inch thick ballast layer, is

$$\delta_P = 0.22 \times (2,272,727)^{0.21} \times \left(\frac{12}{100}\%\right) = 0.57\,\text{in.}$$

Alternatively, assuming that the ballast has been in place for many years, and that the upper 6 in. of ballast has been disturbed by tamping, the ballast settlement one year after tamping is

$$\delta_P = 0.22 \times (2,272,727)^{0.21} \times \left(\frac{6}{100}\%\right) = 0.29\,\text{in.}$$

These rates of ballast deformation will be used together with the subgrade deformation rate to show how much each substructure layer contributes to total track settlement.

4.3.6.2 Subgrade layer deformation

The subgrade deformation model, Equation 4.58, is used for subgrade deformation as follows:

$$\varepsilon_P = a\left(\frac{\sigma_d}{\sigma_s}\right)^m * \left(N^b\right)\%$$

For an assumed soft clayey soil (CH soil type) that will form a progressive shear failure over the top 18 in. of subgrade, with an unconfined soil strength of 10 psi and annual tonnage of 75 MGT, the subgrade settlement rate will be calculated using the soil strain equation. The soil deviator stress for the near surface subgrade should be determined using an analysis method such as GEOTRACK, although a conservative assumption for σ_d for these conditions is 6 psi providing a stress/strength ratio σ_d/σ_s of 0.6. The repeated load N representing the number of axle load cycles per year is calculated as

$$N = [75 * 1,000,000 \text{ tonnes} * (2000 \text{ lb/tonne})]/(4 * 33,000 \text{ lb})$$

$$= 1,136,400 \text{ cycles}$$

Where the term "4 wheels/cycle" pertains to the loading from two axles.

Based on the strain and subgrade layer thickness, one has a soil deformation after the first year of 0.78 in. for newly constructed track:

$$\delta_P = (H) \times a\left(\frac{\sigma_{d1}}{\sigma_s}\right)^m \times \frac{N^b}{100}\% = (18) \times 1.2 \times \left(\frac{6}{10}\right)^{2.4} \times \left[\frac{(1,136,363)^{0.18}}{100\%}\right] = 0.78\,\text{in.}$$

This large initial settlement is characteristic of a subgrade that has been compacted during construction, but not yet experienced traffic loading.

As loading accumulates for larger "N" beyond the initial few years, per the previous relationship, the subgrade deformation rate becomes "flatter" with settlement accumulating much slower. In fact, for subgrade that has been in place for many years, a linear rate of deformation can be assumed to apply with little error. For example, if the same assumed CH soil with a stress ratio of 0.6 has been in place for many years, it has been calculated that the linear deformation rate is $5.02 \times 10^{-7}*(N)$, and the resulting amount of subgrade settlement for one year is

$$\delta_P = (18) * 5.02 \times 10^{-7} * \left(\frac{1,136,363}{100\%} \right) = 0.10 \, \text{in}.$$

Next, these ballast and subgrade settlement rates will be combined to show their relative contribution to total track settlement.

Settlement rates for the case of newly placed ballast, and newly placed and compacted soft subgrade are shown in Figure 4.35, together with the total track settlement. Note that for this case of new track construction, a soft subgrade may provide somewhat more settlement than the ballast in the years following construction.

However, for the more common case, where ballast and subgrade have been compacted by several years of traffic, the opposite trend is true with ballast settlement exceeding that of the subgrade in the initial years following tamping, as shown in Figure 4.36. The large initial amount of ballast layer deformation is due to recompaction with traffic following tamping. The

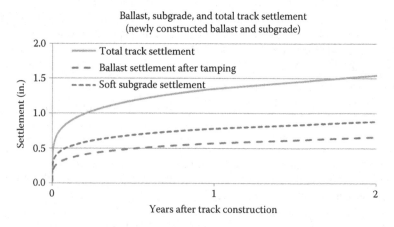

Figure 4.35 Settlement for a newly constructed track.

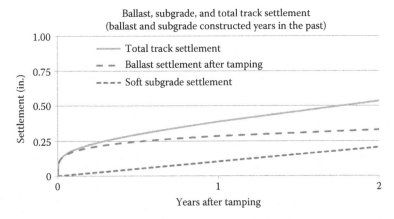

Figure 4.36 Settlement for an existing track following a tamping operation.

tamping disturbance of ballast essentially resets the "N" in Equation 4.48 to zero, providing a large settlement just after every tamping application. On the other hand, subgrade strain is not "reset to zero," but continues to contribute without much higher initial deformation.

This analysis shows that for the older track, ballast settlement typically predominates in the short term following tamping, and subgrade settlement may equal or surpass that of the ballast only over the longer term. If a track transition is constantly in need of tamping to correct geometry, perhaps on a yearly or more frequent basis, most of the track resettlement will typically be due to the ballast.

Yet another observation from field measurements shows further evidence that the source of settlement at track transitions is in the ballast and not in the subgrade. Figure 4.37 shows the typical raise and settlement trends observed at an open-deck bridge transitions near Chester, Pennsylvania. Both rails had maximum dip magnitudes of approximately 2.75 in. referenced from the rail elevation on the bridge deck, and this maximum depth of the dip was located 160 ft away from the bridge transition. Track geometry car profile space curves obtained shortly before and shortly after surfacing at these transitions show that track dips typically return quickly following tamping, but they do not appear to settle further than the dip that existed before tamping. If the source of settlement was a deforming subgrade at these locations, this settlement should continue with traffic rather than become stable shortly after maintenance. Instead, the observed initially rapid settlement rate and subsequent stabilization are consistent with the behavior of ballast compaction under loading, rather than subgrade deformation.

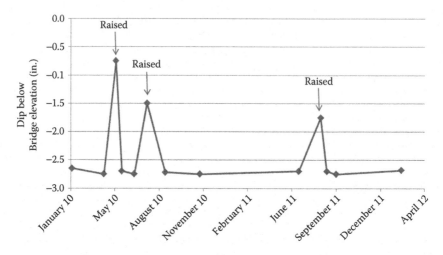

Figure 4.37 Track maintenance intervention at bridge transition.

4.3.6.3 Field investigation and modeling results

Although the method described earlier for ballast and subgrade deformation calculations were developed from years of field testing, a recent comprehensive field study into the causes of poor geometry at track transitions provides further validation of the method and its findings. In an extensive investigation in the United States (Li and Davis 2005), TTCI worked in cooperation with a major Western railroad to install and monitor the settlement of four types of track foundations under the approach track to four ballasted deck concrete bridges. The four bridges, located on the same track route, had a significant amount of HAL (36,000 pound) and approximately 180 MGT of traffic annually. The bridges and approach track were all newly constructed and the existing subgrade from the four locations was mostly lean clay (CL type), but also included clayey sand (SC type) and sandy silt (ML). The measured, unconfined compressive strength of the soil ranged between 14 psi (soft) and 26 psi (medium) that are typical strengths for these soil types at the measured moisture contents. The four track foundations that were placed over the subgrade included a layer of HMA, a geocell layer, a cement-stabilized backfill, and a control section with the railroads' conventional 20 in. of ballast and subballast combined over the compacted subgrade.

Permanent track settlement and vertical deflections under load were monitored on the bridge and in each transition section. As expected, the approaches settled much more than track on the bridges with measurements of differential settlement between bridge and approach of 1 in. and more. However, despite expectations of subgrade providing most of the approach track settlement, evidence began to accumulate that the ballast layer contributed more, including (1) measurements of approach track modulus were well above the values expected for a soft, deformable subgrade; (2) the special track foundations designed and known to reduce subgrade stress and settlement did not reduce the measured track settlement compared with the control; (3) the cement-stabilized backfill approach section has a depth of stabilization of 6.6 ft and yet still experienced significant track settlement; and (4) although vertical displacement rods to monitor subgrade settlement were damaged later in the investigation, early observations did not indicate significant subgrade settlement.

Repeated lifting and tamping was required to restore approach track vertical track geometry, but these did not stabilize the track. Settlement rates in the bridge approach areas for the four bridges decreased after the first 100 MGT, but the rates continued to be excessive requiring track geometry maintenance several times a year. Observations from these extensive tests seem to indicate that ballast contributed more to approach track settlement than the subgrade, and that this is true initially after new track construction and on a continual basis with repeated tamping.

The method described in this section was used to predict the settlement of one particular approach track and ballasted bridge track. This recently constructed bridge and approach track were examined to provide the input shown in Table 4.14. The model output provides a means to verify if the predicted settlement is similar to measured settlement from the case study.

A comparison of predicted and measured settlement for the current track configuration is shown in Table 4.15. The measured settlement value was taken over a period of 70 MGT after the settlement at the site had become stable. The modeled results were taken from the initial 70 MGT after the second tamping. For example, the second tamping occurs at 385 MGT of total traffic, and the settlement values are taken from the following 70 MGT, for total lifetime traffic of 455 MGT.

It is apparent from Table 4.15 that the method provides reasonable agreement between the predicted and measured track settlement values. The method was used to determine the best means to reduce differential settlement and tamping requirements at track transitions. This was accomplished by incorporating results from a parallel study to determine how ballast and subgrade stresses and settlements are affected by changing the parameters in Table 4.14. The goal was to reduce ballast and subgrade stresses on the softer side of the transition to develop a stable track transition with acceptable differential settlement rates. See Chapter 5 for the best practices to improve bridge approach performance.

Based on information from investigations, it is possible to put forward a scenario of how the dip forms. Newly constructed track over a bridge will have a different settlement rate from the track in the approach areas and will eventually produce a dip in elevation profile in the approach.

Table 4.14 Model parameters for track approach settlement

Operation parameters

Annual MGT tonnage	175 MGT
Inspection per year	5
Static wheel load	36 kips
Climate	Wet (25–50 in. precip. per year)
Max allowed differential settlement	0.6 in.
New construction?	Yes
Schedule for track maintenance	5 years

Track parameters

Ballast quality	Good
Ballast fouling index	Moderately fouled
Ballast thickness	12 in.
Tie type	Concrete
Tie spacing	24 in.
Soil classification	CL
Subgrade quality	Poor
Subgrade depth	60 in.
Rail weight	133 lb/yard

Bridge parameters

Ballast quality	Good
Ballast fouling index	Moderately fouled
Ballast thickness	12 in.
Tie type	Concrete
Tie spacing	24 in.
Rail weight	133 lb/yard
Tamping raise	1 in.

Table 4.15 Measured result versus model results

Parameter	Measured	Modeled
MGT since tamping	70	70
Total MGT	Unknown	455
Number of tampings (first five years, 875 MGT)	Unknown	4
	Measured settlement (in.)	**Modeled settlement (in.)**
Approach track	1.31	1.15
Bridge	0.74	0.76
Differential	0.57	0.39

After the ballast settles relative to the bridge by a certain amount, track lifting and tamping at the transitions will eventually become necessary, and this will provide a "ramp" after surfacing on either side of the bridge. If the geometry correction were durable, these transitions would require future tamping applications no more frequently than any other section of the ballasted track. However, due to the decrease in ballast density due to tamping (as explained earlier), these dips will often quickly return. This mechanism is the source for the repeated formation of many dips in transition areas.

Chapter 5

Design

This chapter discusses general principles, considerations, criteria, and methods for the design of track foundation that includes the geotechnical layers under the rails and ties. Three types of track foundations are described in detail in this chapter, including conventional ballasted track, asphalt track, and slab track. Design considerations for track transitions are also discussed in this chapter, because it is often a location where track tends to have more problems.

Regardless of type of track foundation, track is often designed to meet the following criteria under repeated wheel loadings:

- Rail-bending stresses not exceeding the allowable
- Tie-bending stresses not exceeding the allowable
- Rail or track deflections not exceeding the allowable
- Ballast stress not exceeding the allowable
- Subgrade stresses or deformation not exceeding the allowable
- Track modulus or track stiffness within a certain allowable range

In the case of asphalt track or slab track, design criteria also include the allowable stresses or strains for the asphalt layer or the concrete slab. Ideally, design of a track foundation should take into account vehicle/track dynamic interaction, repeated wheel load applications, and cumulative tonnage (Chapters 2 and 4).

What is important to understand, however, is that the design of a track and its foundation is intended to achieve three major objectives for railway operations: to avoid train speed restrictions, avoid frequent track maintenance requirements, and reduce the risk of train derailment, due to excessive deformation and failure.

5.1 BALLASTED TRACK

5.1.1 Background

Ballasted track uses granular layers (ballast and subballast) as the foundation to transfer and reduce train-induced loads to the underlying subgrade. Granular layer thickness is the combined thickness of the ballast and subballast layers between the subgrade surface and the tie bottom, as shown in Figure 5.1. It is also a good practice to design and build a top subgrade layer with better materials (such as granular or soil admixed with lime) than the native subgrade soil (sometimes referred to as the "prepared subgrade" or "subgrade blanket"). In this case, this improved subgrade layer can be considered as part of the granular layer or track foundation.

Adequate granular layer thickness is required to reduce traffic load-induced stresses in the subgrade. With growing trends of railways toward heavier axle loads, higher train speeds, and increasing amount of traffic, higher and more dynamic wheel loads will be exerted on the track structure. Without sufficient granular layer thickness, the subgrade will be subjected to higher repeated stresses, which in turn may lead to excessive subgrade deformation and failure. This will greatly increase the costs of track maintenance. On the other hand, excessive granular layer thickness will result in unnecessary costs and may not be effective in maintaining track stability under repeated traffic loadings.

To reduce train load-induced stresses in the subgrade, several alternative approaches can be used including use of alternative track structures such as a hot mix asphalt (HMA) layer above the subballast/subgrade, geosynthetics, or slab track. Increasing rail weight, increasing tie–ballast bearing area, reducing tie spacing, and so on can also be effective to some degree. In addition, the subgrade can be modified and improved to allow higher

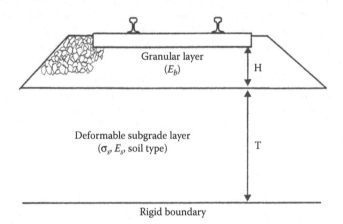

Figure 5.1 Simplified track granular layer and subgrade.

stresses. However, designing an adequate granular layer is likely to be the most effective and economical approach to achieve and maintain good track and subgrade performance under most circumstances.

5.1.2 Design methods for granular layer thickness

A number of design methods are available in the literature for designing granular layer thickness. The following describes four major design methods: the American Railway Engineering and Maintenance Association (AREMA) recommended design equations, the Raymond design methods, the British Rail method, and the Li–Selig method. Table 5.1 gives a comparison of features considered in each of the four design methods.

5.1.2.1 AREMA equations

The manual of the AREMA recommends several design equations, including the following classical Talbot equation:

$$H = 0.24 \left(\frac{P_m}{P_c} \right)^{0.8} \tag{5.1}$$

where:

H is the granular layer thickness (m)

P_c is the allowable subgrade stress (138 kPa recommended by AREMA)

P_m is the vertical stress applied on the ballast surface

This empirical equation was developed based on the field tests conducted during the 1910s and 1920s. Its application, however, represents an oversimplified

Table 5.1 Comparison of four design methods

Feature	AREMA equations	Raymond method	British Rail method	Li–Selig method
Traffic	Single wheel	Single wheel	Single wheel	Multiple wheel loads, total tonnage
Subgrade property	None	Soil type	Threshold strength	Resilient modulus, compressive strength
Ballast property	None	None	None	Resilient modulus
Design criteria	138 kPa allowable	Allowable stress based on soil classification	Progressive shear failure	Progressive shear failure, cumulative deformation
Model for stress	Empirical	Homogeneous	Homogeneous	Multilayer

situation for track under heavier axle loads and higher train speeds. Subgrade soil conditions, effects of repeated dynamic loads, granular layer material quality, and so on are not reflected at all in this equation. Because subgrade varies widely in strength, use of a universal allowable stress of 138 kPa will lead to difficulty in soft soil conditions and will be too conservative under good soil conditions.

5.1.2.2 Raymond method

Raymond (1985) modified the design method recommended by AREMA. Instead of a universal allowable stress of 138 kPa, soil classification was used to relate to allowable bearing pressures for various subgrade soils. Figure 5.2 shows the design chart developed by Raymond for vehicles weighing 70–125 tons. In this chart, vertical stress curves are shown for three different vehicle weights. Safe, allowable pressures for different soil types are superimposed in the chart. Ballast depth is determined by locating the depth at which vertical stress is equal to the safe bearing pressure of a given subgrade soil. Note that the stress calculation was based on an assumption of a homogeneous half-space to represent ballast, subballast, and subgrade

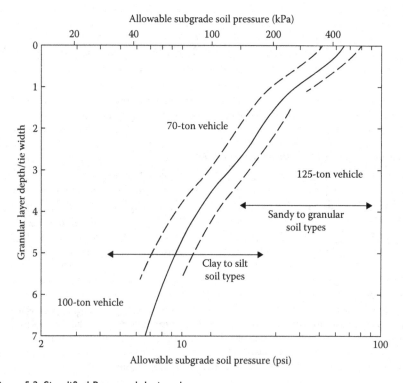

Figure 5.2 Simplified Raymond design chart.

without considering the properties of individual layers. This method does not consider the number of repeated stress applications as a design parameter.

5.1.2.3 British Rail method

British Rail (Heath et al. 1972) developed a "threshold stress" design method for selecting granular layer thickness. In this method, the design is to limit the stress in subgrade soil to less than a "threshold" stress (also called threshold strength) to protect against subgrade progressive shear failure. The threshold stress was determined from repeated load tests in which the soil cumulative strain was measured as a function of the number of loading cycles applied. At stress above the threshold stress level, plastic deformation accumulation would be rapid. At stress below the threshold stress level, deformation accumulation would be small.

Threshold stress (or threshold strength) for cohesive subgrade soil is lower than soil static strength. In practice, threshold strength was often assumed to be equal to 50% of soil static strength. Based on the subgrade soil "threshold stress" concept, design charts were developed for selecting granular layer thickness for various subgrade soil conditions and axle loads. Figure 5.3 shows the design charts, based on design axle load and soil threshold strength/soil modulus.

The "threshold stress" design method considers the effect of repeated stresses on subgrade soil strength. However, design charts were developed using a track model neglecting the effect of much higher stiffness of the granular layer above the subgrade, that is, assuming that ballast, subballast, and subgrade acted together as a single homogeneous layer. As a result, this

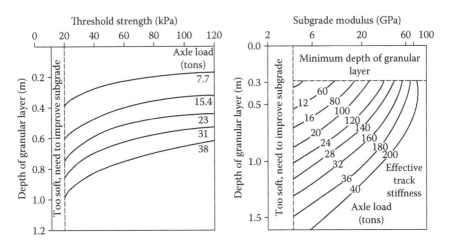

Figure 5.3 British rail design charts.

design method often gives large required granular layer thickness because of high-calculated stresses in the subgrade.

Another major limitation of the "threshold stress" method is that this method uses a single value of axle load without considering the amount of cumulative tonnage, as this is also a major limitation for almost all available design methods in the literature. In other words, by using a method without considering the effect of cumulative tonnage, a track with an annual tonnage of 10 million gross tons (MGT) would be designed with the same granular layer thickness as a track with an annual tonnage of 100 MGT, if the maximum axle loads are the same for both tracks.

5.1.2.4 Li–Selig method

The Li–Selig method (1998) can be used either on the basis of the allowable stress (allowable cumulative plastic strain) at the subgrade surface or on the basis of the allowable subgrade deformation. Figures 5.4 and 5.5 are two examples of design charts (see Section 5.1.3). Figure 5.4 is a chart based on

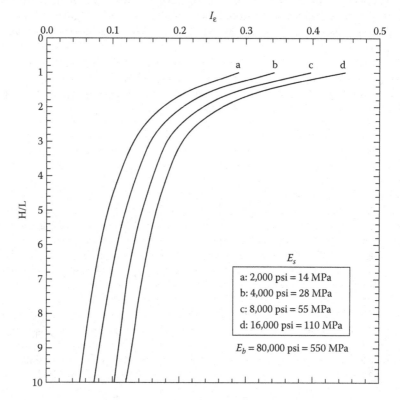

Figure 5.4 Example depth design chart based on allowable stress or plastic strain at subgrade surface.

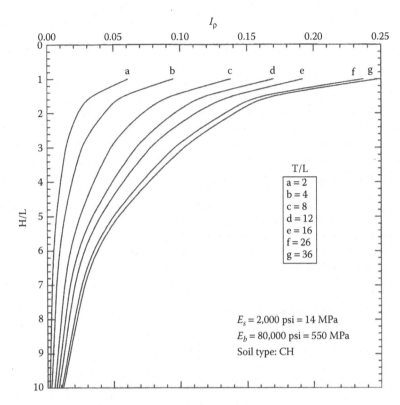

Figure 5.5 Example design chart based on allowable total subgrade deformation.

the use of allowable stress or plastic strain at the subgrade surface, while Figure 5.5 shows a chart based on allowable subgrade deformation.

In these design charts, H (see Figure 5.1) is granular layer thickness. L is a factor used to make charts dimensionless (6 in. in English units and 0.152 m in SI units). I_ε and I_ρ are referred to as the strain influence factor and deformation influence factor, respectively, and are defined by the following two equations:

$$I_\varepsilon = \frac{\sigma_{da}A}{P_d} \tag{5.2}$$

$$I_\rho = \frac{\rho_a/L}{a\left(P_d/\sigma_s A\right)^m N^b} \times 100 \tag{5.3}$$

where:

σ_{da} is the allowable deviator stress at the subgrade surface (note σ_{da} is σ_d or deviator stress when the allowable cumulative plastic strain is used instead of allowable deviator stress, see Section 5.1.3)

ρ_a is the allowable total subgrade plastic deformation for the design period

N is the total equivalent number of load repetitions during the design period

P_d is the design dynamic wheel load

σ_s is the soil compressive strength

$a, m,$ and b are the material parameters dependent on soil type (see Table 5.2)

A is the area factor selected to make deformation influence factor dimensionless (0.645 m^2 or $1,000$ in.^{2})

5.1.3 Development of Li–Selig method

This section gives more details that led to the development of the Li–Selig design method. It describes the typical subgrade failure types that the design is intended to prevent, how these failures are characterized in terms of stress, strain and deformation equations, design criteria, and design chart development.

5.1.3.1 Subgrade failure criteria

In the Li–Selig design method, the design is to prevent two most common subgrade failures due to repeated stress applications: progressive shear failure and excessive plastic deformation (ballast pocket).

Subgrade progressive shear failure is shown in Figure 5.6. This type of subgrade failure was first reported by the European railways (ORE 1970).

Table 5.2 Values of soil parameters a, b, and m for various soil types (Chapter 4)

Soil type	a	b	m
CH (fat clay)	1.2	0.18	2.4
CL (lean clay)	1.1	0.16	2.0
MH (elastic silt)	0.84	0.13	2.0
ML (silt)	0.64	0.10	1.7

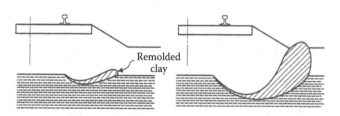

Figure 5.6 Subgrade progressive shear failure.

Many field investigations in the United States, especially for tracks under heavy axle load operations, have shown this type of failure, especially for subgrade consisting of fine-grained soils without adequate granular layer thickness.

Subgrade progressive shear failure develops at the subgrade surface as the soil is gradually sheared and remolded due to repeated overstressing. The surface soil gradually squeezes outward and upward following the path of least resistance. This failure occurs primarily on subgrades of fine-grained soils, particularly on those with high clay content. As shown, a subgrade depression beneath the track is matched by a corresponding heave of soils at the track side. Heaved soils in the ballast shoulder will hinder track drainage. The depression beneath the track traps water entering from the above granular layer, thus aggravating subgrade failure. As progressive shear failure develops at the subgrade surface, any subgrade with a soft layer on the top is susceptible to this failure due to repeated loading.

The second common subgrade failure due to repeated loading for fine-grained soil subgrade is excessive plastic deformation that results in ballast pocket, as shown in Figure 5.7. Although subgrade progressive shear failure is accompanied by progressive shear deformation, ballast pockets grow from the vertical component of progressive shear deformation, deformations caused by progressive compaction, or consolidation of the entire subgrade layer due to repeated loading. As illustrated in Figure 5.7, ballast pocket growth may be accompanied by little subgrade heave at the track side. A subgrade with large ballast pockets often has soft soils extending to a substantial depth.

These two types of subgrade failures are not completely independent. They both occur as a result of soft subgrade soils subjected to large repeated stress applications. However, progressive shear failure can be considered mainly a subgrade surface failure, while excessive plastic deformation can be considered a failure influenced by a substantial depth of subgrade. An adequate granular layer is intended to prevent both types of subgrade failures.

Other types of subgrade failures such as subgrade attrition (by ballast) with mud pumping, massive subgrade shear failure (slope stability failure), and excessive consolidation settlement due to self-weight may also occur.

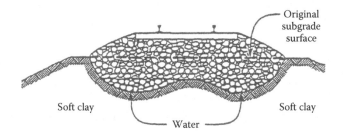

Figure 5.7 Excessive subgrade plastic deformation (ballast pocket).

Increasing granular layer thickness is not the approach to prevent these other types of subgrade failures.

Subgrade progressive shear failure and excessive plastic deformation (ballast pocket) under repeated loading occur mainly in subgrades comprised of fine-grained soils and can be related to subgrade cumulative plastic strain (ε_p) and deformation (ρ), as represented by the following two equations (Chapter 4):

$$\varepsilon_p(\%) = a\left(\frac{\sigma_d}{\sigma_s}\right)^m N^b \qquad (5.4)$$

$$\rho = \int_0^T \varepsilon_p dt \qquad (5.5)$$

where:
    ε_p is the cumulative soil plastic strain
    N is the number of repeated stress applications
    $(\sigma_d = \sigma_1 - \sigma_3)$ is the deviator stress due to train loads
    σ_s is the soil compressive strength
    a, m, and b are the parameters dependent on soil type (see Table 5.2)
    ρ is the cumulative soil plastic deformation
    T is the subgrade layer depth (Figure 5.1)

By reviewing Equations 5.4 and 5.5, it can be seen that deviator stress is the main stress factor influencing performance of fine-grained soil subgrades, rather than vertical stress or confining stress alone. Because shear stress is simply half of deviator stress, deviator stress can be considered to have the physical meaning of shear stress.

Effects of soil physical state such as moisture content and dry density on subgrade performance are indirectly represented by soil static strength, σ_s, which is directly dependent on soil physical state and soil structure. The effect of soil type is reflected by material parameters (a, m, b).

Design for preventing subgrade progressive shear failure is thus to limit the total cumulative plastic strain at the subgrade surface to an allowable level for the design period:

$$\varepsilon_p \leq \varepsilon_{pa} \qquad (5.6)$$

where:
    ε_p is the total cumulative plastic strain at the subgrade surface for the design period
    ε_{pa} is the allowable plastic strain at the subgrade surface for the design period

Because there is a direct relationship between cumulative plastic strain and deviator stress at the subgrade surface (see Equation 5.9), this criterion can be converted into

$$\sigma_d \leq \sigma_{da} \tag{5.7}$$

To prevent excessive subgrade plastic deformation, the design is intended to limit the total cumulative plastic deformation of the subgrade layer to an allowable level for the design period:

$$\rho \leq \rho_a \tag{5.8}$$

where:

   ρ is the total cumulative plastic deformation of the subgrade layer for the design period
   ρ_a is the allowable plastic deformation of the subgrade layer for the design period

From Equations 5.6, 5.7, and 5.8, two approaches can be employed to control cumulative plastic strain at the subgrade surface and cumulative plastic deformation of the subgrade layer. Allowable plastic strain or deformation should not be understood as being related to the cycle of ballast surfacing and tamping. Rather, they are the criteria used to assure that the subgrade would perform adequately within its design period. To achieve this, the first approach is to improve subgrade soil strength, and the second approach is to reduce deviator stress transmitted to the subgrade. Designing granular layer thickness belongs to the second approach and is the focus of this chapter.

5.1.3.2 Effect of granular layer on subgrade stress

A key element in the development of granular layer thickness design method is reasonable calculation of subgrade stresses due to train axle loads. Railway track and subgrade consists of multiple layers, each with different properties. The top granular layer usually possesses much higher stiffness than the underlying subgrade layer. The stiffer granular layer spreads and transmits train loads to the subgrade. A major limitation of most design methods in the literature is the assumption of a homogeneous half-space for all these layers. This assumption neglects different properties of individual layers and often leads to overestimation of subgrade stresses due to train loads.

Several multilayer track models are available for analysis of stresses in the track and subgrade (Chapter 4). A common feature of these multilayer models is the incorporation of all major components of track and subgrade, that is, rails, ties, fasteners, ballast, subballast, and subgrade layers. In addition, a finite element model can adequately characterize different layers of track structure.

A parametric study of deviator stress in the subgrade using GEOTRACK (Li 1994) showed that for a given wheel load, the most important factors reducing deviator stress in the subgrade are the thickness of the granular layer and its resilient modulus.

Figure 5.8 shows the effects of granular layer thickness and resilient modulus on deviator stress at the subgrade surface underneath the tie (half-a-tie length). The difference between Figure 5.8a and b is the granular layer resilient modulus. As can be seen, for the smallest granular layer thickness (0.30 m), deviator stress at the subgrade surface is the highest toward the tie end and is the lowest toward the tie center. This is consistent with observed progressive shear failures in the field, which often exhibited the largest subgrade depressions near the tie ends (Li and Selig 1995a, b). As illustrated in Figure 5.8, an increase in granular layer thickness not only reduces the magnitude of deviator stress but also results in a more uniform distribution of deviator stress across the track (note that the deviator stress becomes constant across the track at a granular layer thickness of 1.07 m in Figure 5.8a and at a granular layer thickness of 0.76 m in Figure 5.8b).

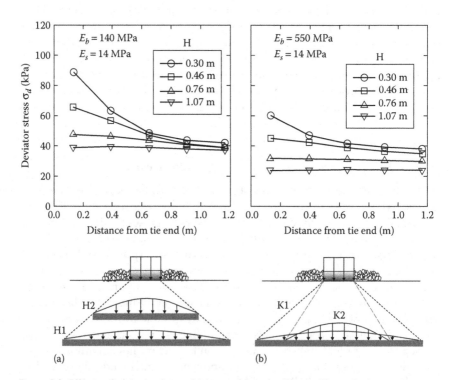

Figure 5.8 Effects of granular layer thickness (a) and stiffness (b) on deviator stress at subgrade surface with illustration on the bottom graph.

For the same granular layer thickness, an increase in granular layer stiffness (i.e., resilient modulus) also significantly reduces the deviator stress at the subgrade surface. This is evident by the comparison of deviator stress results shown in Figure 5.8a and b. This effect of granular layer stiffness clearly indicates the limitation of a homogeneous half-space assumption for the combined granular layer and subgrade.

The effects of granular layer thickness and granular layer stiffness on subgrade stress are further illustrated in the bottom graph of Figure 5.8.

Figure 5.9a shows the effect of granular layer thickness on deviator stress distribution with subgrade depth for a soft subgrade (resilient modulus = 14 MPa). As can be seen, for any of the four granular layer thicknesses selected, deviator stress in the subgrade decreases with subgrade depth. When the granular layer is much stiffer than the subgrade layer as in this case, increasing granular layer thickness reduces deviator stress in the subgrade through two effects. First, the depth of the subgrade surface from the tie bottom is increased. This leads to an automatic reduction of deviator stress in the subgrade because the depth of any point in the subgrade from the tie bottom increases. Second, the stress-spreading capability of the granular layer increases. However, this second effect will diminish if the stiffnesses of the granular layer and subgrade is close to each other, as shown by the results in Figure 5.9b. As can be seen in Figure 5.9b, increasing granular layer thickness for reducing deviator stress in the subgrade is achieved mainly through its first effect, that is, resulting from an increased depth of the subgrade from the tie bottom.

A comparison between Figure 5.9a and b also indicates that, other conditions being the same, a stiffer subgrade will result at a higher deviator stress

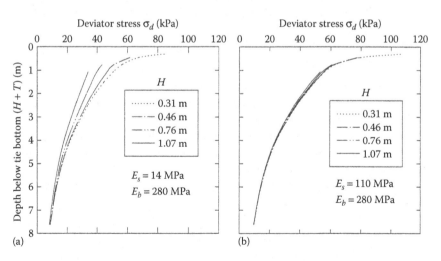

Figure 5.9 Distribution of deviator stress with depth influenced by granular layer thickness: (a) soft subgrade; (b) stiff subgrade.

in the subgrade. This does not mean, however, that a stiffer subgrade would perform worse because it has a higher deviator stress. This is because subgrade performance is influenced not only by its stress state but also by its soil strength. Obviously, a stiffer subgrade will possess a higher strength, and the increase in strength is generally much larger than the increase in deviator stress (Li 1994).

5.1.3.3 Design chart development

In the Li–Selig design method, granular layer includes both ballast and subballast layers, as shown in Figure 5.1, and is characterized by its thickness (H) and resilient modulus (E_b). Subgrade is simplified as one homogeneous deformable layer (T) underlain by a rigid boundary. The properties considered for the subgrade layer include resilient modulus (E_s), soil compressive strength (σ_s), and soil type. Although soil compressive strength is used, design is not based on soil static strength. Instead, the design criteria are based on subgrade cumulative plastic strain (or deviator stress) and total permanent deformation under repeated loading, as represented in Equations 5.6, 5.7, and 5.8.

To prevent progressive shear failure, the first step in the design is to determine the allowable deviator stress at the subgrade surface (σ_{da}) from Equations 5.4, 5.6, and 5.7, or as follows:

$$\sigma_{da} = \left(\frac{\varepsilon_{pa}}{aN^b 100} \right)^{1/m} \sigma_s \tag{5.9}$$

where:
  ε_{pa} is the allowable cumulative plastic strain at the subgrade surface (without %)
  N is the number of load cycles for the design period
  σ_s is the soil compressive strength

Appendix A.1 shows the actual charts that can be used to determine allowable deviator stress from allowable cumulative plastic strain for different soil types and different load cycles.

When one prefers to design the granular layer thickness starting directly from allowable deviator stress, then the previous equation is not needed. However, it is recommended that the design to prevent progressive shear failure start with the allowable cumulative plastic strain at the subgrade surface.

The next step is to determine deviator stress at the subgrade surface as a function of granular layer thickness, and resilient moduli of the granular and subgrade layers. Because stress analysis using GEOTRACK or other models is often done assuming linear elastic track and substructure, the ratio of deviator stress to design dynamic wheel load is constant for a given track structure. This allows the results to be represented by the following dimensionless strain influence factor:

$$I_\varepsilon = \frac{\sigma_d A}{P_d} \tag{5.10}$$

where:
σ_d is the deviator stress at the subgrade surface
P_d is the design dynamic wheel load
Parameter A is an area factor arbitrarily selected to make the strain influence factor dimensionless. Its value is 1,000 in.^{2} when English units are used or 0.645 m^2 when SI units are used

Figure 5.10 shows the design charts developed for a track foundation with given resilient moduli of granular and subgrade layers (also in Appendix A.2).

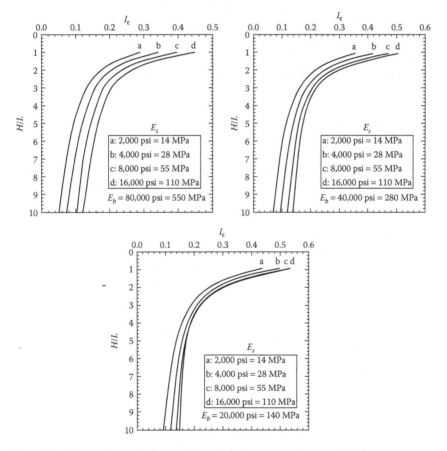

Figure 5.10 Design charts for preventing subgrade progressive shear failure.

As illustrated, each graph is developed for one granular layer modulus but includes four different subgrade moduli. To make the design charts dimensionless, granular layer thickness (H) is also normalized by a length factor (L). The length factor value is arbitrarily chosen to be 6 in. if English units are used or 0.152 m if SI units are used. There is no specific physical meaning associated with this length factor. Its only purpose is to be able to directly use these charts in both SI and English units.

To develop design charts to prevent excessive plastic deformation of the deformable subgrade layer of depth T, cumulative plastic deformation needs to be obtained through the integration of cumulative plastic strain as follows:

$$\rho = \frac{aN^b}{100(\sigma_s)^m} \int_0^T (\sigma_d)^m \, dt \tag{5.11}$$

Equation 5.11 can be rearranged as

$$\rho = I_\rho La \left(\frac{P_d}{\sigma_s A} \right)^n \frac{N^b}{100} \tag{5.12}$$

Where I_ρ is a dimensionless deformation influence defined as follows:

$$I_\rho = \int_0^T \left(\frac{\sigma_d A}{P_d} \right)^m \frac{dt}{L} \tag{5.13}$$

Note that the area and length factors (A and L) previously defined are incorporated to make the deformation influence factor dimensionless.

As indicated by Equation 5.13, deformation influence factor is a function of deviator stress distribution with depth in the subgrade and the subgrade layer depth. Deviator stress distribution in turn is dependent on granular layer thickness and resilient moduli of granular layer and subgrade. Figure 5.11 gives an example of this relationship. As shown, an increase in subgrade layer depth results in an increase in deformation influence factor. This increase can be seen to be significant until at a depth of approximately 4.5–6.0 m. In other words, subgrade at a depth of up to 4.5–6.0 m can significantly contribute to the total plastic deformation.

Deformation influence factor values were calculated using Equation 5.13 for various combinations of subgrade layer depths, granular layer thicknesses, four subgrade soil types (Table 5.2), and various moduli of granular and subgrade layers. Design charts have been developed with deformation influence factor plotted against normalized granular layer thicknesses for various granular and subgrade layer conditions.

Figure 5.12 shows two examples of the design charts. Each curve corresponds to one subgrade layer depth. Each chart is for one soil type, and one

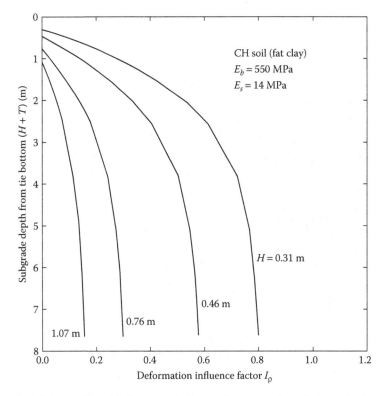

Figure 5.11 Relationship of deformation influence factor to subgrade layer depth.

modulus combination between granular and subgrade layers. A total of 48 design charts have been developed and are given in Appendix A.3 (also in a report by Li et al. 1996).

5.1.4 Application of Li–Selig method

This section describes the actual procedures and examples of using the Li–Selig method for designing granular layer thickness for the ballasted track structure.

5.1.4.1 Design traffic

The Li–Selig method emphasizes the influence of repeated traffic loads on subgrade performance. Thus, the design traffic parameters are

- Static wheel loads
- Train speed
- Total tonnage

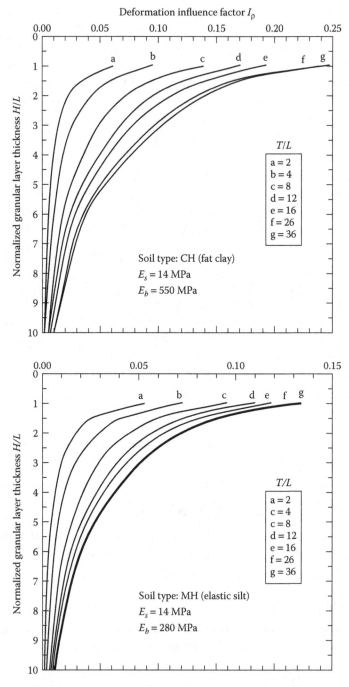

Figure 5.12 Examples of granular layer thickness design charts for preventing excessive plastic deformation.

These parameters are converted into two design variables: the design dynamic wheel load, P_d, and the total equivalent number of design load applications, N, for the design period. The design period is defined as the time for the track to carry the desired amount of traffic without subgrade progressive shear failure or excessive deformation.

A dynamic wheel load, P_{di}, corresponding to a static wheel load, P_{si}, can be determined based on the statistical analysis of past measurement results, based on modeling simulations (Chapter 2), or can simply be calculated based on the AREMA recommended equation as follows:

$$P_{di} = \left(1 + \frac{0.0052V}{D}\right) \tag{5.14}$$

where:
   V is the train speed (km/h)
   D is the wheel diameter (m)

Characteristics of typical freight cars and locomotives in North America are given in Table 5.3. However, the previous equation is valid only for operating speeds of freight vehicles, as it was derived based on lower-speed freight operations. Design examples in this chapter as well as in Chapter 10 include other methods used for determining dynamic wheel loads for design.

The number of repeated load applications, N_i, of any wheel load, P_{si}, during the design period is determined by the following equation, assuming four-axle passes under two adjacent bogies from two adjacent cars as one loading cycle for the subgrade (Chapter 2):

$$N_i = \frac{T_i}{8P_{si}} \tag{5.15}$$

where:
   T_i is the total "tonnage" for the wheel load P_{si}, for the design period but in the same unit as T_i

Table 5.3 Characteristics of typical freight cars and locomotives in North America

Car type	Wheel diameter (m)	Static wheel load (kN)	Axles per vehicle	Axle spacing on a truck (m)
125 tons	0.97	173	4	1.83
110 tons	0.91	160	4	1.78
100 tons	0.91	147	4	1.78
70 tons	0.84	125	4	1.73
Six-axle locomotive	1.02	138	6	3.40
Four-axle locomotive	1.02	142	4	2.84

The axle spacing for the six-axle locomotive is for that between the "outside" axles of a truck. It should be noted that because the spacings between trucks on the ends of passenger cars may be significantly larger than those of freight cars, as shown in Table 5.3, there may be circumstances where the axles of a single truck will constitute one load pulse to the subgrade for passenger car, rather than two trucks combining to form one load pulse, as assumed in Equation 5.15. For such a case, the quantity "8" in Equation 5.15 should be replaced with "4."

To represent the influence of all levels of wheel loads on subgrade performance, the following equation is used to convert N_i cycles of wheel load, P_{di}, to N_i^0 cycles of the design wheel load, P_d (Chapter 4):

$$N_i^0 = N \left(\frac{P_{di}}{P_d} \right)^{m/b} \tag{5.16}$$

Where m, b = parameters dependent on soil type (Table 5.2).

Thus, the total equivalent number of load applications, N, for the design load, P_d, is calculated by

$$N = N_1^0 + ... + N_i^0 + ... \tag{5.17}$$

5.1.4.2 Material properties

As shown in Figure 5.1, the material property considered for the granular layer is resilient modulus, E_b. A stiffer granular layer will result in a lower stress level in the subgrade caused by repeated traffic loading.

Many factors affect resilient modulus of the granular layer, including ballast type, degree of fouling, stress state, and compaction level of the granular layer. In Table 5.4, some typical values of resilient modulus for granular layer are given corresponding to quality of the ballast and subballast layers (Selig and Waters 1994).

Table 5.4 Properties of granular material and subgrade soil

Material	Resilient modulus (MPa)	Compressive strength (kPa)
Granular material		
Good	552	–
Intermediate	276	–
Poor	138	–
Subgrade		
Stiff	69–128	207–344
Medium	28–69	103–207
Soft	6.9–28	34–103

Subgrade is simplified as one deformable layer overlying a rigid boundary. The material properties considered for subgrade include resilient modulus, E_s, soil compressive strength, σ_s, and soil type. In general, other conditions being the same, a stiffer subgrade will result in a higher stress level in the subgrade. However, this higher level of stress will be compensated for by the corresponding higher strength of subgrade soil. For a stiffer subgrade, the increase in soil strength is generally greater than the increase in subgrade stress level. Table 5.4 gives the ranges of resilient modulus and soil compressive strength for various subgrade soil conditions (Li 1994).

5.1.4.3 Design criteria

Two criteria are used to design granular layer thickness for preventing subgrade failure. One design criterion is based on limiting cumulative plastic strain at the subgrade surface and is intended to prevent subgrade progressive shear failure (Equations 5.6 and 5.7). The other criterion is intended to prevent excessive subgrade plastic deformation (Equation 5.8). Both criteria should be evaluated to determine the one that gives larger granular layer thickness in each case.

In the following, the design procedure based on the criterion for preventing subgrade progressive shear failure is referred to as design procedure 1, and the design procedure based on the criterion for preventing excessive subgrade plastic deformation is referred to as design procedure 2.

5.1.4.4 Design procedure I

Figure 5.13 is the flow chart for designing granular layer thickness based on limiting cumulative plastic strain at the subgrade surface. The design procedure consists of the following three steps:

1. Prepare the information required for the design, including:
   a. *Traffic condition*: Dynamic wheel load must be determined based on the statistical analysis of past measurements, dynamic modeling simulation, or using Equation 5.14, and the number of load repetitions for the design period can be determined using Equation 5.15. If there are several major groups of wheel load magnitudes, the corresponding dynamic wheel loads and the numbers of load repetitions should be determined separately. Then, Equations 5.16 and 5.17 can be used to determine the total equivalent number of load repetitions, N, for the design dynamic wheel load, P_d.
   b. *Allowable strain*: Magnitude of the allowable cumulative plastic strain at the subgrade surface, ε_{pa}, must be determined for a certain number of load repetitions (i.e., for the design period).
   c. *Subgrade characteristics*: Subgrade soil type, soil compressive strength, σ_s, and soil resilient modulus, E_s, must be determined.

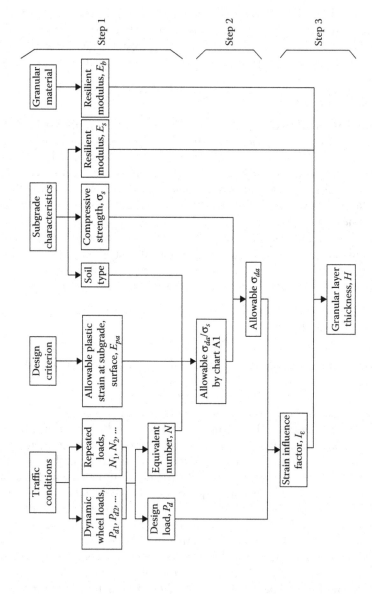

Figure 5.13 Flowchart for design procedure 1.

d. *Granular material*: Resilient modulus, E_b, for granular material must be specified.

2. Determine the allowable deviator stress at the subgrade surface. This can be done using Equation 5.9 or charts in Appendix A.1. This is completed based on the information obtained from Step 1, that is, soil type, the allowable cumulative plastic strain at the subgrade surface for the design period, and soil compressive strength. However, as an alternative to the allowable strain determined from the first step, the allowable deviator stress at the subgrade surface may be selected directly at this step.

3. Select the required granular layer thickness to prevent subgrade progressive shear failure as follows:

   a. Calculate the strain influence factor, I_ε, by Equation 5.2.

   b. Determine the value of H/L corresponding to the strain influence factor, I_ε, using the design charts in Figure 5.10 (or Appendix A.2), based on the values of granular layer resilient modulus, E_b, and the subgrade resilient modulus, E_s.

   c. Multiply H/L by length factor L to get the required granular layer thickness, H. L is equal to 6 in. for English units or 0.152 m for SI units.

5.1.4.5 Design procedure 2

Figure 5.14 is the flow chart for the design procedure based on the criterion of limiting total plastic deformation of the subgrade layer. Design charts for various subgrade and granular layer conditions are given in Appendix (also in a document by Li et al. 1996). Again, granular layer thickness design consists of the following three steps:

1. Prepare all the information required for the design: In addition to the information required in design procedure 1, design procedure 2 requires knowledge of the thickness of the deformable subgrade layer, T. The allowable cumulative plastic strain at the subgrade surface for design procedure 1 is replaced by the allowable total plastic deformation of the subgrade layer for design procedure 2.

2. Calculate deformation influence factor, I_ρ, from Equation 5.3.

3. Select the required granular layer thickness to prevent excessive subgrade plastic deformation as follows:

   a. Select a design chart such as those in Figure 5.12, which best corresponds to soil type, subgrade resilient modulus, and granular layer resilient modulus.

   b. Calculate T/L, and locate the point in the design chart (Appendix A.3) corresponding to I_ρ and T/L.

   c. Obtain the value of H/L for that point, and multiply H/L by the length factor, L, to get granular layer thickness, H.

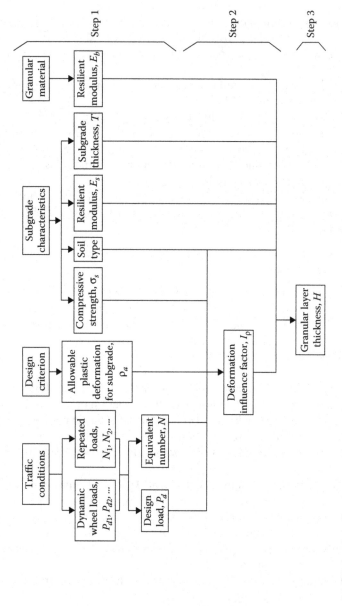

Figure 5.14 Flow chart for design procedure 2.

Table 5.5 Example granular layer thickness design

Design parameter	Value
Traffic condition	$P_s = 173$ kN
	$V = 64$ km/h
	Design tonnage $= 60$ MGT
	$D = 0.97$ m
Subgrade soil	$\sigma_s = 90$ kPa
(Clay. CH)	$E_s = 14$ MPa
	$T = 1.5$ m
Granular layer material	$E_b = 276$ MPa
Design criteria	$\varepsilon_{pa} = 2\%$
	$\rho_a = 25$ mm

Example 5.1

The design procedures are illustrated using an example of a test track built on a soft clay subgrade under repeated heavy axle loadings. The information needed for the granular layer thickness design is given in Table 5.5.

PROCEDURE 1

Step 1: The first step is to obtain the information needed for the design, that is, traffic condition, subgrade soil and granular material properties, and the allowable subgrade strain (see Table 5.5).
The dynamic wheel load is determined by Equation 5.14 to be: $P_d = 236$ kN; and the equivalent number of load repetitions is determined by Equation 5.15 to be: $N = 380,000$.

Step 2: For the allowable cumulative plastic strain, $\varepsilon_{pa} = 2\%$, and $N = 380,000$, the allowable deviator stress from Equation 5.9, $\sigma_{da} = 47$ kPa.

Step 3: The strain influence factor is determined by Equation 5.2 to be: $I_\varepsilon = 0.128$. From design chart Figure 5.10, the required H/L is found to be 4.3 for $E_b = 276$ MPa, $E_s = 14$ MPa, and $I_\varepsilon = 0.128$. Therefore, the required granular layer thickness, $H = 4.3 \times 0.152 = 0.65$ m.

PROCEDURE 2

Step 1: Again, using the information given in Table 5.5, the dynamic wheel load is: $P_d = 236$ kN; and the equivalent number of load repetitions is: $N = 380,000$.

Step 2: By Equation 5.3, the deformation influence factor is determined to be: $I\rho = 0.047$.

Step 3: Using the design chart (Appendix A.3) for $E_b = 276$ MPa, $E_s = 14$ MPa, $T/L = 1.5/0.152 = 10$, and $I_p = 0.047$, the required H/L is found to be 4.6. Therefore, the required granular layer thickness, $H = 4.6 \times 0.152 = 0.70$ m.

For this example, the required granular layer thickness for procedure 2 is larger than the granular layer thickness determined for procedure 1. Thus, procedure 2 governs.

5.1.5 Comparisons with test results

The following gives comparisons between the results using the Li–Selig design method and the actual results from field tests. One field case involved a test track under heavy axle load operations, and another field case involved two revenue service sites under mixed freight and passenger train operations.

5.1.5.1 Low track modulus test track

A low track modulus (LTM) test track 183 m in length was constructed on a soft soil subgrade at the High Tonnage Loop, Pueblo, Colorado. Prior to the LTM track construction, a pilot low track modulus (TLTM) test track 30 m in length was built and tested to evaluate the feasibility of constructing and testing the longer LTM track. The main purpose of the LTM test track was to study the effect of a soft subgrade on track performance under heavy axle loads.

As the natural subgrade soil at the test site is a silty sand and does not represent a soft subgrade soil, a trench 3.66 m in width was excavated in the natural subgrade to a depth of 1.5 m from the ground surface, and backfilled with the Mississippi buckshot clay ($PI = 40 - 45$, $LL = 60 - 70$). The clay was compacted at a specified high moisture content and dry density to achieve a subgrade of low stiffness. Although the specified soil moisture content for both the TLTM and LTM subgrades was 30%, the actual average moisture content was 29% for the TLTM subgrade and 33% for the LTM subgrade. Accordingly, the corresponding soil static strength was approximately 166 kPa for the TLTM track subgrade, and 90 kPa for the LTM subgrade. The corresponding soil resilient modulus was approximately 41–55 MPa for the TLTM subgrade, and 14–21 MPa for the LTM track subgrade. Table 5.6 summarizes the required information for the design and the corresponding design results using design procedure 2.

The Li–Selig design method was not available at the time when the TLTM and LTM tracks were built. However, field measurements of track performance and subgrade conditions were conducted throughout the testing for both TLTM and LTM, which provided excellent opportunities to evaluate this design method. Furthermore, in the phase 2 testing of the LTM track, the design method presented here was applied directly in the reconstruction of the LTM granular layer after failure of the LTM subgrade in phase 1.

Table 5.6 Design results for TLTM and LTM test tracks

Design parameter	Value
Traffic condition	See Table 5.5
Granular material	See Table 5.5
Design criteria	$\rho_a = 25$ mm
Subgrade	Soil type: CH (fat clay), $T = 1.5$ m
	LTM ($\sigma_s = 90$ kPa, $E_s = 14$ MPa)
	TLTM ($\sigma_s = 165$ kPa, $E_s = 41$ MPa)
Design results	LTM: $H = 0.7$ m
	LTLM: $H = 0.3$ m

The original adopted thickness for both the TLTM track and LTM phase 1 track was 0.45 m (0.30 m ballast plus 0.15 m subballast), based on an assumption of 30% soil moisture content and a minimum density of 90% of the standard maximum dry density. The TLTM track with such a granular layer thickness was able to sustain heavy axle loads for approximately 53 MGT of traffic without generating excessive plastic deformation in the subgrade. However, the LTM track with the same 0.45 m granular layer thickness had much more difficulty maintaining the required track surface geometry. The subgrade for the LTM track experienced rapid progressive shear failure and the test track required frequent surfacing. Eventually, the traffic had to be stopped at approximately 62.3 MGT for rebuilding the test track.

A comparison between the design results in Table 5.6 and the originally adopted thickness indicates that the adopted 0.45 m thickness for the TLTM track was larger than the required 0.30 m thickness, but the adopted 0.45 m thickness for the LTM phase 1 track was much smaller than the required 0.70 m thickness. Thus, the TLTM track was able to maintain the required track geometry, while the LTM phase 1 track was not. Obviously, the field test results were consistent with the design results.

After the failure of the LTM subgrade in phase 1, the test section was rebuilt based on the design results using the Li–Selig method. The granular layer thickness was increased to 0.68 m (0.30 m ballast and 0.38 m subballast). Traffic then resumed and measurements showed little track geometry degradation until at 9.3 MGT when the heavy rainfall during train operations caused sudden deterioration of the track geometry. The causes for this track failure were attributed to the loss of granular layer stiffness due to saturation and excessive pore water pressure buildup in the thick subballast layer under train loading (Chapter 6; Read and Li 1995).

5.1.5.2 Mix passenger and freight revenue service sites

Two track sites between Philadelphia and Baltimore required frequent track maintenance. One site was in Edgewood, Maryland. The tracks at this site experienced recurring problem of differential settlement and required ballast tamping and surfacing at least twice a year over a distance of approximately 10 km. Remediation methods such as lime slurry injection and geotextile application had been applied in 1984, but were not successful. The other site was located in Aberdeen, Maryland. At this site one track approximately 60 m long in a multitrack system had a mud pumping problem, while geometry deterioration was not a main concern. A section of track roughly 30 m long was undercut and the fouled ballast was replaced in 1991. However, the pumping problem extended from where it occurred before.

To find out the major causes of the problems and to determine material characteristics for the remedy design, field investigations and laboratory tests were conducted in 1994. For the Edgewood site, the major cause of the differential track settlement was the nonuniform soft subgrade support, resulting from variation in subgrade soil strength aggravated by water trapped along the track. Subgrade progressive shear failure caused significant squeezing and heave of the subgrade soils across the track. Deep ballast pockets were also observed at several locations. Poor track drainage due to fouled ballast also contributed to the problem because the rainfall water could not drain out of the track easily, thus causing softening of the subgrade soils. For the Aberdeen site, ballast breakdown and abrasion was the major reason for the ballast fouling. Mud pumping occurred when the hydraulic actions caused by traffic loading and the free water in the ballast ejected the mud slurry onto the ballast surface. The subgrade was strong and no progressive shear failure and ballast pockets were observed.

Remedial designs included evaluation of the granular layer thicknesses for both the Edgewood and Aberdeen sites. The traffic was a mix of passenger trains with speeds up to 190 km/h and freight trains with the heaviest car being 120 tons. Table 5.7 gives the traffic values used for designs for these two sites. Note that dynamic wheel loads were available from the wayside track instrumentation.

The subgrade soils at the Edgewood site were generally lean clay with unconfined compressive strengths in the range of 48–83 kPa. The subgrade soils at the Aberdeen site were either sandy lean clay or silty sand with gravel. The unconfined compressive strength varied from 97 to 290 kPa. Table 5.8 gives the subgrade soil properties required for design for both sites.

Table 5.7 Traffic parameters for amtrak sites

Traffic type	Annual tonnage	Dynamic loads wheel load (kN) percentage		Static loads wheel load (kN) percentage	
Freight	37 MGT	178	50%	156	40%
		67	50%	44	60%
Passenger	15 MGT	156	20%	67	100%
		89	80%		

Table 5.8 Evaluation of track granular layer thickness for amtrak sites

Parameters	Edgewood site	Aberdeen site
Subgrade	Lean clay (CL)	Sandy lean clay (CL)
	$\sigma_s = 48$–83 kPa	$\sigma_s = 97$–290 kPa
	$E_s = 14$ MPa	$E_s = 28$ MPa
Granular material	$E_b = 280$ MPa	$E_b = 280$ MPa
Design criteria	$\rho_a = 25$ mm	$\rho_a = 25$ mm
Required granular thickness	1.0 m	0.4 m
Actual thickness	0.3–0.5 m	0.7–1.0 m
Actual subgrade performance	Subgrade progressive shear failure and ballast pocket	No subgrade failure

Other required information for design and the corresponding design results are also given in Table 5.8. Based on the design, the required granular layer thickness for the Edgewood site should be 1.0 m, while the required thickness for the Aberdeen site is only 0.40 m. The difference in required thickness between these two sites is a result of the difference in subgrade soil strength between them.

However, the existing granular layer thickness at the Edgewood site was found to be only 0.30–0.50 m from field cross-trench measurements and from cone penetration tests (CPT). Obviously, this existing thickness was not sufficient to reduce traffic load-induced stresses in the subgrade below the allowable value to prevent progressive shear failure. Consequently, significant subgrade progressive shear failure and ballast pockets occurred, causing excessive differential track settlement.

On the other hand, the existing granular layer thickness at the Aberdeen site ranged from 0.70 to 1.0 m, larger than the required design thickness. As discussed previously, there were no subgrade failures at the Aberdeen site, that is, the traffic-induced stresses transmitted to the subgrade were lower than the required to prevent subgrade failure. In other words, mud pumping at this location was not subgrade related; rather it was from ballast degradation and poor track drainage conditions.

5.1.6 Initial granular layer construction thickness

In the parallel discipline of highway pavement design, every new construction needs to comply with local- or national-level design standards or codes. These standards or codes are updated as new research results and new technologies are put forth. In contrast, emphasis has not been adequately given to the proper design of initial granular layer depth during construction of new railway track. A railway track differs from a highway pavement mainly in that the track granular layer is adjustable by means of track surfacing and ballast tamping operations. As such, there exists a common but mistaken belief that a subgrade deformation due to overstressing can be compensated for by the addition of ballast and tamping. However, many investigations into track substructure problems have shown that once a subgrade becomes excessively deformed, resulting track problems can rarely be corrected using conventional maintenance practices and remedies.

When the initial granular layer thickness during construction is less than required, subgrade will deform or fail progressively, which will lead to poor track geometry. As a result of soil shearing and remolding, soil strength will decrease. A deformed subgrade surface will usually trap water, further aggravating subgrade conditions. Repeated ballast tamping used to compensate for the loss of track geometry will cause more ballast to displace the weaker subgrade soil, leading to a notorious subgrade problem called a ballast pocket.

Figure 5.15 shows an example of a large ballast pocket found in a problem track site in revenue service. The result shown in this figure is the tip resistance from a CPT. The track at this site was built over a very soft subgrade. The annual traffic was 50 MGT. This track required tamping every two to 3 months due to excessive track geometry deviations. The soft clay subgrade allowed formation of roughly 2 m deep ballast pockets, due to frequent ballast placement and tamping to compensate for the loss of track geometry.

Due to a distorted subgrade soil profile, water was trapped with the granular layer, further weakening the soil and accelerating its deformation. Because of very low soil strength and the standing water trapped in the pocket, repeatedly adding more ballast was ineffective to fix this problem for long term. In fact, the ballast layer was so large that it adversely imposed a high overburden stress to the soft subgrade (48 kPa in this case).

This example also indicates another important aspect in designing initial granular layer thickness. When the required thickness is too large, other alternatives such as replacing or improving soft subgrade soil should be considered. In many cases, designing and constructing an initial large ballast depth may not be effective. In fact, when a required thickness exceeds a value of roughly 0.9–1.1 m, which usually corresponds to a soft subgrade,

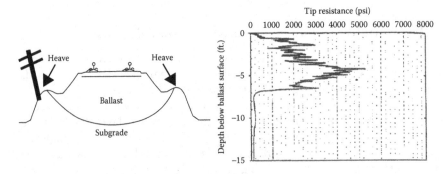

Figure 5.15 Large ballast pocket as a result of repeatedly adding ballast.

a slight increase in load-induced subgrade stress will require a large amount of additional ballast depth. In other words, for a very soft subgrade, the required granular layer thickness is not only large, but also very sensitive to variation of dynamic wheel loads. As can be seen in Figures 5.10 and 5.12, slopes of the design curves increase with increasing normalized granular layer thickness (H/L). When a granular layer thickness becomes too large, a slight decrease in the value of strain or deformation influence factor, which is a direct function of soil strength and dynamic wheel load, will cause a large increase in the required granular layer thickness.

On the other hand, however, as discussed by Hay (1982), a granular layer thickness equal to tie spacing should be considered a desirable minimum. This is because effects of adjacent ties will be superimposed on each other, leading to a relatively uniform stress transmitted to the subgrade.

5.2 ASPHALT TRACK

5.2.1 Asphalt track foundation

An asphalt track foundation is designed to have a layer of asphalt between the ties and the subgrade. This asphalt layer is often constructed when asphalt mix is hot, thus it is called HMA. An asphalt track foundation can be of two different designs: an overlayment HMA trackbed and an underlayment HMA trackbed (see Figure 5.16; Rose 2013). In the underlayment

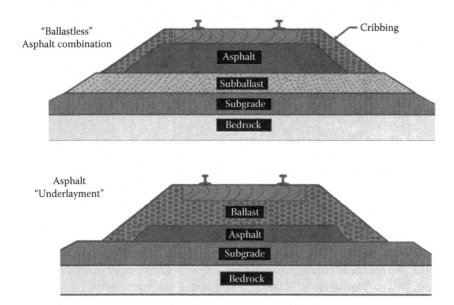

Figure 5.16 Types of asphalt track foundations. (Courtesy of Professor Jerry Rose.)

design, the asphalt is used as a mat or as part of the subballast layer between the ballast and subgrade. In the overlayment or full-depth design, the asphalt mat is placed directly on the subballast and sleepers are placed directly on top of the asphalt. In practice, the underlayment design is preferred, especially for heavy axle load freight operations. This is because this design maintains the ballast within the track structure so that the track geometry can still be easily adjusted when required. An additional benefit of asphalt underlayment is that it is covered by ballast that protects the asphalt from sunlight and keeps temperature variance in a low level.

Asphalt track foundations, particularly the underlayment design, have been used for rehabilitating existing lines with track substructure problems as well as for new track construction projects. It can be used not only in heavy axle load freight track, but also for light rail, commuter rail, and high-speed passenger rail. Use of asphalt track foundation is often based on anticipated track substructure problems if conventional ballasted track is used, or as a means to extend the long-term performance of the track foundation and the life of track components.

As compared to conventional ballasted track, track built with an asphalt underlayment foundation has the following advantages:

- A strengthened track support layer below the ballast to uniformly distribute reduced loading stresses to the subgrade, especially when the subgrade is of marginal quality
- An impermeable layer that diverts rain water to side ditches and prevents water from entering and weakening the underlying subgrade soil
- A separation layer between the ballast and subgrade to reduce the likelihood of subgrade mud pumping
- An all-weather, uniformly stable surface for placing the ballast and track superstructure

Track sites deemed most applicable for an HMA underlayment are those with one or more of the following existing or anticipated conditions:

- Difficulty in establishing and maintaining a sufficiently strong and stable subgrade to adequately support the ballast and track
- Difficulty in establishing and maintaining proper surface drainage to divert water away from the track structure
- Difficulty in preventing groundwater from penetrating upward
- High-impact stress areas such as special trackwork, road crossing, bridge and tunnel approaches, and rail joints

Areas where these conditions exist are likely to show rapid track degradation, ballast fouling, poor drainage, and rough track geometry.

Problems that have been reported with asphalt track foundation include cracking and durability of the HMA layer and development of pore water

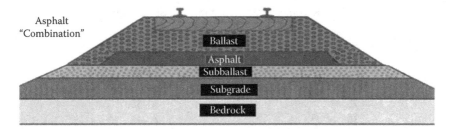

Asphalt
"Combination"

Figure 5.17 Underlayment HMA with subballast. (Courtesy of Professor Jerry Rose.)

Figure 5.18 Underlayment HMA under construction for a heavy axle load freight track.

pressure under HMA if it is built directly on the subgrade where ground water exists. A subballast layer between the HMA and the subgrade should be used when pore water pressure buildup is anticipated, as illustrated in Figure 5.17.

Figure 5.18 shows a picture of the underlayment HMA under construction for a heavy freight track in the United States. A layer of subballast was used between the clay subgrade and the HMA.

5.2.2 Asphalt mix design

The design of an asphalt track foundation includes the asphalt mix design. According to the Asphalt Institute (1998), the asphalt mix that provides stability to the track structure is a low-modulus plastic mix that has design air voids of 1%–3%. This mix will easily compact to <5% air voids. This can be accomplished by specifying the dense-graded highway base mix having a maximum aggregate size of 25–37 mm (1–1.5 in.).

Table 5.9 gives the recommended composition ranges for the asphalt track foundation of railway applications. It is essentially similar to a base course for highway construction and similar to the American Society for

Table 5.9 Recommended asphalt mix by Asphalt Institute

Sieve size (mm)	Amount finer (mass), %
37.5	100
19	70–98
9.5	44–76
4.75	30–58
2.36	21–45
1.18	14–35
0.6	8–25
0.3	5–20
0.075	2–6
Asphalt	3.5–6.5

Table 5.10 Marshall mix design criteria for HMA underlayment

Property	Required range
Compaction	50 blows
Stability (N)	3300 minimum
Flow (mm)	3.8–6.4
Percent air voids	1%–3%
Voids filled with asphalt VFA	80%–90%
In-place density[a]	92%–98%

[a] Percent maximum theoretical density based on ASTM D2041.

Testing and Materials (ASTM) D3515 Mix D-3. This low-void and impermeable mix allows minimum oxidation from the effects of air and water. It provides a layer with reasonably consistent stiffness in warm weather, but slightly resilient in cold weather. Furthermore, the tendencies for the mix to rut, bleed, or crack are significantly reduced, thus ensuring a long life for the HMA layer. For the asphalt content shown in this table, however, authors of this book recommend the higher value (5%–6.5%) to produce higher fatigue strength of asphalt.

The required HMA strength and the ability to reduce the access of water into the subgrade are achieved by meeting the Marshall design criteria (Asphalt Institute 1998). Table 5.10 gives the design criteria and recommended design values.

5.2.3 Asphalt track foundation design

In addition to the asphalt mix, design of asphalt track foundation includes the design of HMA layer thickness and ballast layer thickness based on the known axle load, traffic density, and subgrade soil condition.

In general, design of underlayment HMA is based on two main design criteria: (1) limiting stresses to the top of the subgrade to limit subgrade deformation and (2) limiting the tensile strain at the bottom of the HMA layer to prevent fatigue cracking of HMA (Huang et al. 1987; Chrismer et al. 1996).

Subgrade deformation from repeated loading has been discussed extensively. On the other hand, fatigue cracking of HMA is localized cracking that results from repeated loading causing the buildup of tensile strains at the bottom of the HMA layer. This is a distinction between the bound HMA layer and an unbound ballast layer. "Unbound" material such as ballast has no tensile strength, and therefore tensile stress and strain cannot develop within a granular material layer. On the other hand, "bound" HMA material can develop significant tensile stress and strain.

Fatigue failure of HMA begins with the initiation of microcracks caused by repeated cycles of tensile strain in the bottom of the HMA layer. Once a microcrack has formed, it then propagates through the HMA layer with the rate and extent of propagation governed by the tensile stress at the tip of the crack. As fatigue cracks become more prevalent and widespread in the HMA layer, the layer ceases to act as a bound material and the stiffness of the layer becomes significantly reduced. The reduction in stiffness results in a significant reduction in load distribution and a corresponding increase in stress to the subgrade. Additionally, once the crack propagates and becomes widespread, water can penetrate the cracks and then soften the subgrade. The combination of increased stress in the subgrade and the subgrade's reduced strength due to water-softening or high water pressure, can then result in severe track foundation deterioration. Unlike highway pavements, HMA track underlayment is shielded from view so early detection of signs of HMA cracking is often impossible.

As with minimizing excessive subgrade deformation, a high fatigue resistance of HMA is achieved by a design that reduces tensile strain and stress in the HMA layer. This is accomplished by (1) reducing the applied stress to the HMA by increasing ballast thickness, (2) increasing the stiffness of the subgrade layer under the HMA, and (3) increasing the thickness of the HMA layer.

The specified thickness for an underlayment HMA varies depending on the quality of subgrade support, traffic load, and type of installation. A 125–150 mm (5–6 in.) HMA is usually specified for average subgrade conditions, and at least 200 mm (8 in.) is specified for conditions of poor subgrade support and/or high-impact area. Ballast thickness normally ranges from 200 to 300 mm (8 to 12 in.). These HMA and ballast thickness values are slightly less for transit and light freight lines.

Most of the models described in Chapter 4, such as GEOTRACK and Finite Element Analysis (FEA), can be used for the analysis and design of asphalt track foundations. KENTRACK was particularly developed for the asphalt track foundation.

With KENTRACK, design is based on two equations: one for determining the number of allowable repetitions for the subgrade layer before failure occurs due to excessive vertical compressive stress, and is computed by Equation 5.18 (Huang et al. 1984), and the other for determining the number of allowable repetitions for HMA before the failure occurs, and is calculated using Equation 5.19.

$$N_d = 4.837 \times 10^{-5} \sigma_c^{-3.734} E_s^{3.583} \tag{5.18}$$

$$N_a = 0.0795 \varepsilon_t^{-3.291} E_a^{-0.853} \tag{5.19}$$

where:
   N_d is the number of allowable repetitions for the subgrade layer before failure occurs due to excessive vertical compressive stress
   N_a is the number of allowable repetitions for HMA before failure occurs
   σ_c is the vertical compressive stress on the top of subgrade in psi
   E_s is the subgrade modulus in psi
   ε_t is the horizontal tensile strain at the bottom of asphalt
   E_a is the elastic modulus of asphalt in psi

In other words, Equation 5.18 is intended to control subgrade failure from excessive deformation by limiting compressive stress on the subgrade surface, whereas Equation 5.19 is intended to control HMA failure from fatigue cracking by limiting tensile strain in the bottom of the HMA layer.

From these two equations, one can calculate the service life of the subgrade as well as the service life of the HMA by dividing the allowable number of repetitions for the subgrade layer or allowable number of repetitions for HMA by the actual number of load applications per year (Chapter 2).

Design using KENTRACK does not actually calculate required thickness of the ballast and HMA. Instead, this method allows a design engineer to first select the cross section of asphalt track foundation, and then KENTRACK can calculate service life for the subgrade or HMA. For HMA underlayment, the thickness of HMA can vary between 4 and 8 in., with a minimum 8-in. ballast thickness.

The failure criteria used for the railway trackbeds was developed based on highway loading conditions and environments. Rose contends that the service life predicted by KENTRACK is conservative because the stress levels experienced by railway subgrades are lower than that of highways. Also, settlement in railroad trackbeds is not as significant as in highways and can be rectified easily. It has also been found that unlike highways, where the settlement generally occurs due to water infiltrating into the subgrade, the

HMA acts as a waterproofing layer in the trackbed preventing the infiltration of water into the subgrade.

GEOTRACK has also been used for HMA track foundation design and analysis. Figure 5.19 shows the importance of having and maintaining a good subgrade under the HMA based on GEOTRACK modeling. The analyses

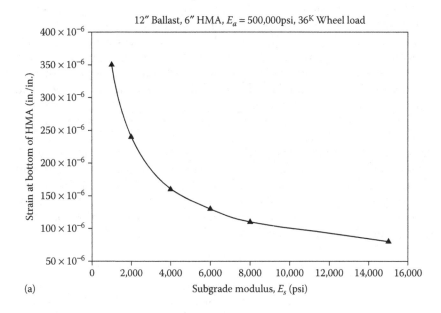

(a)

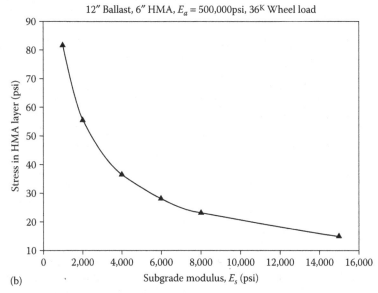

(b)

Figure 5.19 Strain (a) and stress (b) in HMA layer for various subgrade moduli.

were performed for a typical case of HMA track underlayment, with 12-in. of ballast, 6-in. layer of moderately stiff HMA (E_a = 500,000 psi), concrete ties, and 36-ton axle loads. The top figure shows the level of horizontal strain in the bottom of an HMA layer for various subgrade moduli, and the bottom figure shows the level of stress within the HMA for various subgrade moduli. As shown, if the subgrade softens under the HMA from a moderate subgrade (6000 psi) to a soft subgrade (2000 psi), then the strain at the bottom of the HMA increases by 85% and the stress on the HMA increases by 100%.

Subgrade stress levels under an HMA layer will be less than the subgrade stress under an equal thickness of granular material. This is because HMA is a "bound" layer, as opposed to an "unbound" granular layer, and thereby has tensile strength and relatively high stiffness. The higher strength and stiffness of a layer of HMA compared with the same layer thickness of ballast or subballast is particularly useful in situations where the amount of granular layer thickness that can be used is limited due to overhead or undergrade structures or grade crossings. Figure 5.20 (Chrismer et al. 1996) shows a comparison of the stress reduction capabilities of different layer thicknesses of moderately stiff HMA (500,000 psi in modulus) and conventional granular-based substructure. As shown, for 12-in. of ballast and 6-in. of HMA, the stress on the subgrade is the same as that for 29-in. of combined ballast and subballast.

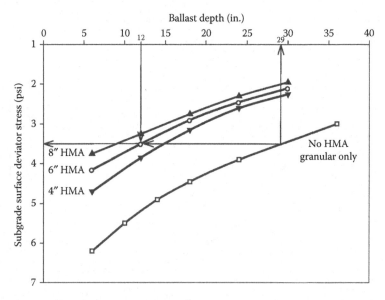

Figure 5.20 Comparison of subgrade surface stress for HMA and granular material. (Data from Chrismer, S. M., Terrill, V., and Read, D. (July 1996). *Hot Mix Asphalt Underlayment Test on Burlington Northern Railroad*, R-892. Association of American Railroads, Transportation Technology Center, Pueblo, Colorado.)

5.2.4 Use of asphalt track for drainage

Use of underlayment HMA can have a key benefit of improving track drainage by shedding water due to the relatively impermeable nature of HMA (Chapter 6). In many instances, the root cause of track substructure problems stems from water remaining in the substructure for prolonged periods of time. This is often the result of the water being unable to effectively drain out of the track because the water elevation is below any effective drainage paths. HMA can be used effectively to raise the drainage surface within the track substructure and thereby provide effective gravity drainage away from the track. Figure 5.21 illustrates this application.

Grade crossings often have very little lateral drainage; therefore, water drainage out of grade crossings is typically longitudinal along the track and out of the crossing ends. Good drainage is a key component of a well-performing grade crossing, and it is commonly observed that crossings contain excessive amounts of water. Chapter 6 provides more details showing potential application of HMA in a crossing to raise the drainage surface so that water can effectively escape.

A potential problem with the use of HMA underlayment is the possibility of trapping excess water beneath the HMA. Water trapped below the HMA can reduce soil strength and cause local failure of the subgrade (see Section 5.2.5). Also, if cracks develop in the HMA, concentrated hydraulic action may occur through the openings causing local soil erosion. The impermeability of HMA could result in pore water pressure build up in the subgrade under the HMA if it is constructed on fine-grained soil that does not drain rapidly. This could result in a loss of strength of the material. In this case, it is recommended to place a permeable subballast layer under the HMA to provide for dissipation of pore pressure.

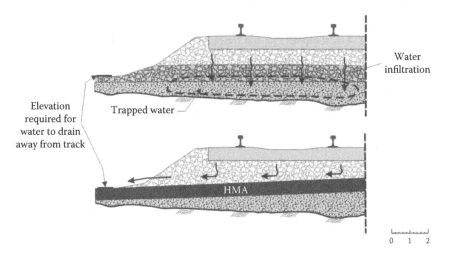

Figure 5.21 Typical application of HMA to provide drainage within track.

5.2.5 Asphalt track foundation test under heavy axle loads

In the summer of 1999, two HMA underlayments were placed as a course under the ballast but above the soft subgrade test section (Section 5.1.5—LTM Test Track). Each segment is about 107 m long. One segment has a 100 mm HMA, and the other has a 200 mm HMA. Figure 5.22 illustrates the longitudinal cross section of these two segments. For the entire test section, a 100 mm subballast layer was used between the subgrade and the two HMA underlayments. At construction, the ballast thickness above the HMA was 300 mm over the 100 mm HMA but was 200 mm over the 200 mm HMA. For both segments, the total granular/HMA thickness was therefore 500 mm.

The asphalt mix design was based on the guidelines recommended by the Asphalt Institute (see Table 5.9). Table 5.11 (left two columns) gives the actual compositions for the HMA mix. The required HMA strength was

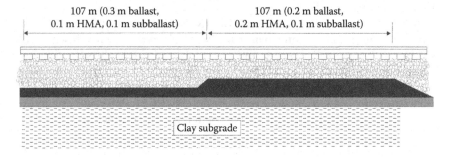

107 m (0.3 m ballast, 0.1 m HMA, 0.1 m subballast)   107 m (0.2 m ballast, 0.2 m HMA, 0.1 m subballast)

Clay subgrade

Figure 5.22 Longitudinal cross section of HMA test track. (Courtesy of Transportation Technology Center Inc., Pueblo, Colorado.)

Table 5.11 Actual asphalt mix and strength test results for HMA test track

Sieve size (mm)	Amount finer (mass), %	Property	Result
37.5	100	Compaction	50 blows
19	76	Stability (N)	7700
9.5	52	Flow (mm)	6.1
4.75	41	Percent air voids	2%
2.36	30	Voids filled with asphalt	86%
1.18	23	In-place density	94%[a]
0.6	17		
300 μm	11		
75 μm	4.5		
Asphalt	6.4		

[a] Average nuclear density test results.

achieved by meeting the Marshall design criteria (see Table 5.10). Table 5.11 (right two columns) gives the test results of the actual mix.

During construction, the HMA was placed in one lift for the 100 mm HMA, but in two lifts for the 200 mm HMA. To achieve the desired HMA density listed in Table 5.11, a steel-wheeled, vibratory roller was used to compact the HMA layer while the mix was still between 85°C and 150°C. Following compaction, a nuclear density gage was used to obtain the final *in situ* HMA density results. In addition, a number of HMA core samples were obtained for further laboratory testing.

This test was the first to apply HMA underlayment over a soft subgrade under 36 metric ton (350 kN) axle loads. Since its installation, the performance of this test track has been evaluated in terms of track geometry degradation with traffic as well as the amounts of track modulus increase and subgrade stress reduction compared with conventional granular layer construction.

Figure 5.23 gives the track modulus test results obtained at 83 MGT and the subgrade stress test results under a static wheel load of 18 metric tons

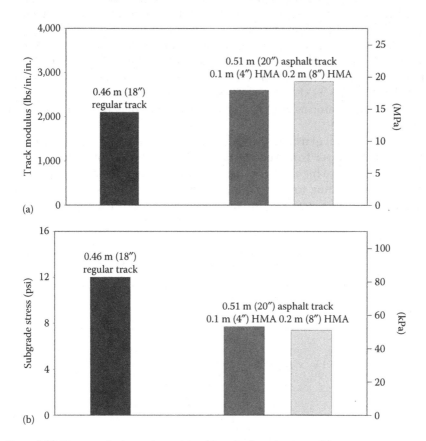

Figure 5.23 Test results in track modulus (a) and subgrade stress (b).

(175 kN). As shown, the average modulus values for the two HMA segments are 20 and 23 MPa for the fully compacted ballast (increased from 18 to 19 MPa, respectively, at 0 MGT). Obviously, the HMA underlayment application significantly increased track modulus from the 450 mm granular track with an average track modulus of 14 MPa. As a result, the measured subgrade stresses were lower for the asphalt trackbeds than for the 450 mm granular track. Under 18 metric ton (175 kN) static wheel load, only 50–55 kPa of subgrade stress was generated under the HMA underlayments, compared to 83 kPa under the 450 mm granular track structure.

To show how subgrade stresses induced by wheel loads are reduced by the presence of HMA over the subgrade, Figure 5.24 shows the dynamic stresses under an actual train operation (64 km/h) measured on the 200 mm HMA surface as well as on the subgrade surface. As illustrated, use of a 200 mm HMA underlayment reduced the subgrade stress approximately by half.

Figure 5.25 shows the results of average track settlement as a function of traffic for both the segments. As illustrated, after the initial higher rate due to early ballast compaction, the settlements became gradual, characteristic of typical and normal track deformation. After almost 91 MGT, about 37 mm of total settlement was accumulated for the 100 mm HMA segment, while about 33 mm of total settlement was observed for the 200 mm HMA segment. Nevertheless, the settlements (mainly due to ballast deformation) have been uniform along and across the test track.

Another benefit of using HMA underlayment beneath ballast is insulation of the asphalt layer from the weather. This keeps the asphalt less susceptible (compared to highway construction) to the oxidation and temperature effects, thus leading to longer asphalt life without weathering and cracking. In Figure 5.26, temperature measurements of the HMA underlayment

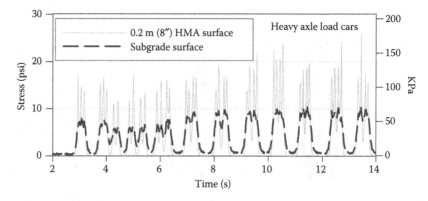

Figure 5.24 Reduction of dynamic stresses from 200 mm HMA to subgrade.

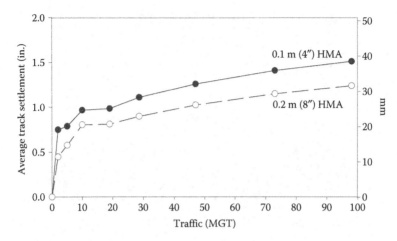

Figure 5.25 Track settlement of HMA track as a function of traffic.

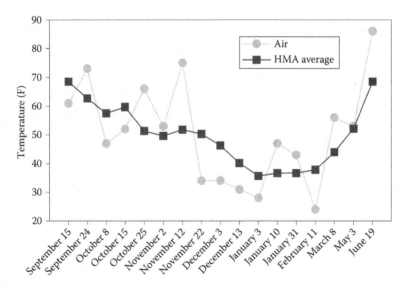

Figure 5.26 HMA temperature versus air temperature.

and the air made for more than approximately 1 year show that the HMA experienced much less temperature variation compared with that of the air.

Since its construction in 1999, this test track has been performing well until 342 MGT when a slight cross-level deviation was noticed in the 4-in. HMA section. Little change in track geometry was noted in an additional 17 MGT; however, one rail then began to settle more rapidly. The first surfacing was needed in February 2005 with 371 MGT on the HMA test track when cross-level exceeded 1.75 in. Sixty-five feet of track were surfaced.

Figure 5.27 Early indication of asphalt failure.

Similar maintenance was needed several more times in the next 35 MGT. As the rate of degradation increased, it became evident that the asphalt had failed. Figure 5.27 shows early evidence that the asphalt pushed up and outward from the ends of the ties.

At 406 MGT, the track was removed for investigation and repair. It was found that approximately 80 ft of asphalt cracked longitudinally. The asphalt that had been pushed up was about 2 ft higher than the asphalt that had not. There was water under and along the asphalt that failed. This was at a low spot in the track, and where track geometry problems had occurred earlier. This low spot had historically collected water from rain and snowfall in the surrounding area.

Apparently, the permeable subballast provided an entry path for the water to get under the asphalt, and water contributed to the start of the failure by weakening the subballast and clay subgrade, and accelerated the degradation following failure initiation. The underlying cause of the asphalt failure was progressive shear failure and the clay subgrade, which also pushed upward and outward. This was consistent with the previous subgrade failure in the same test section (Section 5.1).

Figure 5.28 (left) shows the HMA where the failure occurred. For comparison, Figure 5.28 (right) shows undamaged asphalt nearby.

Asphalt cores were taken from damaged and undamaged asphalt, in the 4-in. and 8-in. segments. Clay samples were also taken to determine whether moisture content had changed, and to measure the current strength of the clay. The core taken from the track and cores that had been taken at the time of HMA construction were tested. The following tests were conducted: bulk specific gravity, washed sieve analysis, kinematic viscosity at 135°C, absolute viscosity at 60°C, and dynamic shear rheometer.

Figure 5.28 Left—Damaged asphalt and standing water. Right—Nearby undamaged asphalt.

The tests were to determine whether the properties of the asphalt had changed since the asphalt was installed. Of primary importance and most likely to change were the viscosity and dynamic shear values, as they provide an indication of the strength of the asphalt. In summary, there was no indication of any significant deterioration of the asphalt. The 1999 and 2005 cores did not differ significantly in gradation, asphalt contents, or volumetrics.

The clay samples were tested for moisture content. On average, moisture content was 31%, slightly lower than 33%, which was the average moisture content when the clay subgrade was built in 1991. The California Bearing Ratio (CBR) at 0.5% penetration was 1.0%, which relates to a resilient modulus of approximately 10.3 MPa (1500 psi) and is indicative of very low strength of the clay soil in the subgrade. The failure can be related to Figure 5.19 where the subgrade resilient modulus decreased from 14 MPa (2000 psi) when the clay was first placed to approximately 10 MPa with increased moisture, causing an exponential increase in HMA stress and strain.

The damaged and displaced asphalt and subballast were removed. Asphalt was then placed directly on the clay, which differed from the original installation when there was a layer of subballast. The primary reason for placing asphalt directly on the clay was to better seal the surface water from the clay by removing the permeable entry path. However, a subballast layer between the HMA and fine-grained subgrade is preferable in most situations. If the clay is too soft to support the HMA-paving operation, a layer of subballast can provide the necessary support and load distribution for the paving equipment. In addition, if ground water is present, the subballast can allow water to escape under the asphalt and also prevent potential buildup of pore water pressure.

After the repair of the damaged HMA section, drainage in the area was improved with the installation of a trench to intercept and divert surface water from the clay (see Figure 5.29). The trench was filled with coarse gravel.

Figure 5.29 Trench prior to being filled with coarse gravel.

These repair measures addressed the problem and strengthened the damaged track. Since then, no maintenance had been required for the underlayment HMA built on the clay subgrade until September 2013 when another similar failure occurred in the 4-in. HMA section. As a comparison, the adjacent 8-in. HMA section has not failed since the beginning of the installation.

5.3 SLAB TRACK FOUNDATION

5.3.1 Overall requirements

Slab track is a ballastless track structure. Although it can be designed and constructed in a variety of types, a slab track often consists of a concrete slab built on a subbase supported by the soil subgrade. Compared with a conventional ballasted track structure, slab track requires less maintenance and provides better track geometry, if designed and constructed properly. This is due in part to the fact that in most instances the predominant source of settlement in conventional ballasted track comes from ballast settlement or breakdown and abrasion of the ballast. High-speed passenger trains can tolerate only small amounts of track geometry degradation. For instance, for a single deviation from uniform vertical profile, the tolerance level is twice as stringent for a 160 mph passenger track in the United States compared to a 60 mph freight track. Therefore, use of slab track has become a viable and often advanced alternative to the conventional ballasted track for high-speed operations.

In general, the design of slab track foundation includes designing the slab (concrete mix and reinforcement), the pad/fastening system, and the subbase and subgrade. By its nature, a slab track is a stiff system as the resiliency provided by ballast is no longer available. However, track requires a certain amount of resiliency, and this is provided by the pad/fastening system. Design of the pad/fastening system needs to consider dynamic responses of slab track including considerations for noise and vibration. To control noise and vibration, the fastening system must have a resilience that is designed to be within a certain range, meaning that both minimum and maximum limits on track resiliency must be considered. Proper concrete mix and reinforcement design is essential for a concrete slab to perform as a structured layer, and withstand dynamic wheel loads and temperature changes.

Another requirement is the proper design and construction of the subbase and subgrade as the foundation of slab track. The subbase under the concrete slab allows a uniform distribution of wheel load stresses, dissipates these stresses, and may also provide frost protection. The subbase also provides a platform for construction of the concrete slab.

The subgrade should be capable of providing a suitable working base for installation of the subbase and concrete slab, supporting the slab track, and accommodating the stresses due to traffic loads without failure or excessive deformation. A stable support must be maintained over time that is not excessively affected by environmental conditions (e.g., wet/dry, freeze/thaw). Subgrade should also be designed so that surface water and ground water drain away from the subgrade.

If a slab track is built on a poorly designed or constructed subgrade, maintenance costs for slab track may exceed those costs required for a conventional track structure. The relative ease with which the geometry error of a conventional ballasted track may be corrected with ballast tamping and surfacing does not apply to slab track. Therefore, uniform and adequate subgrade support is an essential requirement for slab track.

5.3.2 Failure modes and design criteria

There are no well-documented sources of information on typical failure modes of slab tracks. However, if not designed and constructed properly, concrete slab may exhibit failure modes similar to highway concrete pavement, including cracking, faulting (for jointed slab), and pumping due to poor subgrade. It is therefore the role of design to ensure that anticipated failure conditions will not develop during its design life (typically 50 years), or if certain progressive failure conditions do develop, these conditions will be manageable through routine corrective maintenance and repair.

In addition to the methods used for designing many proprietary slab track systems used in Germany, Japan, and China, AREMA Manual for

Railway Engineering (2012) contains the recommendations for design of the slab track for North American applications, and includes the following design criteria:

- Allowable rail-bending stress = 11,000 psi (77 MPa)
- Allowable rail deflection (elastic) = 0.25 in. (6 mm)
- Minimum slab width = 10.5 ft (3.2 m)
- Allowable stress on the subbase = 30 psi (0.2 MPa)
- Allowable stress on the subgrade = 20 psi (0.14 MPa)
- Concrete compressive strength (28-day) = 4,000 psi (28 MPa)
- Minimum modulus of subgrade reaction = 350 pci (0.09 N/mm^3)
- Allowable crack width of concrete slab = 0.012 in. (0.3 mm)
- Minimum fastener insert pullout load = 14,000 lb (62,300 N)
- Toe load for fastening clips = 2,200–3,200 lb (9,800–14,240 N)
- Vertical spring rate for fastening system = 90,000–300,000 lb/in. (15.8–52.5 kN/mm)
- An impact factor of 2.0 should be used for design of the slab track. Note that for the design of slab track for high speed, an impact factor of 2.5 and 3.0 are recommended for 250 and 300 km/h, respectively (MOR, China 2009).

Apparently, depending on the loading and operating environments, these design criteria may need to be expanded and their values adjusted.

5.3.3 Design of concrete slab

Concrete slab is usually designed with longitudinal and transverse reinforcement. Longitudinal reinforcement in the slab is designed for longitudinal bending moments, and to keep shrinkage cracks very small and uniformly distributed. Transverse reinforcement in the slab is provided to resist transverse bending moment. The reinforcement in both directions is provided by mats of reinforcing bars placed near the top and bottom of the slab layer. Figure 5.30 shows examples of reinforcement for two types of slab tracks designed for shared high-speed and heavy freight operations (Lotfi and Oesterle 2005; Li 2010).

Slab thickness should be selected to withstand warping stresses (temperature differential between the top and bottom of the slab), bending stresses from wheel loads, and longitudinal stresses induced as a result of anchoring rail to the slab. In general, the selection of slab thickness is based on past experience as well as on numerical analysis to ensure that the design criteria (such as those listed in Section 5.3.2) are met.

Concrete mix design can also follow from past experience, as long as it meets the 28-day strength and other construction criteria. For the two examples shown in Figure 5.30, the mix design was similar to that typically used in highway pavement applications, which was a 6-bag mix consisting

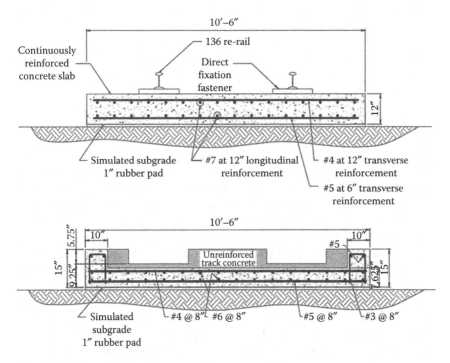

Figure 5.30 Examples of reinforcement in concrete slab.

of 560 lb of cement, 150 lb of fly ash, 964 lb of fine aggregate (FA-1), 1,987 lb of coarse aggregate, 32 gallons of water, a water-reducing gent, and an air-entraining agent. Also specified was a water–cement (w/c) ratio of 0.38, a slump of 3 ± 1 in., air content between 4% and 7%, a maximum aggregate size of 1 in., and an unconfined compression strength of 5,000 psi at 28 days.

5.3.4 Subgrade and subbase

Allowable subgrade stress is the criterion most commonly used in the design for subgrade. Subgrade stiffness can also be used in the design, which can be defined in terms of Young's modulus, resilient modulus, CBR, or modulus of subgrade reaction (k value). It is also possible to design the slab track based on the allowable deformation of the subgrade. Two approaches can be used to achieve this objective (Section 5.1). One approach is based on a criterion to limit the maximum allowable settlement for the design life, and the other is to limit the maximum allowable cumulative plastic strain for the design life.

Ideally, the allowable subgrade stress and stiffness is determined from a field *in situ* test or a laboratory test on high-quality soil samples collected in the field. In the absence of such a test, published values for various soil types can be used for preliminary design (Chapter 3).

The subgrade shall preferably be a sandy/gravel material. As mentioned in Section 5.3.2, subgrade for slab track should have an allowable stress of 20 pci and a minimum modulus of subgrade reaction of 350 pci (0.09 N/mm³).

The subbase is the layer of material that lies directly below the concrete slab. Soil–cement, cement-treated subbase, stabilized asphalt, and lean concrete are all considered suitable. The subbase layer thickness should be a minimum of 4 in. (100 mm), although a geotechnical evaluation should be performed to determine if a thicker layer is needed. The subbase should project 2 ft (610 mm) beyond each side of the concrete slab.

Figure 5.31 shows the example of two-slab track sections built in a 5° curve under heavy axle load train-operating environment. In this example, the subgrade soil is silty sand. Soil moisture content varies between 5% and 12%. The surface of the subgrade was recompacted at an optimum moisture content of 10.5%. The resilient modulus of the soil was approximately 10,000 psi, with a compressive strength of approximately 50 psi.

The subbase is a 6-in. soil–cement layer. This layer was installed by mixing 5% of cement (based on weight) with the soil and compacting at an optimum moisture content of 12.5%. Compaction was specified to be 98% of the maximum (modified Proctor). The target compressive strength was 700 psi, with subsequent cores showing actual strength between 780 and 840 psi.

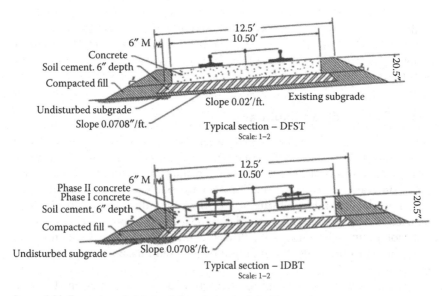

Figure 5.31 Examples of slab track foundation built in 5° curve.

5.3.5 Analysis of slab track foundation

Because there is no well-documented procedure for step-by-step slab track design, modeling of slab track is often necessary to assure that it will meet the end-result design requirements. Although the beam on elastic foundation model (Chapter 4) can be used for analysis of slab track foundations, the most popular analytical approach is to use multilayer computer models such as GEOTRACK or FEA model, whether in 2D or 3D.

At a minimum, the output from modeling should include rail stresses and deflections, fastener loading and compression, slab deflections and bending moment, and subbase and subgrade stresses.

5.3.6 Slab track design based on subgrade deformation

The strength and resistance of subgrade layers to deformation is of great importance in any track foundation design. However, because correcting slab track geometry often requires more effort, cost, and time than with conventional ballasted track, the control of slab track settlement and roughness is more critical. To justify the higher initial construction cost of slab track, maintenance requirements must be low. The following gives a design approach that can be used to determine the expected subgrade stress or settlement and if it is within tolerable limits (Chrismer and Li 2000). This approach considers the performance of slab track over subgrades of varying strength and stiffness, and of several cases of passenger and freight traffic loading.

Three vehicle-loading cases are considered: light rail vehicle (LRV), commuter car, and freight shared with passenger traffic. The loading characteristics from these cases are shown in Table 5.12. The yearly tonnage is based on the typical number of cars over the track in a year for that vehicle type. For example, a two-commuter car consist is assumed to pass over the track every 45 minutes each day, every day of the year, providing 32 trains per day. The yearly tonnage for this case is then 32 consists/day ·x 18 ton/axle ·x 8 axles/consist × 365 days/year = 1.7 MGT/year. The

Table 5.12 Vehicle types and loading characteristics used in slab track analysis

Vehicle type	Number of axles per vehicle	Wheel load (lb)	Axle spacing (in.)	MGT per year
140,000 lb light rail vehicle	6 (3 trucks)	12,000	72	3.8
140,000 lb commuter car	4	18,000	102	1.7
263,000 lb freight car	4	33,000	70	10.0

annual tonnage calculation for LRV traffic assumes a single articulated car every 20 minutes, the increased frequency being typical for this vehicle type. Annual tonnage of 10 MGT for freight is assumed to represent the case of track with shared freight and passenger traffic, where the amount of freight traffic is not large (by freight track standards), but its larger loads determine the track design. It is assumed that the much lighter passenger loads on this shared track do not affect the slab track foundation design, but the passenger traffic dictates that the track geometry must be maintained to FRA (Federal Railroad Administration) Class 6 or better.

At the heart of this analysis are two essential tools. One is the GEOTRACK model to predict the imposed subgrade stresses from traffic loading. The other tool is the Li–Selig design procedures developed for conventional track foundation (Section 5.1.3).

The cases considered are shown in Table 5.13, including four subgrade conditions. The compressive strength range shown is typical for the modulus, although this stiffness–strength relation is only a general approximation. In a design process, the actual subgrade strength should be obtained by testing. The lower compressive strength value for each modulus is used in this analysis in terms of design criterion 1 (Equations 5.6 and 5.7) to prevent subgrade surface progressive shear failure, where water may more readily accumulate and reduce the strength. The second design criterion (Equation 5.8) in terms of preventing subgrade settlement uses the higher strength found at deeper locations where moisture fluctuations are less pronounced. The commonly used CBR is also included as a reference.

GEOTRACK is used to determine the deviator stresses in the subgrade. The deviator stress is the major principle stress minus the minor principle stress on a subgrade element, and its magnitude relates to the amount of shearing stress and deformation of the subgrade. The track will deform less when these deviator stresses are relatively low and uniform under the track. Deviator stresses are calculated at the subgrade surface to determine the stability under design criterion 1 (Equation 5.7) to prevent the failure shown in Figure 5.6. Analysis under design criterion 2 (Equation 5.8) to prevent the failure shown in Figure 5.7 is conducted by breaking up the 120-in. thick subgrade layer into four equal 30-in. thick sublayers and

Table 5.13 Stiffness and strength of fine-grained subgrade soil used in slab track analysis

Resilient modulus (psi)	Compressive strength (psi)	CBR
2,000	7–9	1.3
3,000	10–12	2
5,000	15–18	4
10,000	25–30	10

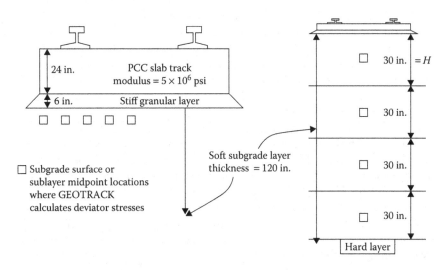

Figure 5.32 Slab track used in analysis.

calculating the deviator stress, and resulting deformation, at the midpoint of each sublayer (see Figure 5.32 for the slab track analysis example).

As an example to illustrate the design process, the LRV loading case will be considered with slab track over a subgrade with a modulus of 2,000 psi. With design criterion 1, the number of repeated load cycles, N, of any static wheel load, P_s, during the design period is determined by the following equation, where T is the total traffic tonnage for the wheel load P_s, for the design period in the same unit as P_s, and W is the number of wheels contributing to one load cycle for the subgrade.

$$N = \frac{T}{(W \cdot P_s)}$$

This equation will give the "N" for one year, unless the number of years is entered. Assume that the design life period is 15 years. If T is in MGT/year (tons) and P is in pounds, the 2,000 lb/ton conversion factor shown next must be used. For the 3.8 MGT/year assumed with the LRV loading, the preceding equation gives

$$N = 3.8 \cdot (15 \, \text{years}) \cdot (1 \, \text{M}) \cdot (2{,}000)/4/12{,}000 = 2.4 \, \text{M}$$

Note that the "4" relates to four wheels of one truck that give one load cycle. The distance between adjacent LRV trucks is typically so large that there is no interaction of the load pulses from each truck. For freight cars, the end trucks are usually close enough that the total of eight wheels makes up one load pulse on the subgrade as assumed in Equation 5.15.

Next, Equation 5.9 is used to determine the allowable deviator stress based on the N value and the allowable soil plastic strain for a CL type (lean clay) soil. Commonly, the allowable strain of 1%–3% has been used, with the upper range used for freight track. Because passenger traffic is being considered (assume FRA track Class 6), the smaller value of 1% will be used. For a 1% strain, $N = 2.4$ M, and using lower strength of 7 psi, the value of the allowable deviator stress is 2.1 psi.

From GEOTRACK, the average deviator stress along the subgrade surface is determined to be 1.0 psi, below the 2.1 psi allowable. Therefore, design criterion 1 is satisfied. The next step is to determine if criterion 2 is also satisfied by calculating the subgrade settlement for the design period of 15 years.

For criterion 2, it considers the settlement over the entire deforming layer due to the imposed stresses and the number of load cycles that occur over the design life. The limit of settlement for this criterion is 1 in. In general, a slab settlement in the order of 1 in. may be tolerable and can be compensated for by shimming or other means. The generalized settlement equation is

$$\rho = \sum_{i=1}^{4} \left(\frac{H}{100} \right) \cdot a \cdot (N)^b \left(\frac{\sigma_{di}}{\sigma_s} \right)^m$$

Where ρ is the settlement in inches; the number of soil sublayers is four, H is the sublayer thickness; a, m, and b are soil type parameters (see Table 5.2). For the assumed CL type soil over the entire subgrade layer thickness, a is 1.1, b is 0.16, and m is 2. The GEOTRACK predicted deviator stresses for criterion 2 at the midpoint of the sublayers are 1.3, 1.0, 0.8, and 0.3 psi in order from the top layer down. The subgrade strength is taken to be 9 psi, the upper value for the 2,000 psi resilient modulus soil. The predicted settlement over 15 years is then

$$\rho = \left(\frac{30}{100} \right) \cdot 1.1 \cdot (2.4M)^{0.16} \left[\left(\frac{1.3}{9} \right)^2 + \left(\frac{1.0}{9} \right)^2 + \left(\frac{0.8}{9} \right)^2 + \left(\frac{0.3}{9} \right)^2 \right] = 0.15 \, \text{in.}$$

Which is <1 in. and therefore acceptable.

Therefore, both criteria are met, and the slab track should perform satisfactorily with LRV loading over 15 years on a subgrade with a resilient modulus of 2,000 psi, and compressive strengths of 7 psi at the surface and 9 psi at depth.

Next, consider the heavier commuter car loading of the 18,000-pound wheel load and use the same track conditions that are acceptable with the lighter LRV loading. The N value for this case is

$$N = 1.7 \cdot (15 \, \text{years}) \cdot (1M) \cdot (2000) / 4 / 18,000 = 708,300 = 0.71 M$$

From Equation 5.9, the allowable deviator stress is determined to be 2.3 psi. GEOTRACK predicts an average subgrade surface deviator stress of 1.7 psi. Because the allowable stress is greater than the predicted 1.7 psi, it meets the criterion to prevent progressive shear failure.

The predicted deviator stresses at the midpoints of the four subdivided soft soil layers are 1.9, 1.7, 1.3, and 0.8 psi. Therefore, the settlement for this case is

$$\rho = \left(\frac{30}{100}\right) \cdot 1.1 \cdot (0.71\,\text{M})^{0.16} \left[\left(\frac{1.9}{9}\right)^2 + \left(\frac{1.7}{9}\right)^2 + \left(\frac{1.3}{9}\right)^2 + \left(\frac{0.8}{9}\right)^2\right] = 0.31\,\text{in.}$$

The predicted settlement for this case of 0.31 in. is less than the allowable 1.0 in., acceptable for and is therefore commuter car loading. Next, the adequacy of the slab track subgrade support will be evaluated for freight car loading.

As stated previously, it is assumed for simplicity that the freight loading dominates, and the much lighter passenger loading is considered to have no bearing on the slab track design. In actual practice, this should not be assumed at the outset. The equivalent load cycle "N" value calculation should include the load magnitudes and number of axles for both traffic types. If passenger traffic is found to contribute very little to the total "N" value, the contribution from passenger traffic can be disregarded in the subsequent analysis.

As a start, assume a 3,000 psi modulus subgrade is required for freight car loads. The associated soil compressive strength is then assumed to be 10 psi (Table 5.11). The "N" value is calculated as

$$N = 10 \cdot (15\,\text{years}) \cdot (1\,\text{M}) \cdot (2,000)/8/33,000 = 1,136,400 = 1.1\,\text{M}$$

From this, the allowable subgrade surface deviator stress can be determined from Equation 5.9 to be 3.1 psi. However, this is a failing condition because GEOTRACK predicts an applied average deviator stress of 3.2 psi at the subgrade surface.

If the subgrade modulus is 5,000 psi the soil strength is assumed to be 15 psi and the allowable subgrade surface deviator stress is now determined to be 4.7 psi, which is acceptable because GEOTRACK predicts 3.8 psi subgrade surface stress.

Next, check the adequacy of the 5,000 psi modulus subgrade with criterion 2. The GEOTRACK predicted stresses at the sublayer midpoints are 4.9, 4.5, 3.6, and 2.0 psi from top down. Note that two trucks (eight wheels) are assumed to contribute to one load cycle in this loading case. Also a compressive strength of 18 psi is assumed for the 5,000 psi soil. Therefore, the settlement is

$$\rho = \left(\frac{30}{100}\right) \cdot 1.1 \cdot (1.1\,\text{M})^{0.16} \left[\left(\frac{4.9}{18}\right)^2 + \left(\frac{4.5}{18}\right)^2 + \left(\frac{3.6}{18}\right)^2 + \left(\frac{2}{18}\right)^2\right] = 0.58\,\text{in.}$$

The predicted settlement of 0.58 in. is less than the allowable 1.0 in. and is therefore acceptable.

Therefore, subgrade with a modulus of 5,000 psi and the related compressive strength is acceptable for the 10 MGT/year freight loading. It is often found that the required subgrade strength increases exponentially in relation to wheel load magnitudes.

5.4 TRACK TRANSITION

5.4.1 Design principles/best practices

Typical problems and their root causes associated with track transitions are described in Chapter 3. This section discusses the general principles that should be considered in the design of track transitions. These principles also apply to remediation of problems associated with existing track transitions.

Because the main root cause of most problems at track transitions is the abrupt change of track structure and foundation, design of a track transition should seek to make this change as small as possible. As change of track structure or foundation is inevitable at such track transitions as bridge approaches, ballasted track to slab track, grade crossings, and special trackwork, design of transitions should aim to provide a consistent support and resistance to deformation under train loading. In other words, track transitions should be designed with similar track strength, stiffness, and damping characteristics in both vertical and lateral directions to ensure that under train operations, track will not accumulate significant differential displacement in either the vertical or lateral direction.

The following gives the best practices for track transition design, which were developed based on the authors' many years of research, testing, and consulting experience in dealing with transitions at bridge approaches, slab track to ballasted track, special trackwork, and grade crossings.

5.4.1.1 Bridge approach

For bridge approaches, design or remediation should be focused on controlling deformation in the approach where track tends to have lower stiffness, as well as on minimizing differential settlement where plastic deformation tends to accumulate more rapidly. If weak subgrade exists in the approach it should be removed and replaced or treated so that the approach track does not accumulate large plastic deformation under loading. However, if ballast settlement in the transition is the main source of the track dip, a means to stabilize the ballast should be sought. Another method is to reduce the ballast stress at the tie–ballast interface.

The design of a stable bridge transition must consider the bridge type. A ballasted deck bridge will likely require a different solution from an open deck girder bridge. The fixed elevation points and rigid structure of the open deck type of bridge are especially challenging to the control of deformation and settlement in the approach.

From open track to bridge, change of track structure in some cases and change of track foundation in all cases are inevitable; therefore, creating a true transition rather than an abrupt change in properties is essentially to reduce track stiffness and increase track damping for the track structure on bridge.

In addition, facilitating drainage both on the bridge as well as in the approach is one of the most important principles for bridge approach design.

5.4.1.2 Slab track to ballasted track

By its nature, a concrete slab is much stiffer than a granular ballast layer. As such, it was often considered that the focus of track transition design was to provide a stiffness transition between the slab track and the ballasted track. This is actually a misconception, as a slab track can be designed and built with proper resilient and damping materials installed between the rails and the slab and the resulting track stiffness and damping can be very similar to the adjacent ballasted track. In other words, there is usually no need to design a "stiffness transition" between the slab track and the adjacent ballasted track when stiffness reduction can be provided by slab track design.

Rather, the focus of track transition design between slab track and ballasted track should be on the settlement of the latter. This is because ballasted track will inherently settle from the compaction of ballast particles over time, whereas slab track will not. If track transition design is not done properly, differential track settlement would grow under traffic, which in turn will lead to larger dynamic wheel loads, further accelerating track settlement.

Therefore, the best practices for track transition design between slab and ballasted track are similar to those for bridge approaches to control track settlement in the ballasted track that results from repeated wheel load applications over time. This can be achieved by reducing ballast stress at the tie–ballast interface, stabilizing ballast, and "design lifting" ballast section (Chapter 8), all of which are intended to reduce and control ballast settlement in the transition. Concurrently, design of the fastening and pad system for the slab track will need to achieve track stiffness and damping characteristics similar to the adjacent ballasted track.

5.4.1.3 Special trackwork

At special trackwork locations such as diamond crossings, abrupt change occurs mainly as a result of change of track superstructure components, involving changes of rail cross section, rail fastening, and tie plate. However, what makes this type of track transition especially challenging is the change of rail running surface, as gaps and discontinuities cause large impact forces locally.

Best practices for track transition design at special trackwork locations include the following: (1) If weak subgrade soil is identified, it needs to be removed or treated; (2) because the ballast in these areas is difficult to tamp, it is especially important to design and maintain good drainage so that water would not stay in the track; (3) because of large impact forces exerted on the track from rail running surface discontinuities, the track foundation should be strengthened using methods such as underlayment HMA; and (4) design and install resilient and damping materials, such as rubber pads, under the platework or ballast mats between the ballast layer and the HMA underlayment to reduce and attenuate dynamic impact forces.

5.4.1.4 Grade crossing

Although grade crossings are not typically allowed for high-speed passenger track, they can be found at many locations in freight railway track as well as in transit track. The cause of many problems in grade crossings are associated with the change of track superstructure at the crossing, as well as poor drainage conditions.

Best practices for this type of track transition design include the following: (1) design and maintain good drainage so that water will not accumulate and stay in the grade crossing as well as in the adjacent track; and (2) if weak subgrade soil is identified, it needs to be removed or treated not only under the grade crossing but also under the adjacent track.

As stated for other transition types, it is also a common practice to reduce tie spacing and increase tie footprint in the track that approaches the grade crossing to reduce ballast and subgrade stress to minimize the formation of a dip. Also, stabilization of the ballast should be considered.

5.4.2 Examples of design and remediation

The following are several track transition design and remediation examples that have been implemented in the field: most have been proven effective while one has not as will be described. These examples are chosen to illustrate how design and remediation can meet the design principles described earlier.

5.4.2.1 Reduce stiffness and increase damping for track on bridge

At the Facility for Accelerated Service Testing (FAST) in Pueblo, Colorado of the United States, two multispan concrete bridges were built in 2003, both with ballast decks. Standard concrete ties were used on the bridges and in the approaches. Track roughness started to develop at the approaches soon after the bridges went into service, and ballast tamping and surfacing operations were required every 2–8 MGT. This situation is found on many railways' ballast deck concrete and steel bridges, where these bridges and their approaches require frequent track maintenance work due to rough track geometry, mud pumping, and tie cracking.

Investigations at FAST and also at other revenue service test sites showed (Li et al. 2010) that track with standard concrete ties on ballast deck concrete bridges had high track stiffness and insufficient track damping, contributing to high wheel/rail impact forces.

To address this issue, several methods were selected and tested on the two bridges at FAST, which included the following: concrete ties with rubber pads at the bottom surface (see Figure 5.33), plastic composite ties, wood ties, and the use of a ballast mat under the ballast layer, all of which were intended to reduce track stiffness and increase track damping for the track on the bridges.

Test results at FAST have shown that track modulus on the bridges was lowered significantly, that is, reduced to a level similar to that of the approach tracks (roughly 34 N/mm/mm or 5,000 lb/in./in.). See Figure 5.34 for track modulus test results. Surfacing requirements were reduced significantly (approximately by a factor of 10), as a result of those remedies.

Figure 5.33 Concrete ties fitted with rubber pads on the bottom.

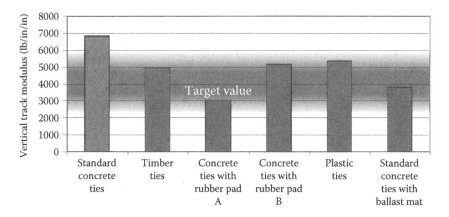

Figure 5.34 Track modulus reduction from various methods.

Figure 5.35 shows reductions in impact produced by these remedies, as measured by strain gages on the bottoms of the bridge beams. Average impact was reduced by 20%–40%.

At the western US heavy axle load high-tonnage site, concrete ties with rubber pads were also installed on a ballast deck concrete bridge in September 2007, and test results have shown a 30% reduction of dynamic bending strains measured at the bottom of the bridge beam, as Figure 5.36 shows.

Figure 5.37 shows the reduction of dynamic wheel/rail forces and vibration as a result of reduced track stiffness and increased track damping for

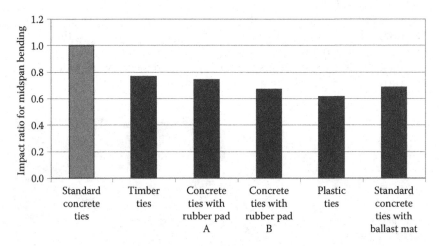

Figure 5.35 Reduction of impact from various methods of reducing track stiffness.

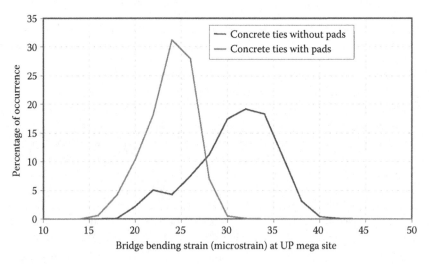

Figure 5.36 Effect of reducing track stiffness for ballast deck concrete bridge in revenue service.

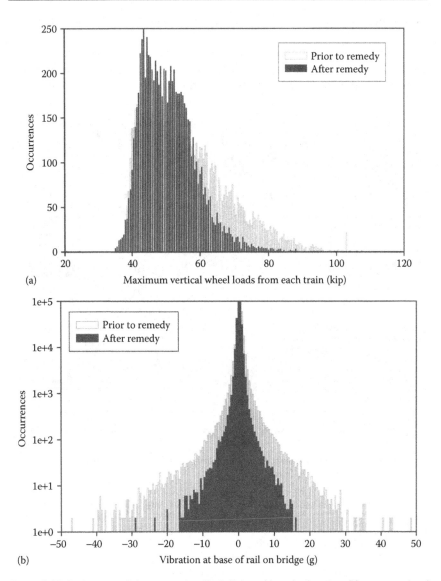

Figure 5.37 Reduction of dynamic wheel/rail forces (a) and vibration (b) as a result of reduced track stiffness and increased track damping.

the track on this bridge. Since the installation of ties fitted with rubber pads plus accommodation of track drainage at this location, no track maintenance was needed for at least 1,000 MGT, an increase of maintenance cycle by at least 20 times.

At another bridge approach location, ballast mats were installed in July 2009 for the same purpose of reducing track stiffness and increasing track damping for the track on the bridge. Track drainage was also improved from prevention of clogging of drainage holes on the bridge as well as the drainage paths provided in the two approaches.

Figure 5.38 shows photographs of the installation of ballast mats on this bridge. Figure 5.39 shows the conditions of track before and after the remediation. In addition, track condition for this bridge was compared with a

Figure 5.38 Installation of ballast mats in a heavy axle load and high-tonnage bridge in western United States.

• Before remediation   • 500 MGT after remediation

Figure 5.39 Improvement of track conditions from reducing track stiffness, increasing track damping, and track drainage accommodation.

- MP 80.5, 2 months after conventional undercutting maintenance

- MP 78.2, 1.5 years after ballast mat/ drainage installation

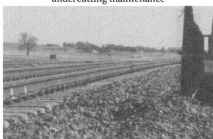

Figure 5.40 Comparison of track geometry condition for two nearby bridges with different maintenance practices.

nearby bridge where no materials were installed nor any measures taken to reduce track stiffness and increase track damping for the track on the bridge. Figure 5.40 shows a comparison of track geometry conditions at these two bridge locations. From the photographs shown in Figures 5.39 and 5.40, it is apparent that this design of reducing track stiffness is also effective.

Figure 5.41 gives a summary of track modulus and track damping test results from various tests conducted by Transportation Technology Center, Inc. (TTCI) for the heavy axle load freight railroads in the United States. These results can be used as reference values for modeling and design of bridge approach projects.

5.4.2.2 Consistent track strength/restraint

For some bridges, especially those located in curves, it is preferable to design the track on the bridge to match the strength or restraint characteristics with that of the adjacent open track. It has been found that when open deck steel bridges located in sharp curves are changed to ballasted deck, the similar lateral track resistance provides a more uniform alignment and profile geometry.

Figure 5.42 shows two photographs of a bridge before and after such a change. As a result lateral track strength, including gage strength at the rail–tie interface and panel shift strength at the tie–ballast interface becomes more consistent. When the track in this sharp curve moves laterally due to temperature change, it now moves in a more uniform manner, reducing or eliminating the misalignment problem. Changing from open deck to ballasted deck bridges allows ballast tamping operations to be continuous from the approaches to the bridge, which eliminates one key source of differential settlement: the discontinuity of a ballast layer that allows track movement and a rigid bridge structure that does not.

Figure 5.43 gives the lateral gage strength test results prior to (left) and after (right) the deck replacement for the bridge shown in Figure 5.42.

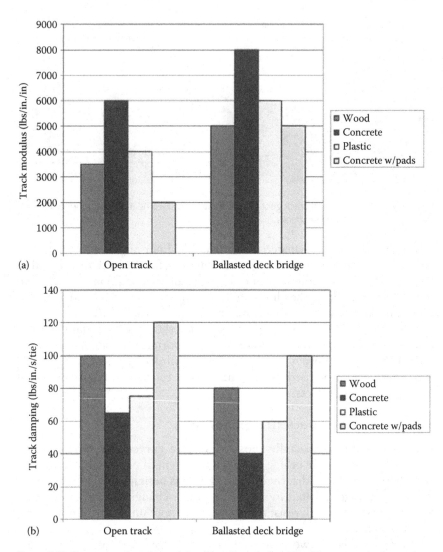

Figure 5.41 Summary of track modulus (a) and track damping (b) test results for various track conditions.

Comparing the test results in these two figures shows that gage strength is more uniform for the bridge with ballasted deck.

Before this remedy was applied, the track required frequent surfacing and alignment maintenance on a monthly basis. Since the replacement installed in September 2008, this bridge and its approaches have required little maintenance.

Note that conversion to ballasted deck requires sufficient bridge load capacity to carry additional dead load, and sufficient distance to run off the change in top of rail elevation that results.

- Before—track was "fixed" to the deck, different fastenings on bridge

- After—track on bridge moves and deforms the same manner as approach track

Figure 5.42 Change of open deck to ballast deck to achieve consistent track strength and restraint.

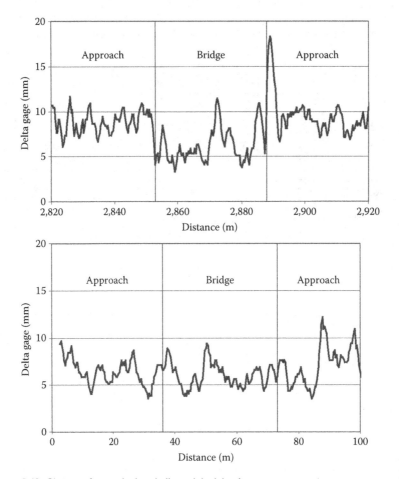

Figure 5.43 Change of open deck to ballasted deck leading to more consistent gage strength.

5.4.2.3 Soil improvement

As discussed, weak subgrade soil can cause instability at bridge approaches, leading to excessive track deformation under train operations. One of the effective remedies to address this is to strengthen subgrade soil. Figure 5.44 shows one of the strengthening methods, using stone columns in the weak subgrade.

This method was used at a site in the western United States to address bridge approach problems caused by weak subgrade. Since the installation in 2003, this bridge approach has required little maintenance.

5.4.2.4 Unnecessary stiffness transition from slab track to ballasted track

A 25-ft-long transition zone was designed and built for a slab track section at FAST. The transition was designed to have a 1-ft-thick reinforced concrete slab with upturned curbs to retain the ballast, a 12 in. ballast layer, and concrete ties. Figure 5.45 shows the 25-ft base slab built in the transition and the final view of the transition track to the ballasted track with wood ties.

Figure 5.44 Use of stone columns to strengthen weak subgrade in the bridge approach.

(a)   (b)

Figure 5.45 Unnecessary base slab for "stiffness transition" between slab (a) and ballasted (b) track.

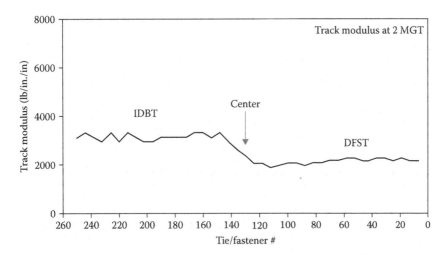

Figure 5.46 Track modulus for two types of slab tracks at FAST.

The 25-ft base slab was designed to provide a transition of track modulus between the ballasted track and the slab track. However, this was found to be unnecessary because the slab track resiliency was significantly greater than first thought, that is, the slab track was equally or even more resilient than the ballasted track at FAST.

Figure 5.46 shows the actual track modulus test results for the two types of slabs placed next to each other at FAST. Because of the resilient rubber pads used under the tie plates for the direct fixation slab track (DFST), and the rubber pads and boots used under the dual block ties for the independent dual block track (IDBT), the slab tracks as built were very resilient, with track modulus measured to be 2,100 and 3,000 lb/in./in., respectively, for two types of slab tracks. In fact, these slab tracks were more resilient than the adjacent wood tie ballasted track, which had track modulus measured between 3,000 and 4,000 lb/in./in.

The measured track modulus for the transition built on the base slab was found to be between 4,600 and 5,000 lb/in./in., which is significantly higher than the modulus of the slab tracks and also higher than the track modulus for the ballasted track, completely defeating the purpose of this "stiffness transition."

Figure 5.47 shows vertical track deflection test results, obtained using the Track Loading Vehicle (TLV; see Chapter 8). It shows a continuous vertical track deflection profile along the slab track and the adjacent ballasted track, obtained under a constant wheel load of 40 kips. As illustrated, the DFST showed an average deflection of 0.2 in., while the IDBT showed an average deflection of 0.15 in. The results shown in Figure 5.47 are consistent with the results shown in Figure 5.46, that is, the DFST exhibited higher deflection or lower track modulus than the IDBT.

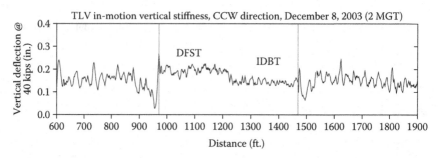

Figure 5.47 Track defection measurement over slab track and adjacent ballasted track.

Figure 5.47 also shows that deflections generated in the DFST were higher than those in the adjacent ballasted track, indicating that the DFST was more resilient than the adjacent ballasted track. The IDBT showed track deflections similar to the adjacent ballasted track. Nevertheless, both slab tracks (DFST or IDBT) showed more uniform track stiffness than the adjacent ballasted track.

Figure 5.47 shows that immediately next to the DFST or the IDBT (or 25-ft transition area), track deflections were significantly lower than those of the slab tracks or the ballasted track nearby. In other words, the design and installation of the 25-ft transitions did not achieve their objective to provide a gradual stiffness transition. Because the slab track was actually either more resilient than, or equally resilient to the ballasted track, the installation of these two transitions became unnecessary in terms of track stiffness.

Chapter 6

Drainage

Drainage of the track is the most important consideration in track design and maintenance. Track must have both internal and external drainage, where internal drainage maximizes the flow of water out of the track and external drainage controls the sources of water having access to the track and ensures that water exiting the track is carried away. Further, both internal and external track drainage must work together. Water has such a dominant effect on the track substructure that poor drainage increases every aspect of track maintenance cost. Despite its fundamental importance, drainage is often given insufficient attention in the design and maintenance of track. This chapter is intended to provide an understanding of the principles of drainage in relation to railway track, and present methodologies for designing and maintaining track drainage systems.

6.1 SOURCES OF WATER IN THE TRACK

There are three main sources of water into the track substructure. The first is *direct water* in the form of rain or snow falling directly onto the track. The second is *runoff water*, which is surface water from rain or snowmelt that begins on higher ground adjacent to the track and then flows into the track along the ground surface. The third is *ground water*, which flows up into the track substructure from below due to either a pressure gradient or a capillary action. Figure 6.1 illustrates these main sources of water into the track. Water is also sometimes transported within the track by longitudinal flow within the ballast/subballast layer.

6.1.1 Direct water

For ballasted track, the access of direct water into the track structure is inevitable except in covered areas such as tunnels and continuous overhead structures. Direct water from precipitation varies widely from place to place and time of year. The drainage measures for track and right-of-way should be based on the magnitude and duration of precipitation for a particular

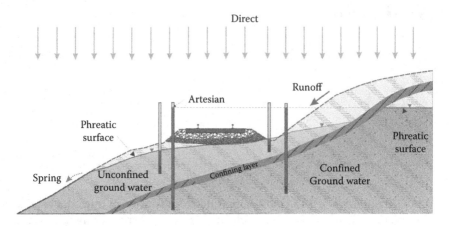

Figure 6.1 Source of water into the track substructure.

area. The impact of direct water on track performance is minimized by good internal drainage, which essentially is free-draining ballast and a sub-ballast layer that promotes lateral drainage out of the track. Once out of the track, direct water is conveyed further away by good external drainage.

6.1.2 Runoff water

The amount of water from runoff that has access to the track is a function of the rain intensity and the surface area of the contributing drainage basin. Often, features that divert surface water away before it can enter the track can also function as external drainage and thereby serve to remove direct water from the right-of-way. Runoff water from overland flow adjacent to the track should be intercepted to prevent it from entering the track substructure by creating diversion ditches and swales.

6.1.3 Ground water

Ground water develops from infiltration of precipitation and surface water features, such as streams and lakes. Ground water is essentially any water that resides within the subsurface that saturates all soil (or rock) pore spaces. The unconfined water surface at the top of the saturated layer is called the phreatic surface, which is sometimes referred to as the "water table." The elevation of the water table will typically fluctuate with time at a given location. Ground water problems are often difficult to diagnose and control.

Ground water flows through the ground and into the track substructure due to a pressure gradient induced by the difference in phreatic surface elevation. This type of flow is typically lateral flow from high phreatic elevation to low phreatic surface. Ground water flow can be artesian, where the ground water pressure exceeds hydrostatic, and if unconfined, rises above the ground surface. If a pipe is inserted into the ground to the depth of the artesian aquifer, the existence of artesian water pressure can be easily observed by the fact that the water will rise to an elevation above the unconfined water level, or even above the ground surface, as illustrated in Figure 6.1.

Ground water is typically only a problem in cuts because the surrounding higher ground allows groundwater to flow toward the track. Excavation in a cut often extends into the bedrock. If the bedrock contains permeable seams or fissures extending upward to the surface, and if these fissures are filled with ground water, the weight of the train passing over this section can force the fissures to open and close, and in some cases can cause water to be pumped through the fissures. When the bedrock is erodible, such as clay shale, the pumping can produce fine particles which mix with water to form mud, which can be forced up into the ballast (Selig oral communication 1998).

6.1.4 Capillary action of water in soil

Soil can become partially or fully saturated above the water table due to capillary action. Surface tension of the water causes water to be drawn into small continuous voids. Capillary rise can be demonstrated by placing a clean glass tube vertically into water (open to atmosphere) and observing that the water rises in the tube to a certain height above the free water surface, as shown in Figure 6.2. The smaller the opening of the tube, the higher the water rises.

Capillary action enables soil to draw water to elevations above the phreatic surface. In contrast to the thin tubes depicted in Figure 6.2, soil has many pores of varying sizes that are interconnected in an intricate network, as depicted in Figure 6.3. Soil immediately above the phreatic water surface will be saturated, and the zone above that will have gradually decreasing saturation extending to a height to which capillary action allows the water to rise. This height will vary based on the size and interconnectivity of the voids. The variable height of capillary action in soil results from the variable sizes of the void spaces, with water rising highest where voids are narrowest. But if the water encounters large pores as shown in the right side of

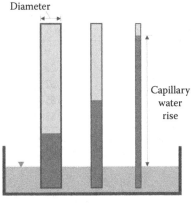

Diameter

Capillary
water
rise

Figure 6.2 Example of capillary rise in tubes showing increasing rise with smaller tube diameters.

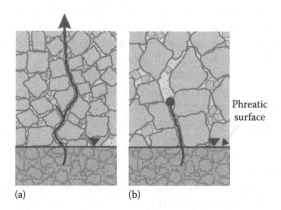

(a)   (b)

Figure 6.3 (a,b) Illustration of capillary rise in soil with different size particles and void spaces.

Figure 6.3, it will not rise above this level. Therefore, the natural variation of a soil deposit will provide a variable amount of capillary rise within it.

Capillary action also enables soil that is draining under the effects of gravity to retain water above the phreatic line. This is known as capillary head, which is usually higher than capillary rise. Capillary head is often a more important consideration for railway track because precipitation leads to saturation of the track substructure from the top downward.

The height of water in soil due to capillary action depends on many factors, including: void ratio, packing density of soil particles, interconnectivity of void spaces, as well as grain shape and roughness. The capillary head and capillary rise values in Table 6.1 are based on data from Lambe and Whitman (1969). The capillary head values in Table 6.1 represent the heights for 100% saturation above the phreatic surface, whereas the capillary rise values in Table 6.1 represent the maximum height that water will rise, but not necessarily causing complete (100%) saturation at the soil.

Table 6.1 Capillary rise and capillary head of soil

Soil	Effective particle size, d_{10} (mm)	Void ratio, e	Capillary head (cm)[a]	Capillary rise (cm)	Capillary rise from Equation 6.1 (cm)[b]
Coarse gravel	0.82	0.27	6	5	9
Fine gravel	0.3	0.29	20	20	23
Coarse sand	0.11	0.27	60	82	67
Fine sand	0.03	0.36	112	166	185
Silt	0.006	0.94	180	360	355

Sources: Lambe, T.W. and Whitman, R.V., *Soil Mechanics*, John Wiley and Sons, New York, 1969; Terzaghi, K. et al., *Soil Mechanics in Engineering Practice*, 3rd Edition, John Wiley & Sons, New York, 1996.

[a] For 100% saturation
[b] Based on empirical coefficient, $C = 0.02$.

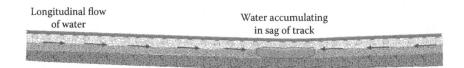

Longitudinal flow
of water

Water accumulating
in sag of track

Figure 6.4 Illustration of longitudinal flow of water along the track.

Table 6.1 presents capillary head and capillary rise for a few different soil types as a function of d_{10}, which is the size (diameter) of soil particles for which 10% of the soil particles by weight are smaller.

Capillary rise in meters can be estimated using Equation 6.1, as suggested by Terzaghi et al. (1996). Equation 6.1 uses an empirical coefficient C varying from 0.01 to 0.05, which is a function of the grain shape and surface impurities. Equation 6.1 also requires the void ratio, e, and the pore diameter d_{10} particle size in millimeters. The right column in Table 6.1 is based on Equation 6.1 using the empirical coefficient of $C = 0.02$, and the void ratio and d_{10} value from the Lambe and Whitman results.

$$h_{cr} = \frac{C}{ed_{10}} \tag{6.1}$$

6.1.5 Longitudinal flow of water in track

The track substructure, especially with a relatively clean ballast layer, can be the most permeable part of the right-of-way and can be a conduit for groundwater flow longitudinally along the track. Water can follow the track grade and collect in a sag in the track grade, as illustrated in Figure 6.4. Water from longitudinal flow can also collect adjacent to features such as at-grade road crossings or undergrade bridges.

6.2 WATER EFFECTS ON TRACK SUBSTRUCTURE

In many instances, the root cause of track substructure problems stems from water remaining in the substructure for prolonged periods of time. In general, excessive moisture in the track substructure layers results in higher rates and magnitudes of deformation, lower resilient modulus, higher plastic strain, and lower shear strength. These characteristics are true for both granular noncohesive soil and fine-grained cohesive soil, albeit the mechanisms of behavior for granular soil and cohesive soil are different.

6.2.1 Granular soil

The strength and deformation characteristics of granular soil are governed by effective stress, σ', which is equal to total stress, σ_T minus the effective pore water pressure, u, that is

$$\sigma' = \sigma_T - u \tag{6.2}$$

The shear strength, s, of granular, noncohesive soil is a function of the effective stress, σ', and the internal angle of friction, ϕ.

$$s = \sigma' \tan\phi \tag{6.3}$$

Pore pressure weakens the strength of soil by reducing the effective stresses within the soil. The greater the pore water pressure, the smaller the effective stress, and consequently, the lower the soil strength. If pore pressure is equal to or greater than the total stress, then the material has zero strength. Under saturated conditions where pore water pressure can develop, certain soil gradations subjected to cyclic loading can develop increased pore pressure that exceeds the total stress, resulting in severe loss of strength and high amounts of deformation. This is known as liquefaction.

Figure 6.5 shows an example of track geometry degradation due to liquefaction of a poorly draining subballast layer. In this actual case, the granular layer was designed with 0.3 m of ballast plus 0.45 m of subballast to reduce heavy axle load-induced stresses in the low-strength subgrade. With this granular layer, subgrade performance was acceptable and the track had limited track geometry degradation, as indicated by the near-flat profile roughness (standard deviation) from 0 to 9.3 MGT. At this point, track geometry rapidly deteriorated immediately following a heavy rainfall event. This track geometry degradation became so severe that the train operation was suspended (Read and Li 1995).

In a similar fashion, localized liquefaction of fouling material in a saturated, highly fouled ballast layer could also result in sudden deformation of the ballast layer, resulting in an abrupt settlement of the rail. However, the conditions for such an event in ballast would be very uncommon due to the larger particles playing a dominant role in ballast mechanical behavior and resisting liquefaction of the layer.

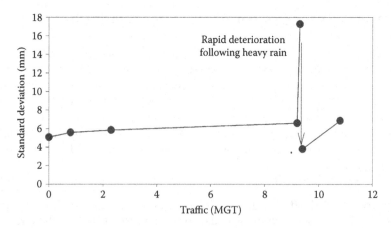

Figure 6.5 Track vertical profile degradation from liquefaction of thick subballast layer.

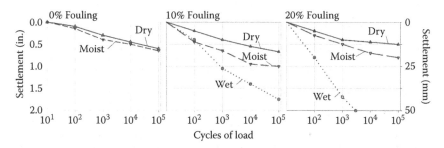

Figure 6.6 Effects of water saturation on the settlement of ballast with different degrees of fouling. (Derived from Han, X. and Selig, E., Effects of fouling on ballast settlement. *Proceedings of the 6th International Heavy Haul Railway Conference,* Cape Town, South Africa, pp. 257–268, 1997.)

Even in partially saturated conditions where the pore pressure is essentially zero, water has an effect on strength and deformation characteristics of granular soil. Figure 6.6 shows the effects of water on the settlement of ballast at different degrees of fouling in the laboratory box test over repetitive cycles of load (Han and Selig 1997). This figure illustrates the greatly increased settlement that occurs due to the presence of water in the ballast. The settlement of the wet, unsaturated granular material is greater than the dry or moist material. Also, Figure 6.6 shows that the greater the amount of fouling material, the more effect water has. This further underscores the importance of maintaining good drainage within the track substructure.

Figure 6.7 further illustrates the effects of moisture/drainage on granular material. Here, the moisture level of the soil is characterized by the degree of saturation, S, and the figure shows that the greater the level of saturation, the greater the settlement due to cyclic loading. The degree of

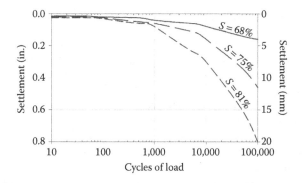

Figure 6.7 Effects of water saturation (S) on the settlement of granular material. (From Haynes, J.H. and Yoder, E.J., *Effects of Repeated Loading on Gravel and Crushed Stone Base Course Materials Used at the AASHO Road Test,* Highway Research Record 39, National Academy of Science, Washington, DC, 1963.)

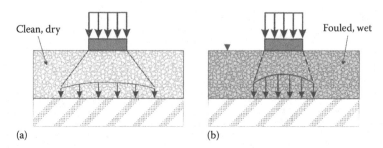

Figure 6.8 Idealized concept of stress transmission from well-drained (a) and saturated granular (b) layers. (Inspired by Cedergren, H.R., *Seepage, Drainage and Flow Nets*, 3rd Edition, John Wiley and Sons, New York, 1989.)

saturation is expressed as the percentage of available void space that is filled with water, that is, 100% degree of saturation means all of the available pore spaces in the soil are filled with water. Depending on the soil gradation and packing density, the moisture content for 100% degree of saturation can vary widely.

Figure 6.8 compares how the stress from train loading is transmitted vertically through, a well-drained, clean ballast layer with that of a saturated ballast layer that has an abundance of fouling material in the ballast. Stress distribution within a granular medium depends on development of intergranular stress between particles, and if the stress is solely transmitted at these strong ballast particle contacts, a large ballast layer stiffness results that significantly reduces the stresses transmitted to the underlying layer as shown in Figure 6.8a. However, if the ballast voids are filled with saturated fine material, the ballast layer effectively softens (becomes less stiff). This softer ballast layer will result in the stress distribution shown in Figure 6.8b, with a greater amount of stress passed along to the layer under the ballast.

6.2.2 Cohesive soil

The strength and deformation characteristics of fine-grained cohesive soil is governed by the effective stress condition, similar to granular soil, but fine-grained cohesive soil is also greatly influenced by the mineralogical make-up of the clay soil that controls the way that clay particles interact with water.

Clay particles have electrical charges on their surfaces that dictate how these very small particles interact. The amount of electrical charge depends on the mineralogical make-up of the clay, which influences the size and shape of the clay particles. Clay particles are typically flat, or "platy," and have net negative electrical charge on their surfaces and a net positive charge of their edges, as depicted in Figure 6.9. Water is a polar molecule with charged sides, also as illustrated in Figure 6.9. The two charged entities, clay and water, therefore tend to combine and become attached. The degree of attraction of water and clay is related to the clay's degree of plasticity.

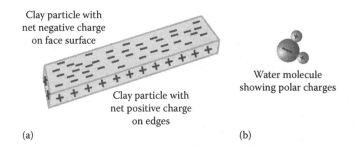

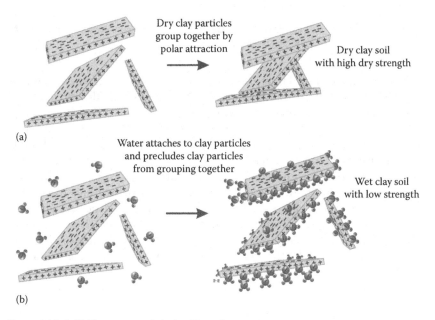

Figure 6.9 Illustration of clay (a) and water (b) as charged particles.

Figure 6.10 (a,b) Illustration of clay's affinity for water.

Figure 6.10 illustrates how water and clay particles can interact to reduce the strength of fine-grained cohesive soil. Dry clay particles are attracted closely to one another, thereby resulting in relatively strong attractive forces and subsequently high dry strength. This "pulling together" of particles also results in the shrinking of the mass of clay particles. When water is present, the polar water molecules are attracted to the clay particles, creating a layer of water between individual clay particles. The presence of the water layer between clay particles inhibit the clay particles from bonding to each other, resulting in low strength of the clay soil mass. The more highly charged the clay, the more water molecules attach to the individual particles, resulting in the development of thicker interparticle water layers. Therefore, more highly charged clay produces more expansive soil, and more clay softening

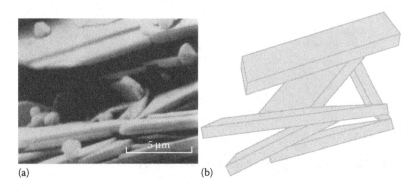

(a)   (b)

Figure 6.11 (a) Photo and (b) diagram of clay particles with *flocculated* structure. (SEM courtesy of Prof. Ernie Selig.)

in the presence of water. High-plasticity, "fat" clays have greater charges and this type of soil (e.g., bentonite and "gumbo") has a high dry strength but expands and softens dramatically with water.

The orientation and distribution of the mass of clay particles can also affect the reaction of clay to water. The structure of clay can be either flocculated or dispersed. In a flocculated structure, like that shown in Figure 6.11, the clay particles are in an edge-to-face orientation. In a dispersed structure, as illustrated in Figure 6.12, the clay particles are oriented face-to-face, or parallel. The clay's structure, being either flocculated or dispersed, not only affects its strength, compressibility, and permeability but also affects the behavior in the presence of water. A dispersed clay structure tends to be lower in strength with high water content and is more susceptible to volume changes with changes in moisture.

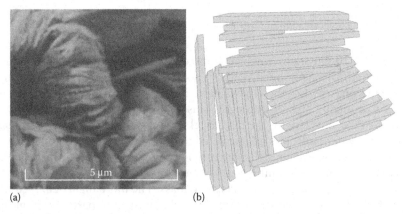

(a)   (b)

Figure 6.12 (a) Photo and (b) diagram of clay particles with *dispersed* structure. (SEM courtesy of Prof. Ernie Selig.)

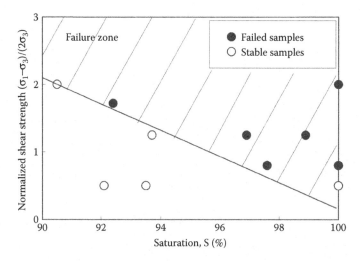

Figure 6.13 Effect of degree of saturation on shear strength of clay. (Data from Miller, G.A. et al., *J. Geotech. Eng.*, ASCE, 126, 139–147, 2000.)

Figure 6.13 shows results of cyclic load triaxial tests on clay and the significant effects that the degree of saturation has on clay strength (Miller et al. 2000). In this figure, soil shear strength is normalized by the confining stress and is plotted as a function of degree of saturation. For a total of eleven cyclic tests conducted, a boundary was drawn to divide the failed and stable soil samples under repeated loading. This boundary represents soil strength under various conditions. Figure 6.13 illustrates the large sensitivity of clay strength with moisture when it is nearly saturated.

The significant effect of moisture on clay soil strength can also be seen in the test results of unconfined compressive strength versus moisture content (Figure 6.14). The clay represented in Figure 6.14 is buckshot clay from Mississippi that was used at the Facility for Accelerated Service Testing (FAST) facility at Transportation Technology Center (TTC) in the Low Track Modulus test section. Again, a small increase in soil moisture content is observed to significantly reduce unconfined soil compressive strength.

6.2.3 Internal erosion/piping

Piping is a process whereby soil is internally eroded by water as it flows through the soil mass. Under certain gradations and water conditions, soil particles can migrate through the void spaces of the soil, and eventually out to the ground surface. Silts are very prone to piping not only because the particle sizes are small but also because they are cohesionless. Graded aggregate filters and geotextiles are often used to prevent the piping of soil due to ground water movement.

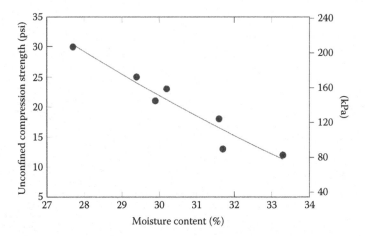

Figure 6.14 Effect of moisture on unconfined compressive strength of clay.

6.2.4 Erosion from surface water

Water flowing along the ground surface can cause erosion of soil when the velocity becomes too high. The amount of erosion caused by high velocity increases with the increase in the steepness of the slope, the reduction in particle size, and the lack of surface protection such as vegetation or paving. The velocity of flow required to initiate particle transport (erosion) at the ground surface depends primarily on the size of the soil particles. Knowing the velocity of flow and particle size, a prediction can be made as to whether the particle moves or not.

Figure 6.15 is based on empirical observations by Lane (1953) of the inception of sediment transport at relatively large particle sizes, including those of ballast-size particles (up to 100 mm diameter). This figure presents the minimum velocity of particle movement as a function of particle size. Figure 6.15 does not reflect the effects of slope or particle shape, and particles on a slope or rounded particles will tend to move at lower velocities than those shown in Figure 6.15. The data in Figure 6.15 match the "zone of transportation" of soil, discussed by Holtz et al. (2011).

6.2.5 Frost heave

Frost heaving of soil is due to the development of ice lenses in the soil. Ice lenses form due to capillary rise of water. In order for ice lenses to develop, and for frost heaving to occur, three factors must be present: freezing temperatures, a source of water, and soil that is capable of sustaining capillary rise. These three factors must all occur to produce frost heave, and when they do occur in the track substructure, track geometry will deteriorate. Soil with high potential for capillary rise is often referred to as frost-susceptible soil.

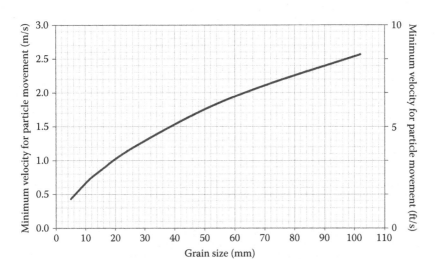

Figure 6.15 Minimum velocity of particle movement as a function of particle size. (After Lane, E.W., Progress report on studies on the design of stable channels of the bureau of reclamation. *Proceedings of the ASCE*, Vol. 79, Reston, VA, 1953; Simmons, D. and Sentürk, F., *Sediment Transport Technology*, Water Resources Publications, Littleton, CO, 1992.)

The most frost susceptible soil are those with potential for high capillary rise and low plasticity, such as silt, fine-grained sand, well-graded sand, and clay with low plasticity (*PI* < 12). Soils are frost susceptible when they are sufficiently fine grained to have a high capillary rise, and sufficiently permeable to have a relatively high rate of hydraulic conductivity. Gravel and sand with 5%–20% fines, which can include certain gradations of fouled ballast, can be frost susceptible (Holtz et al. 2011).

Frost heave occurs not simply from the expansion of freezing water, but from the growth of ice lenses. As illustrated in Figure 6.16, ice lenses will begin to form as the ground freezes from the top downward and water flows upward from capillary action. Capillary water will continue to feed the growth of the ice lens until the ground freezes below the ice lens and the capillary rise stops due to the impermeability of the frozen soil. New lens can continue to form and layers of lenses can develop. Once the frost-penetration stops, the lenses at the interface of frozen and unfrozen soil can grow quite large.

Thawing of ice lenses and frozen ground can create worse track problems than the track heave. This is often referred to as thaw softening. As illustrated in Figure 6.17, as the ice lenses melt, from the top downward, excess free water is present, and this free water is typically unable to drain from the track due to the remaining frozen soil below that is relatively impermeable. The high water content that develops during thawing causes a softening of the soil and can result in a drastic reduction in strength due to loss of effective stress.

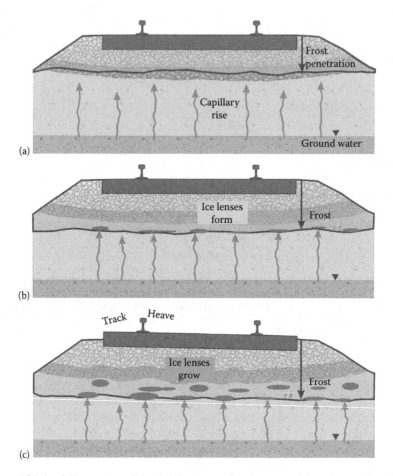

Figure 6.16 (a–c) Illustration of the development of ice lenses and frost heave of track.

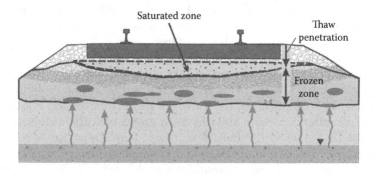

Figure 6.17 Illustration of saturated substructure due to thaw softening.

Frost heave control measures include overexcavation of the soil susceptible to ice lens formation and replacement with select material that is resistant to capillary action. It may not be necessary to replace all of the soil with this select material. A freely draining layer installed at the bottom of the excavation can work to intercept water from below and block further capillary rise. An alternative to providing a cutoff to the water from below is the placement of insulation material to provide a thermal break to freezing temperatures from above. Materials such as extruded polystyrene mats or injected foam can be placed in the substructure to act as thermal barriers.

The application of minerals such as salts to lower the freezing temperature of water, and thereby deter the formation of ice lenses is another frost control measure used for track. These minerals can be placed on the track surface in the hope of sufficient penetration into the track substructure, or they could be injected into the substructure with the same machines used to inject grout. However, sufficient dispersal of these de-icing agents can be difficult to achieve.

Shimming of the rails can often be sufficient to correct for the differential track heave resulting from ice lens formation, although this tactic is aimed at maintenance rather than remediation. In the same way that tamping can provide a durable geometry correction over an actively deforming subgrade, even though it does not correct the subgrade instability, so can shimming provide a reasonably durable correction though it does not address the cause of track heave. As thaw softening can happen rather quickly, the removal of the rail shims must be precisely timed with the track thawing so that track geometry roughness is maintained to acceptable levels.

6.3 WATER PRINCIPLES FOR TRACK

Track drainage includes both subsurface flow of water through soil and overland flow of water along the ground surface. Designing effective internal and external drainage for track requires an understanding of both surface and subsurface water flow.

6.3.1 Water flow through soil

The quantity and velocity of water flowing through soil is affected by the soil's hydraulic conductivity (also referred to as the soil permeability) and the amount of hydraulic head. Greater soil permeability allows easier flow of water through it. This permeability depends mainly on the soil's grain size distribution, porosity, specific surface, shape, and tortuosity (Judge 2013). Specific surface is the ratio of surface area to volume of the soil, and tortuosity is the ratio of the meandering distance traveled by water through soil pores to the straight line distance traveled.

The flow of water through porous material is illustrated in Figure 6.18. The U-shaped channel is shown with the right leg shorter than the left leg

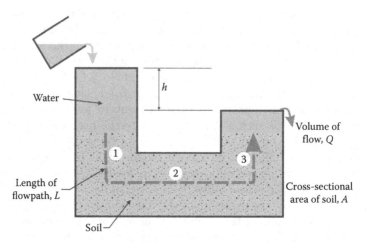

Figure 6.18 Schematic diagram illustrating flow through soil.

and with soil filled to the same elevation in both legs. Water is poured into the left leg to maintain water elevation at the top of the left leg. The water flows down through the soil across the bottom and up the other leg until it overflows at the top of the short leg. At position 1, the water is flowing downward through the soil; at position 2, it is flowing horizontally through the soil; and at position 3, it is flowing upward through the soil. This illustrates the fact that water can flow through the ground in any direction depending on the gradient and layers of soil.

Water flowing through soil, as shown in Figure 6.18, is defined by Darcy's law, which states that the volume of water (Q) is equal to the product of hydraulic conductivity (k) of the soil, the hydraulic gradient (i) within the flow regime, and the cross-sectional area (A) of the flow channel. Darcy's Law is expressed as

$$Q = kiA \tag{6.4}$$

The hydraulic gradient is defined as the hydraulic head difference, h, divided by the length of the flowpath, L, that is

$$i = \frac{h}{L} \tag{6.5}$$

For track drainage, the hydraulic conductivity of the ballast is a significant aspect of internal drainage of the track. The hydraulic conductivity for fouled ballast is based on the amount of fouling, density of ballast layer, and type of fouling. Table 6.2 presents typical ranges of hydraulic conductivity values for ballast with different fouling levels. For clean ballast, the range of hydraulic conductivity values from the work by Parson (1990) are in reasonable

Table 6.2 Hydraulic conductivity for ballast at different fouling categories

Fouling category	Hydraulic conductivity, k (cm/s)
Clean	5.1–2.5
Moderately clean	2.5–0.25
Moderately fouled	0.25–0.15
Fouled	0.15–0.0005
Highly fouled	<0.0005

Source: Parsons, B.K., *Hydraulic Conductivity of Railroad Ballast and Track Substructure Drainage*. Master's thesis, University of Massachusetts, Amherst, MA, 1990.

agreement with testing results from another study by Wnek (2013), which reports clean ballast permeability in the range of 6.1–4.3 cm/s.

Figure 6.19 presents hydraulic conductivity values for various gradations of granular material in yet another study (FHWA 1980; Ridgeway 1982). This figure serves to further illustrate the large control that the amount of fine material in the gradation has over hydraulic conductivity. As ballast becomes increasingly fouled, its permeability decreases drastically.

Soil permeability can be anisotropic, meaning that the horizontal hydraulic conductivity is different than the vertical hydraulic conductivity. This occurs in layered soil, where a continuous layer of high-permeability soil is sandwiched

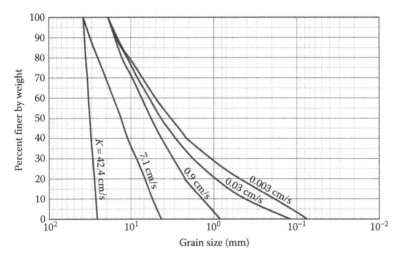

Figure 6.19 Hydraulic conductivity for typical gradation of coarse-grained soil. (Adapted from FHWA, *Highway Subdrainage Design*, Report No. FHWA-TS-80-244, Federal Highway Administration, Offices of Research and Development, Washington, DC, 1980; Ridgeway, H.H., *Pavement Subsurface Drainage Systems*, NCHRP Synthesis of Highway Practice, Report 96, National Academy of Sciences, Washington, DC, 1982.)

between layers of much lower permeability. More groundwater volume flows through the layers of higher hydraulic conductivity that can act as channels. Cedergren (1989) showed an example where lenses of coarse gravel with a hydraulic conductivity of 30 cm/s were located within a layer of silt with a hydraulic conductivity of 10^{-6} cm/s. This resulted in a horizontal seepage that was over five orders of magnitude more than the seepage in the vertical direction.

The hydraulic conductivity of coarse-grained soils can be determined by constant-head or falling-head permeability tests (ASTM 2002). Figure 6.18, which illustrates the flow of water through soil, is actually a schematic of a constant-head test. A large-scale constant-head permeameter was developed at the University of Illinois to measure the hydraulic conductivity of ballast (Figure 6.20).

Coarse aggregate permeability in the range of 1 cm/s to >100 cm/s can also be measured using a laboratory test apparatus, as shown in Figure 6.21, that is based on pneumatic underdamped slug test in monitoring wells (Judge 2014). This test is based on measuring the oscillatory response of water level as it flows through soil.

6.3.2 Surface water flow

The amount of surface water flow that drains to a particular area along the track is governed by the local area's rainfall and ground characteristics. The ground characteristics are defined by the size, shape, soil and ground cover type, land use, and water storage features (ponds, lakes) within the watershed. These conditions can be determined by aerial or ground-based surveys. Rainfall is characterized by its intensity, duration, and return period

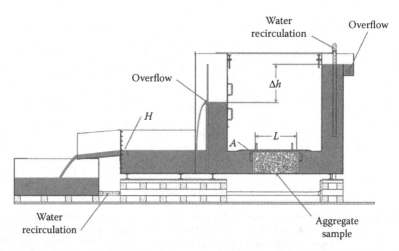

Figure 6.20 Large-scale constant-head permeameter. (Data from Wnek, M., *Investigation of Aggregate Properties Influencing Railroad Ballast Performance*, Master's thesis, University of Illinois at Urbana-Champaign, Urbana, IL, 2013.)

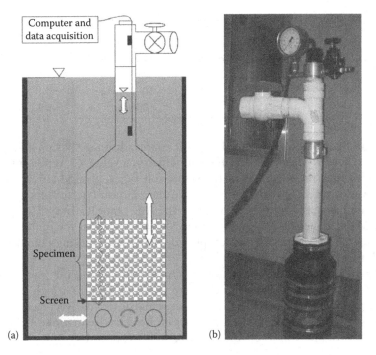

Figure 6.21 Schematic diagram (a) and photograph (b) of the permeameter.

frequency. These factors are typically represented for a specific site based on actual measured rainfall data, which are available from public sources such as the National Weather Service in the United States, or can be obtained directly from local public or private weather stations.

The rainfall storm characteristics that are appropriate for design are based on the sensitivity and importance of the feature being designed. The return period (recurrence interval) is the average time interval between the occurrence of a rainfall event of specified magnitude and an equal or larger rainfall event. The return period frequency is expressed in years and is the maximum storm expected for that particular time period. For instance, a 50-year return period frequency storm is a storm that is expected only every 50 years. Naturally, a 2-year return period storm has a much lower intensity than a 100-year storm. Typical storm return periods that can be used for design are shown in Table 6.3.

Common methods to calculate surface water flow are the Rational method and the Soil Conservation Service (SCS) method. Details of these methods are found in many hydrology textbooks and design manuals, notably Haestad and Durrans (2003) for the Rational method and the TR-55 Guide from the United States SCS (Soil Conservation Service 1986). These models calculate the quantity of surface water runoff by the product of the design

Table 6.3 Storm return periods for design

Application	Storm return period
Track internal drainage	1–2 year
Surface ditch	5–10 year
Culverts	50–100 year
Bridges	100 year

rainfall intensity, the projected area for runoff (drainage basin), and a runoff coefficient. The runoff coefficient, which is a value <1, represents variables such as the steepness of the slope, soil types, ground cover, and the amount of water absorbed into the ground. Runoff coefficients are readily found in many references, notably the American Railway Engineering and Maintenance-of-Way Association (AREMA) Manual for Railway Engineering (AREMA 2012), which provide values for varying conditions in both rural and urban settings.

The Rational method is relatively simple and is good for track design, because track drainage design does not typically require a high degree of refinement. The Rational method is an empirically derived method that estimates peak runoff (Q) as a product of drainage basin area (A), rainfall intensity (i), and runoff coefficient (C) and is expressed as

$$Q = AiC \tag{6.6}$$

The intensity, i, is determined from rainfall records that have been developed into Intensity, Duration, and Frequency (IDF) plots; an example of which is shown in Figure 6.22. The Rational method is based on the assumption that the rainfall lasts long enough for runoff water to travel from the farthest point in the drainage basin to the point of interest. The time for rain water to travel along the ground from the most distant point to the location of interest is called the time of concentration, t_c. From an IDF plot, like Figure 6.22, the design rainfall intensity is determined by selecting an appropriate return-period frequency and using a storm duration equal to the time of concentration of specific drainage basin.

The SCS method uses 24-h rainfall frequency distribution, and characterizes the basin's soil type, land use, and slope. The SCS method is generally more complicated than the Rational method, but it also tends to be more exacting. This method also considers storage within the basin from features such as ponds and lakes, and uses hydrographs for water flow through these features. SCS methods subdivide the basin into sub-basins of similar characteristics to be more accurate.

There are various computer models that can be used by those with training and experience in hydrology and hydraulics as refinements to the Rational and SCS methods, when greater levels of accuracy and certainty are required. The computer models often combine both hydrology and hydraulics to model the

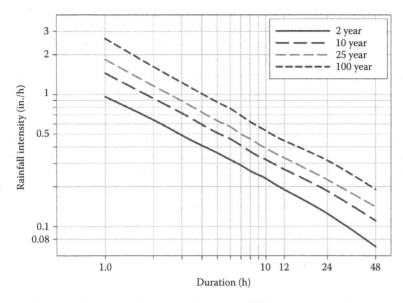

Figure 6.22 Rainfall intensity, duration, and frequency (IDF) curves for use in the Rational method. (Data for Northampton, Massachusetts.)

runoff, and they can take into account physical elements such as culverts, bridges, open channels, storm sewers, and detention basins (storage areas). The computer models allow for determination of flow from numerous and varied watershed areas, and catchments (i.e., individual drainage subareas within larger basins). For accurate determination of flow, the basin must be subdivided to accurately represent all of the runoff conditions, storage features, and so on.

6.4 DRAINAGE MATERIALS

A wide variety of materials are available to provide track drainage. There are various types of pipes and conduits available to be used as subsurface drains. The designer can choose basic materials such as graded stone aggregate to provide simple but effective internal drainage, or more advanced materials such as geosynthetics. Hot mix asphalt (HMA) is another material that can be integrated into internal drainage. These materials are discussed in this section.

6.4.1 Pipes

Pipes are used to convey water longitudinally along the track and laterally under the track. Pipes for drainage applications can be made of steel (galvanized), iron, reinforced concrete, and plastic (high-density polyethylene

[HDPE], polyvinyl chloride [PVC]). Historically, vitrified clay pipes were common and can still be found in the right-of-way. Depending on their application, pipes can be perforated, that is, containing slots or holes that allow ingress of water, or can be nonperforated to provide conveyance of water. Pipes are typically round, however, elongated pipes, such as fin-drains also exist. Pipes are often corrugated to provide added strength at the cost of somewhat reduced flow characteristics.

Pipes need to be of suitable size for the anticipated water quantity, and also need to be of suitable strength to eliminate the possibility of deformation or collapse. In addition, to ensure good flow characteristics, pipes should be installed on slopes of at least 1% for smooth pipes and at least 2% for corrugated pipes. In some situations, it may be necessary to install pipes with as little as a 0.5% slope, and in these cases the pipe should not be corrugated.

6.4.2 Graded aggregate

Using coarse aggregate with its inherent high permeability is often the most economical and effective subsurface drainage material. Design considerations for graded aggregate drainage include material type and gradation, as well as the thickness and configuration of the aggregate layer. One of the key design requirements for aggregate drains is to ensure its long-term performance by limiting particle migration from adjacent soil layers into the void space of the aggregate layer.

To assess the separation effectiveness of the aggregate drain to preclude particle migration from the adjacent soil, both of the following criteria need to be met (Cedergren 1989; Selig and Waters 1994):

$$\frac{D_{15}\left(\text{Aggregate Drain}\right)}{D_{85}\left(\text{Adjacent Soil}\right)} < 5 \tag{6.7}$$

$$\frac{D_{50}\left(\text{Aggregate Drain}\right)}{D_{50}\left(\text{Adjacent Soil}\right)} < 25 \tag{6.8}$$

where D_{xx} (e.g., D_{15}, D_{50}, D_{85}) is the particle size at which xx percent of the material is smaller.

In most instances, high-permeability aggregate, in particular clean ballast, does not meet these criteria for sandy or fine-grained subgrade soil. Therefore, geotextile filter fabrics are usually needed to eliminate particle migration and subsequent clogging of the aggregate layers. As an example, Figure 6.23 shows the grain size distribution of a common sandy subgrade soil along with typical aggregate drain material.

Table 6.4 summarizes the pertinent particle sizes from Figure 6.23 for use in evaluating the potential for particle migration.

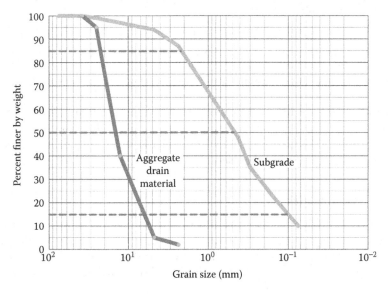

Figure 6.23 Example grain size distribution plots for common sandy subgrade and typical aggregate drain material.

Table 6.4 Summary of pertinent particle sizes of aggregate drain material and adjacent subgrade

Material	$D_{15}{}^a$ (mm)	D_{50} (mm)	D_{85} (mm)
Aggregate drain	5	15	21
Subgrade	0.1	0.5	2

Note: $^aD_{xx}$ = is the particle size at which *xx* percent of the material is smaller.

Using the values in Table 6.4 in the separation criteria, Equations 6.7 and 6.8 yield:

$$\frac{D_{15}(\text{Drain})}{D_{85}(\text{Soil})} = \frac{5}{2} = 2.5 < 5 \Rightarrow \text{OK}$$

$$\frac{D_{50}(\text{Drain})}{D_{50}(\text{Soil})} = \frac{15}{0.5} = 30 > 25 \Rightarrow \text{N.G.}$$

Therefore, the aggregate drain material gradation is not adequate to maintain separation from the adjacent soil, due to intermixing in the mid-range particle sizes. Often a subgrade soil has a greater amount of silt and clay-sized particles than the soil gradation in Figure 6.23, and such a fine-grained soil would be even more likely to infiltrate into the aggregate drain material per these two criteria.

6.4.3 Geosynthetics

The family of products known as geosynthetics is manufactured from synthetic polymers and has a wide range of application in civil engineering works. These geosynthetic products include geotextiles, geogrid, geomembranes, geocell, and geocomposites. In railway track applications, geotextiles are commonly used with the intention of providing separation and retention between layers of geotechnical materials, while other products such as geogrid and geocell are used to provide strength through reinforcement. If the goal is to establish an impermeable layer, then geomembranes or geosynthetic liners can be used.

6.4.3.1 Geotextiles

A geotextile is a permeable geosynthetic that is either woven or nonwoven, as shown in Figure 6.24. Geotextiles are often used to provide filtration and separation of differently graded layers so that water can escape from the fine-grained layer without allowing small soil particles to pass through into the voids of the coarse-grained layer.

The Apparent Opening Size (AOS), which is also referred to as the Equivalent Opening Size (EOS), of the geotextile is used to determine how effective it is at providing separation or soil-retention capability. The AOS is equivalent to the D_{95} of the geotextile (Cedergren 1989). Those with too large an opening size may allow fine soil particles to intrude and possibly pass through so that the intended separation function is lost. The AOS should be less than the D_{85} particle size of the fine-grained soil for the fabric to provide separation (Selig and Waters 1994).

Figure 6.24 Examples of woven and nonwoven geotextiles.

An AOS that is too low may indicate insufficient permeability of a geotextile and possible failure of the drainage function. A geotextile with a small AOS when new can become even less permeable in use and can act more like an impervious membrane than a permeable media, which can lead to the buildup of a very weak soil layer under the geotextile and eventual track instability. Such an example is described by Chrismer (1985), in which the lateral sliding of the track was induced by a woven, thermal-bonded geotextile with an AOS of 200 (opening size of a #200 sieve). This particular geotextile had a relatively low initial permeability, which was reduced further in the track after it was laid directly over a compacted silty clay subgrade, and the fine soil particles began to infiltrate into the fabric as loading accumulated. As the traffic tonnage accumulated to approximately 5 MGT, lateral track strength eventually reduced to the point that the track slid laterally under a locomotive, resulting in a derailment. A postincident investigation revealed that a thin layer of very weak clay slurry had developed immediately under this fabric along its entire length along the track.

This occurrence is a cautionary example that applies to any case where a relatively impervious material is placed over fine-grained soil and thereby prohibits drainage. For the same reason, a layer of asphalt, Portland cement concrete, geomembrane, or any other impermeable layer placed over a fine-grained soil with potential for high moisture content can produce the same kind of track instability, unless free-draining material is placed between the soil and the impervious material.

Geotextiles have the capability of providing drainage in the plane of the geotextile; however, the amount of drainage in the planar direction is typically very small. Koerner (1990) notes that, aside from geocomposites, nonwoven geotextiles have the highest in-place permeability. However, the permeability of nonwoven, needle-punched geotextiles is only in the order of 0.04 cm/s, with a transmissivity (volume of water per time per unit length of geotextile) of 2×10^{-6} m³/s-m (Koerner 1990). When placed horizontally in the track substructure with proper slope (2%) and outlet, the lateral drainage capability is only in the order of about 5 L/h for 1 m of length.

6.4.3.2 Geomembranes

Geomembranes are impermeable, flexible, geosynthetic sheets typically made from neoprene, PVC, chlorinated polyethylene, or bitumen. The primary function of geomembranes in track substructure application is to prevent water from passing from one side of the geomembrane to the other. Membranes can be used to prevent the infiltration of water from the track surface into the substructure, and thereby minimize water softening of the subgrade, shrink/swell volume changes in expansive-type soil, and detrimental water effects on the performance characteristics of the granular ballast and subballast layers. Geomembranes can also provide the isolating function of keeping water from entering the track laterally from the sides.

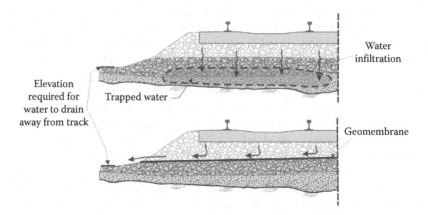

Figure 6.25 Typical application of geomembrane to raise drainage surface within substructure.

Geomembranes can be used to raise the drainage surface within the track substructure and thereby provide effective gravity drainage away from the track. Figure 6.25 illustrates this application of raising the drainage surface and shedding of water.

A potential problem with the use of a geomembrane is the possibility of trapping excess water beneath the membrane. Water trapped below the membrane can create a slipping surface at the soil–membrane interface and cause local failure of the subgrade due to a soil strength reduction as described previously for a geotextile with low permeability. Also, if seams or holes develop in the membrane, concentrated hydraulic action may occur through the openings causing local soil erosion. Therefore, these issues need to be addressed in any application by ascertaining if the potential for trapping water exists, and if it does, by providing a suitable drainage layer beneath the membrane.

Resistance to lateral sliding must also be considered for each specific application where a geomembrane is used. The membrane may have a relatively low friction between it and the ballast placed above it, and so the possibility of lateral ballast movement on the membrane should be considered. However, lateral instability is more likely to arise if a weak soil layer forms beneath the membrane due to lack of pore water pressure release and buildup of weak slurry, as described in the previous section. To avoid this, the membrane should not be in direct contact with a fine-grained soil but should instead be placed on an intermediate granular layer of subballast or similarly graded material that will provide an escape for pore water pressure release under loading.

6.4.3.3 Geocomposites

Geocomposites are the result of combining two geosynthetic materials (or sometimes a nonsynthetic material with a geosynthetic) with different material properties to provide a function in the track substructure that a single geosynthetic material cannot. A common example is a high-tensile strength geogrid that is combined with a filtration geotextile to provide both strength and separation. Another common examples of a geocomposite is a sandwich drain, which is a geotextile that is wrapped around a perforated rigid polymer core that allows water, and not soil, to pass through the geotextile and into the core.

6.4.4 Hot mix asphalt

HMA underlayment can provide improved shedding of water from above it because HMA is essentially impermeable. Therefore, HMA can help reduce the access of water to moisture-sensitive subgrade soils. However, this benefit is limited to the extent that water may gain access to the subgrade from other sources. Figure 6.26 illustrates the application of HMA, similar to a geomembrane, being used to raise the drainage surface within the track substructure, and thereby provide effective gravity drainage away from the track.

Grade crossings often have very little lateral drainage. Therefore, water drainage out of grade crossings is typically longitudinal along the track and out of the crossing ends. Good drainage is a key component of a well-performing grade crossing, and it is commonly observed that crossings contain excessive amounts of water. Figure 6.27 shows an application of HMA in a crossing to raise the drainage surface so that water can effectively escape.

A potential problem with the use of HMA underlayment is the possibility of trapping excess water beneath the HMA. Water softening of subgrade soil under HMA layer can dramatically reduce the life of HMA through accelerated fatigue failure. Water trapped below the HMA can cause local failure of the subgrade due to a soil strength reduction (see LoPresti and Li 2005). Also, if cracks develop in the HMA, concentrated

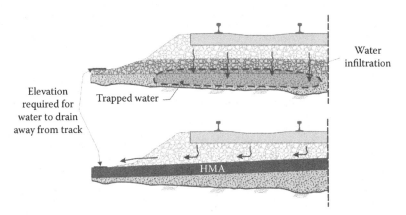

Figure 6.26 Application of HMA to raise drainage surface within substructure.

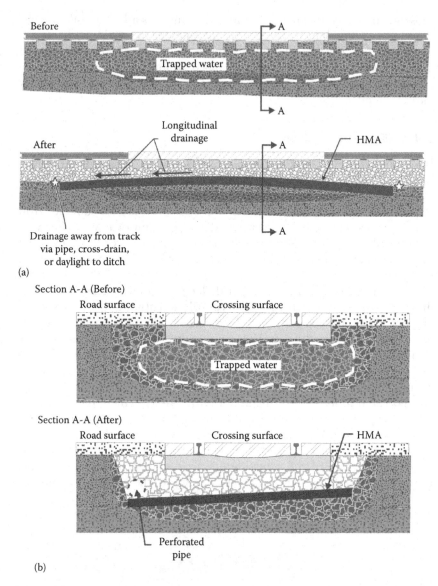

Figure 6.27 Longitudinal (a) and cross-sectional (b) views of HMA application in grade crossing.

hydraulic action may occur through the openings causing local soil erosion. The impermeability of HMA could result in pore water pressure buildup under the HMA slab if it is constructed on fine-grained soil that does not drain rapidly, potentially producing a loss of soil strength and lateral track instability. Therefore, these issues need to be addressed in any application by ascertaining if the potential for trapping water exists, and if it does, by providing a suitable drainage layer beneath the HMA for water to escape laterally.

6.5 EXTERNAL TRACK DRAINAGE DESIGN

The goals of an external track drainage design are to minimize the access of water to the track and to remove water from the track right-of-way. All sources of water must be considered when designing a drainage system, including direct precipitation falling onto the track, surface water from adjacent areas, and groundwater flow. Good external track drainage requires that an unobstructed and downwardly sloping path exists from the ballast shoulder to the ditch, and then from the ditch to the right-of-way egress point, so that water is effectively and rapidly removed from the track vicinity.

6.5.1 Diverting water away from track

Surface water flow must be intercepted and diverted away from the track. Where such interception and diversion does not occur, the infiltrating surface runoff places an added burden on the track drainage system, which is typically only relied upon to remove rain water falling on the track. The basic approach to divert water away from the track is to construct a cutoff ditch, trench, or drain system that allows the water to flow to a discharge point so that it does not enter the right-of-way or the track substructure. If the amount of surface water is small enough, the cutoff ditch function can be provided by the track-side drainage ditch.

Ground water flow can be managed by designing drains that reduce the phreatic surface by gravity. In extreme cases, when gravity flow is not possible, pumps can be used to draw down the water table. Subsurface ballast drains (SBD) using clean ballast are often an effective and economical drainage solution. Figure 6.28 presents a typical SBD. The flow capacity of an SBD is very large and typically does not require a pipe to supply additional capacity. However, perforated pipe is sometimes installed with periodic "clean-outs" to flush the pipe in case it gets clogged.

6.5.2 Right-of-way mapping

Accurate topographic mapping of the right-of-way and adjacent land is important for external drainage design. In addition to ground topography, the location and elevation of the top of rail for all tracks is important. Figure 6.29 provides an example of a three-dimensional, computer-generated digital

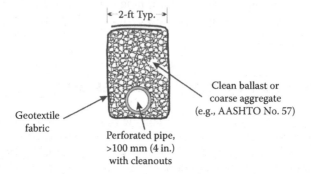

Figure 6.28 Subsurface ballast drain (SBD).

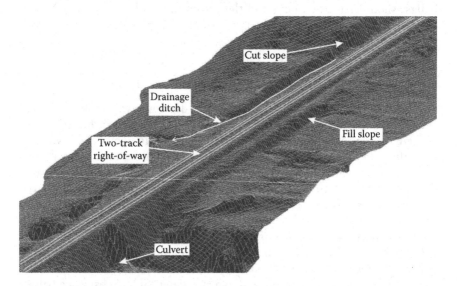

Figure 6.29 Example of digital terrain model from aerial LIDAR data. (Courtesy of M. LaValley, HyGround Engineering, Williamsburg, Massachusetts.)

terrain model, where the ditch is in the cut and the outlet is shown at the cut-to-fill transition.

6.5.3 Ditch and slope

Good external drainage is usually easy to establish when the track is on an embankment and the ground surface slopes away from the track. But for track on flat ground or in a cut, open-channel side ditches are preferred. The ditch bottom must be sloped or inclined along the track to cause the water to flow longitudinally. The minimum ditch depth is the lowest point of the substructure to be drained, usually at the bottom of the subballast at the edge of the track structure, so that the ditch will drain both the ballast and the subballast.

Heyns (2000) presents a useful design procedure for external ditch design that consists of the following steps:

1. Collect right-of-way information
2. Select ditch parameters
3. Determine optimum ditch profile
4. Check ditch capacity
5. Check feasibility of ditch

The first step in designing a ditch system is to collect information on the right-of-way and identify any constraints to ditch construction, such as grade crossings, utilities, and catenary poles. The outlets are the locations where water can discharge from the ditch and away from the track by gravity. In planning for ditches, the highest point between the two outlets should be selected to minimize the amount of earth to be excavated.

Ditch parameters can be defined based on the right-of-way information. These include type of ditch, minimum and maximum ditch slope, cross-section shape, and minimum ditch elevation below the track (Heyns 2000). In addition, at this point in the design process, the need for SBD should be evaluated. SBD should be considered for locations where the required depth of the ditch is large, resulting in excessively wide ditches or in interference with structures. Also, SBDs can be used in lieu of open ditches, where open ditches are prone to clogging from raveling side slopes or from trash, such as in an urban environment. The minimum elevation of the ditch should be based on the internal drainage characteristics of the track. Ground penetrating radar (GPR), discussed more in Chapter 8, can be used to reveal the below-ground features that are needed to establish the ditch elevation. GPR can indicate the presence of a deformed soft subgrade layer that may be holding back water in the track substructure that would otherwise drain to the side ditch. The low elevation point of the deformed subgrade surface is the elevation that must be drained by remedial work, and this establishes the minimum ditch elevation because it must be lower than this point.

Determining the optimum ditch profile typically means minimizing the required excavations by considering the outlet points, grade of the track, and minimum elevation requirements. The minimum longitudinal slope of the ditch bottom is typically 0.5%, although a 1%–2% slope is preferable. The maximum longitudinal slope is typically 2% to prevent soil erosion from high-velocity flow. In the case of earth ditches, the side slopes must also be flat enough to prevent erosion from water flowing into the ditch from both sides.

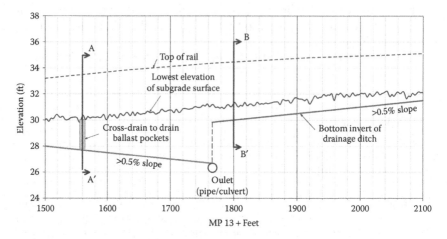

Figure 6.30 Example of external drainage design.

Figure 6.30 shows an example that illustrates an external drainage design. The frame of reference provided by the top of rail indicates the grade of the track. The lowest elevation of the subgrade surface, as revealed by GPR, shows the depth to which any lateral drains (if needed) must be excavated under the track. The cross-drain must slope from the track to where it intersects with the invert of the drainage ditch. The invert of the ditch, shown in Figure 6.30 by the two straight lines terminating at the outlet, must be excavated to an elevation low enough to compliment the track's internal drainage. Note that the slope of the ditch can be similar to that of the top of rail as shown on the right of Figure 6.30, but sometimes the drainage design requires that it slope in the opposite as shown on the left.

The two cross sections AA′ and BB′ in Figure 6.30 relate to the respective cross-sectional profiles shown in Figure 6.31. The lower subgrade surface in the middle of the track in Section AA′ can hold water in the track. It is important that the lowest point of the subgrade surface in Section AA′ be effectively drained, which is often accomplished with a subsurface ballast drain.

Following Heyns' (2000) approach, the next step is calculating the required capacity of the ditch, which is accomplished in accordance with the surface flow analysis. Once the flow quantity is established, open-channel flow analysis is used to determine the size of the ditch and any ancillary piping. The required capacity of the ditch can be determined by the Rational method or SCS method, as discussed in Section 6.3.2.

The final step is to check the feasibility of the ditch by developing right-of-way topographic cross sections with the new ditch superimposed, followed by site reconnaissance.

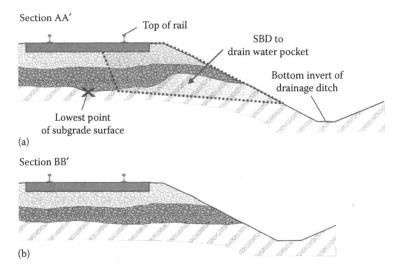

Section AA'

Top of rail

SBD to
drain water pocket

Bottom invert of
drainage ditch

Lowest point
of subgrade surface

(a)

Section BB'

(b)

Figure 6.31 Sample cross section AA' (a) and BB' (b) for Figure 6.30.

6.6 INTERNAL TRACK DRAINAGE DESIGN

The internal drainage characteristics of the track must be designed to provide ballast with sufficient lateral drainage to allow water to escape and prevent saturation. The most important element for good internal track drainage is a highly permeable ballast layer. Once the ballast layer has become fouled, its permeability is adversely affected and the track no longer has good internal drainage and no amount of drainage appurtenances or external drainage will solve the problem. For the internal track drainage design presented here, the underlying assumption is that the ballast under the track drains freely and that the shoulder ballast drains freely from top to bottom. Figure 6.32 presents an ideal track cross section for good internal drainage.

6.6.1 Internal drainage flow nets

The examination of flow nets is useful in understanding internal track drainage. A flow net is a graphical representation of a two-dimensional Laplace equation for flow of water through a porous media (Cedergren 1989). A flow net is comprised of flow lines, which represent channels of equal flow, and equipotential lines, which represent lines of equal head loss.

Figure 6.33 shows flow lines for ballast and subballast, assuming that the ballast is saturated. The upper drawing in Figure 6.33 is a situation where the ballast is more permeable than the subballast layer. The bottom

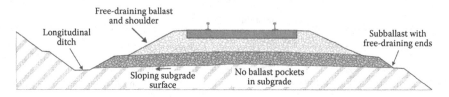

Figure 6.32 Components of good track drainage.

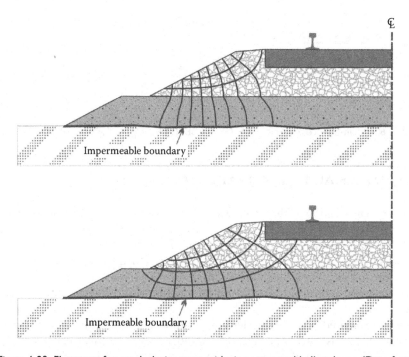

Figure 6.33 Flow nets for track drainage considering saturated ballast layer. (Data from Selig, E. et al., Drainage of railway ballast, *Proceedings of the 5th International Heavy Haul Railway Conference*, Beijing, China, pp. 200–206, 1993.)

drawing in Figure 6.33 is for ballast and subballast with the same permeability. However, seldom does ballast have a low enough permeability where saturation takes place, and the flow nets depicted in Figure 6.33 are rarely achieved.

The ballast layer rarely becomes saturated; instead the rainfall enters the subballast directly from the ballast above. For internal track drainage design, ballast is assumed to have a high permeability (clean to moderately fouled) so that saturation does not take place. Therefore, the actual flow

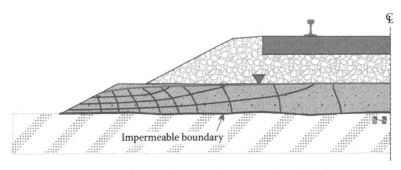

Figure 6.34 Flow net drawings for steady state flow through subballast.

nets for internal track drainage considers that the ballast does not become saturated and the water flow out of the track is through the subballast layer, as presented in Figure 6.34.

The design capacity for internal track drainage can be determined by using the same type of IDF plots that are used for the Rational method. For track internal drainage design, a return period of 2 years can be used as the design rainfall, and instead of determining a time of concentration, a duration of 1 h is used. Rainfall rates that are higher or last a longer period of time will cause more saturation in the track than designed. For instance, using the IDF plot shown in Figure 6.22, for a 2-year storm with 1-h duration, the design drainage capacity is 0.95 in./h.

6.6.2 Ballast and subballast layers

Good internal track drainage depends on the ballast and subballast layers draining freely and minimizing the time water is in the track substructure. Good drainage requires establishing a clean ballast layer in the cribs, shoulder, and under the ties, along with good external drainage. The suggested minimum hydraulic conductivity value for track drainage design is 0.2 cm/s (300 in./h), which is for moderately fouled ballast with a fouling index of about 15. More highly fouled ballast, with correspondingly lower values of hydraulic conductivity, will retain water in the track for longer periods of time.

The drainage of subballast is a very important issue for good internal drainage. Most water that enters the track from precipitation is drained laterally out of the track by the subballast layer. Only for high-intensity rainfall will rain water be shed off the top of the subballast layer, without entering the subballast layer, and then only after the subballast becomes saturated. Even for relatively low subballast permeability, the subballast layer typically does not shed water across the subballast/ballast interface, rather all downward migrating water infiltrates the subballast.

Relatively low-permeability subballast tends to be a track drainage detriment because of the slow drainage rate and the potential for worsening of subgrade saturation. It is ideal that the hydraulic conductivity of the subballast be considerably greater than that of subgrade to allow the water to drain. The need for the subballast to be relatively permeable conflicts with the function of shedding water from downward infiltrating rain water. A solution to this issue is to use free-draining subballast with a geomembrane on the subballast surface for shedding water. Heyns (2000) found that the slope of subballast/subgrade interface does not play a significant role in avoiding subballast saturation, but it does affect the time for drainage of the subballast once the rainfall stops.

It is important to establish the shortest possible lateral drainage path out of the subballast. At a minimum, this means establishing and maintaining a free-draining end of the subballast layer immediately at the toe of the ballast shoulder. Figure 6.35 shows water draining out of the track one day after a rainstorm. The slow drainage at this location is due to the subballast layer under the track extending out of the track and becoming the right-of-way access road. The subballast layer thus has a long drainage path and a corresponding long time needed for drainage. Instead of the water freely draining from the subballast, the water is forced to slowly percolate through access road. In the example photo in Figure 6.35, the subballast under the track stays saturated longer than if the subballast layer terminated at the toe of the ballast shoulder. The internal track drainage would be improved by lowering the right-of-way access road and ensuring that the bottom of the subballast layer under the track drains freely to the top of the access road.

In double-track territory, each track should drain laterally to the field side of the tracks. In triple-track territory, a longitudinal drainage collection trench is needed due to the long drainage path through the subballast.

Figure 6.35 Poor internal drainage due to the long drainage path of a subballast that extends through the right-of-way road.

6.6.3 Design approach

A design approach for internal track drainage has been developed by Heyns (2000), and this approach is based on limiting the saturation of the sub-ballast layer, which prevents excessive deformation due to cyclic loading under saturated conditions. Figure 6.36 shows the condition of the subballast layer with water draining laterally from the location of the maximum saturation height ("H_{max}") toward both ends. Limiting H_{max} to less than the thickness of the subballast layer ensures that the subballast does not fully saturate. This approach assumes that the subgrade is impermeable and that there is no vertical drainage of the subballast into the subgrade, which can yield conservative results for sections of track that have a permeable subgrade that allows downward drainage of the subballast layer.

Equation 6.9 can be used to determine the desired H_{max} for use in the design (Heyns 2000), by specifying a Percentage of Drainage for a given thickness of subballast, H. For design, a Percentage of Drainage of 50%–80% is reasonable.

$$\text{Percentage of Drainage} = 100 \times \left(1 - \frac{H_{max}}{H} \right) \tag{6.9}$$

Once H_{max} has been selected using Equation 6.9, Equation 6.10 (Heyns 2000) can be used to determine the maximum length of the flowpath (L_{flow}), based on a given rainfall intensity and subballast hydraulic conductivity. Or alternatively, if the length of the flowpath is fixed, then the required subballast permeability to handle the design rainfall can be determined.

$$\frac{q_i}{k} = \frac{0.75 H_{max}}{L_{flow}} \left(S_{dL} + \frac{H_{max}}{L_{flow}} \right) \tag{6.10}$$

where:
  q_i is the rainfall intensity infiltrating the subballast, in./h
  k is the hydraulic conductivity of the subballast, in./h

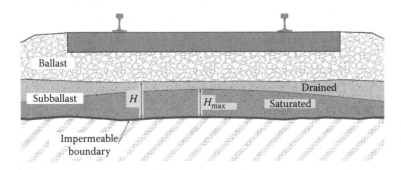

Figure 6.36 Double-sided drainage.

S_{dL} is the slope of drainage layer, in./in.

H_{max} is the saturation height at the no-flow boundary, in.

L_{flow} is the length of flowpath, in.

For a given design rainfall intensity (see Section 6.3.2) and an H_{max} determined from Equation 6.9, the length of subballast flowpath (L_{flow}) from centerline of the track to the edge of subballast can be determined for a given material permeability and cross-slope. Suppose, for example, that the nominal subballast thickness is assumed to be $H = 6$ in., the slope $S_{dL} = 0.5\%$, $H_{max} = 3$ in., and $k = 50$ in./h. For these conditions, using the above equation, the drainage path, L_{flow}, must be shorter than 16.5 ft to preclude subballast saturation under the design rainfall.

Double-sided drainage of subballast often applies for single track or for multiple tracks where there is no lateral obstruction on either side for water to drain. However, if there is a track adjacent to the track being considered, the adjacent track often limits flow of water in that direction, and therefore single-sided drainage conditions may apply for the track under consideration. The analysis for single-side drainage is more complicated and beyond the scope of this book. However, more information on this can be obtained from Heyns (2000).

From Equation 6.10, design charts can be developed, like the one shown in Figure 6.37 from Heyns (2000). For a given rainfall intensity, q_i, and a given subballast permeability, k, the amount of drainage of the subballast can be determined for various combinations of subballast thickness, H, and drainage path length, L_{flow}.

Heyns (2000) also presents an approach to evaluate the time for subballast to drain after a rainstorm stops. For time-dependent drainage of a saturated layer, a sloping subballast surface only provides a significant benefit for less-permeable subballast. However, low-permeability subballast, used where rainfall is frequent, results in the bottom portion of the subballast being saturated most of the time. This is an important consideration when the retained water in the bottom of the subballast is over a moisture-sensitive subgrade, which argues for avoiding low-permeability subballast. Lastly, although a sloping surface provides a limited benefit allowing the subballast layer to drain somewhat faster, it will not reduce the possibility of this layer saturating under a rainstorm. Table 6.5 gives an example of the effects of slope of subballast layer on its drainage time. Table 6.5 is for a 6-in. thick subballast and drainage path (L_{flow}) of 8 ft.

6.6.4 Ballast pockets

Ballast pockets present a unique internal drainage challenge. Ballast pockets originate from the progressive deformation of the subgrade surface directly under the track. As downward-infiltrating water ponds in the depression that is created, the fill continues to soften and further deformation occurs.

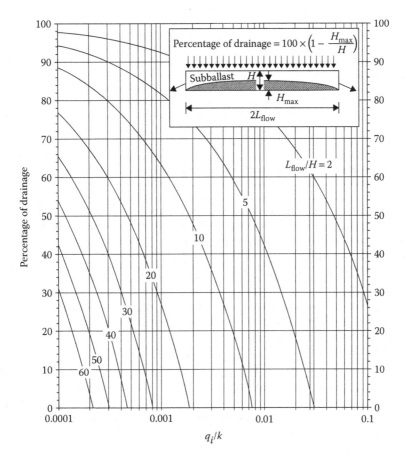

Figure 6.37 Design chart for a percentage of the subballast drainage for a layer slope $S_{dL} = 0\%$. (From Heyns, F.J., *Railway Track Drainage Design Techniques*, Doctoral dissertation, University of Massachusetts, Amherst, MA, 2000. With permission.)

Table 6.5 Drainage time of a 6-in.-thick saturated subballast for a single track after rainfall stops

	Time for complete drainage (hours)			
	50% drainage		90% drainage	
Subballast permeability cm/sec (ft/day)	$S_{dL} = 0\%$	$S_{dL} = 2\%$	$S_{dL} = 0\%$	$S_{dL} = 2\%$
7×10^{-2} (200)	0.8	0.6	6.9	3.5
7×10^{-3} (20)	4.9	3.7	30	22
7×10^{-4} (2)	28	18	230	115

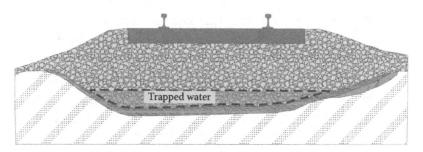

Figure 6.38 Illustration of ballast pocket drainage problem.

As the track settles due to the fill deformation, additional ballast is added and tamped under the ties to raise the track, resulting in a thick ballast pocket, as shown in Figure 6.38. Left untreated, ballast pockets can lead to shear failure within the embankment slope.

The most common approach to correcting ballast pockets is to use SBD (Figure 6.28) excavated at the deepest point of the pocket. The location with the greatest dip from the track geometry data may not necessarily coincide with the deepest part of the ballast pocket. GPR is a good tool for determining the lowest point of the ballast pocket.

Ballast pockets can also develop to one side of the track due to progressive movement and plastic deformation of the clay subgrade. As shown in Figure 6.39, the subgrade deforms under the track and the heave or "push" into the ballast shoulder results in a "bathtub" effect in the track that traps water, which in turn accelerates the softening and deformation of the clay.

Corrections of ballast pocket problems can vary in scope, cost, and effectiveness. For example, the most effective, albeit most expensive, is in-place reinforcement or improvement of the embankment soil. There are many soil improvement and reinforcement techniques that can be used to stabilize embankments and foundation soils (see Chapter 9). The specific site location, extent of problem, and economics will drive the selection of a particular soil-improvement technique.

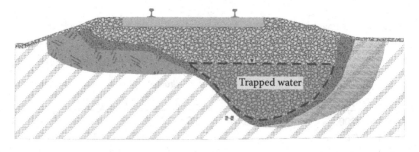

Figure 6.39 Illustration of ballast pocket from progressive movement and plastic deformation.

Providing drainage of the ballast pocket can also provide some measure of remediation. Cross-drains installed at the lowest point of the ballast pocket will remove the most amount of water. Although removing the water reduces the driving weight within the embankment and reduces the chance of further water softening of the embankment soil, it does not improve the soil that has already weakened.

6.7 DRAINAGE IMPROVEMENT AND REHABILITATION

To ensure good long-term track substructure performance, drainage must continue to function well. There are many aspects of track drainage that must work together, including such things as drainage between the tracks in multitrack territory, drainage ditches beyond the ballast being clean and with a minimum 0.5% slope to an outlet, and special considerations for drainage in station areas and other difficult locations. Establishing a clean ballast section is the most important component of track drainage. However, even with clean ballast, problems can occur without proper subballast and external drainage conditions.

The track cross sections shown in Figure 6.40 illustrate features of good and poor internal and external track drainage. The idealized profile has clean ballast from under tie to top of subballast, a uniform subballast layer thickness with a slightly sloping interface with the subgrade, and an ability for the subballast to drain laterally to the side ditch or slope. The lower drawing in Figure 6.40 shows a number of poor drainage characteristics. The ballast is heavily fouled under the tie and may retain excess water. The subballast layer does not terminate near the toe of the ballast but instead extends to the cut slope, which also make the lateral drainage path too long. Due to this, and the fact that no ditch exists on the left, water draining laterally in the

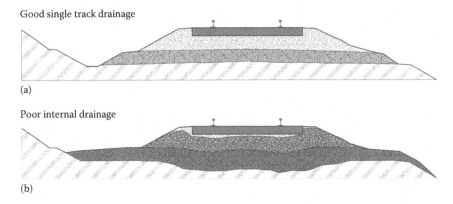

Good single track drainage

(a)

Poor internal drainage

(b)

Figure 6.40 Illustration of good (a) and poor (b) track drainage.

subballast has no escape. Also, the subgrade layer is deformed and will retain water in the ballast pockets.

6.7.1 Ballast cleaning

A clean ballast layer can be established under the track by various means such as track undercutting, track vacuuming, large raising of the track, or complete removal and replacement of the track. These alternatives must be evaluated in terms of cost, track possession requirements, and total work duration time.

Of the various methods available to remove existing heavily fouled ballast and replace it with clean well-draining material, the most common are vacuuming, shoulder cleaning, and track undercutting. Ballast removal by vacuuming is usually reserved for limited sections of track due to its relatively slow rate of progress, but the shoulder cleaning and undercutting methods are mechanized and have a relatively high-production rate. Figure 6.41 illustrates the pre-existing fouled ballast condition (upper drawing) and the cleaned ballast section after using shoulder cleaning and undercutting (middle and lower drawings, respectively).

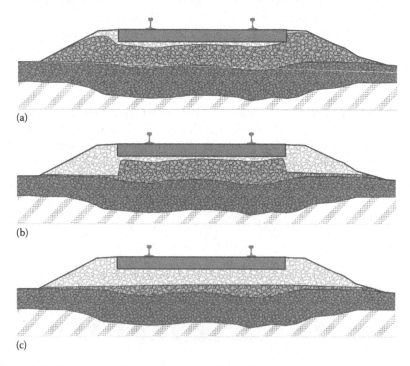

(a)

(b)

(c)

Figure 6.41 Illustration of ballast cleaning alternatives: (a) poor internal drainage; (b) shoulder cleaning; and (c) undercutting.

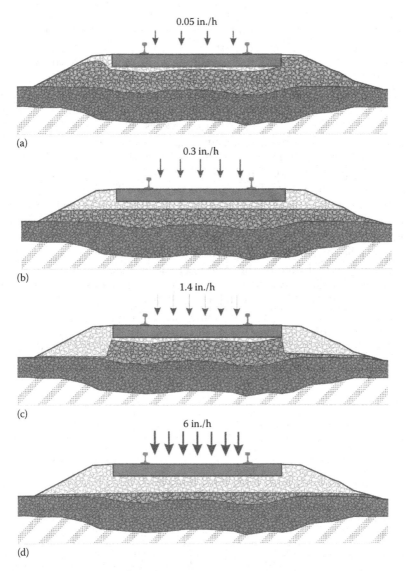

Figure 6.42 Illustration of effects of various ballast-cleaning approaches: (a) fouled; (b) partial depth clean; (c) shoulder clean; and (d) full depth clean. (Based on Selig, E. et al., Drainage of railway ballast. *Proceedings of 5th International Heavy Haul Railway Conference*, Beijing, China, pp. 200–206, 1993.)

Figure 6.42 illustrates the effects of different ballast-cleaning approaches, as based on Selig et al. (1993). In the first case, where the ballast is completely fouled to its full depth, the rainfall rate that saturates the ballast near the center of the track was determined by flow net analysis to be 0.05 in./h. For the second case, it is assumed that undercutting and cleaning of the upper part of the ballast layer (partial depth undercutting) was carried out, extending to the edge of the shoulder. By calculation, it was determined that the resulting rainfall rate required to saturate the center portion of the ballast was increased to about 0.3 in./h, resulting in an improved drainage rate. Shoulder cleaning has an increased drainage rate over partial-depth undercutting. Full-depth undercutting has the greatest improvement of internal track drainage rate. An important reason for the large amount of drainage improvement provided when the entire thickness of the shoulder ballast is cleaned is the shortening of the drainage path laterally through the fouled ballast compared with the length of the fouled section. Figure 6.43 shows a photo of a shoulder cleaner in action.

It is worth noting that when replacing ties with a Track Laying Machine (TLM) a shoulder cleaner is often used first. The TLM by itself produces a thin layer of clean ballast under the tie, and so the amount of drainage improvement would be comparable to that shown by the "Partial Clean" case. But, the further drainage improvement produced by full-depth shoulder cleaning is indicated by the "Shoulder Clean" case.

The undercutting machine shown in Figure 6.44 can separate fouling material from the ballast (through on-board sieving) and can return cleaned ballast to the skeletonized track. Some undercutting machines allow the operator the ability to either waste all of the old ballast, or to attempt to recover at least a portion of the larger particles sizes to the track by diverting some or all of the old ballast to the on-board sieves. Whether or not

Figure 6.43 Picture of a shoulder cleaner. (Courtesy of Loram Maintenance of Way, Hamel, Minnesota.)

Figure 6.44 Excavating chain of ballast undercutter. (Courtesy of R. Wenty, Plasser & Theurer, Vienna, Austria.)

cleaning and returning some of the old ballast to the track is attempted, new clean ballast can be added to the track along with undercutting operation. The length of the undercutting bar should be at least 12 ft where the track centers are 13 ft or less to allow the area between the two tracks to be cleaned. This provides drainage between tracks by overlapping the excavation from each track. In other areas with wider track centers, other means will be needed to provide unblocked drainage between the two tracks. The ballast shoulders on the field side of the tracks must be cleaned to allow water to flow out of the track to a drainage outlet.

Control of the undercutting depth and cutting bar inclination determines the minimum invert depth of the ditch below the longitudinal track profile so that water may drain laterally out of the ballast and subballast and flow into the ditch. Planning the undercutting depth is particularly important in a multiple track system. Lateral discharge of water out of the track is impeded in a multitrack system if the undercutting depth of a middle track is lower than that of the outer tracks. In such areas, the undercutting depth of the outside track should be adjusted to match that of the middle track to allow water to drain laterally out of the tracks, as shown in Figure 6.45. Around a curve, the tracks can have significant super elevation, and it may be necessary to undercut the tracks so that water discharges only to one side, as shown in Figures 6.45 and 6.46. Undercutting depths should therefore be adjusted to allow slope toward the side.

A large vacuum machine can be used in lieu of an undercutter to remove fouled ballast in cribs and shoulders, and to some extent the

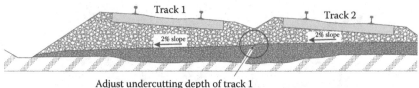

Adjust undercutting depth of track 1
to match undercutting depth of track 2

Figure 6.45 Example of a center track lower than outside track.

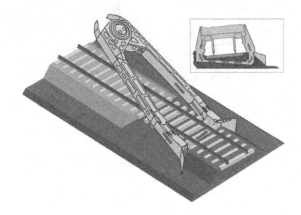

Figure 6.46 Example of tracks undercut to one side around a curve. (Courtesy of R. Wenty, Plasser & Theurer, Vienna, Austria.)

ballast under the tie, over a limited length of track. Once the fouled ballast has been removed from the cribs, clean ballast is placed and the new ballast can be forced under the tie by tamping. The vacuum machine can be used to excavate fouled ballast from the center tracks, straight through the outer tracks, to an outlet point at the perimeter of the right-of-way to provide a lateral escape for water to the ditch in multiple track locations. It can also be used to remove fouled material and provide a path for water to drainage features such as culverts and drainage pipes, as shown in Figure 6.47.

Typically, drainage improvement along the right-of-way is first performed on the external drainage, with the internal drainage of the track performed after the external drainage is established.

6.7.2 Subballast drainage

Water drains out of the track through the shortest possible lateral path out of the subballast. At a minimum, this means establishing and maintaining a

Figure 6.47 Ballast vacuum train. (Courtesy of Loram Maintenance of Way, Hamel, Minnesota.)

free-draining end of the subballast layer immediately at the toe of the ballast shoulder. Further drainage enhancement includes using longitudinal drainage collectors placed in the subballast nearer to the tie. Longitudinal collectors or free-draining zones may be necessary in multiple track territory. Extending the subballast layer across an adjacent right-of-way access road is not a good practice because of the increased drainage length, as shown in Figure 6.48. A photograph of this condition is shown in Figure 6.35.

Substructure problems can be related to the drainage of the subballast, as illustrated by the case shown in Figure 6.48. The deformation of the subballast layer is due in part to water retention in the subballast, as well as the

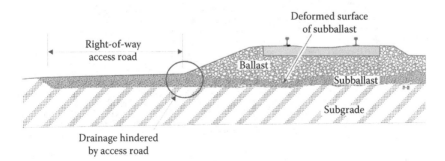

Figure 6.48 Substructure problem related to insufficient subballast drainage.

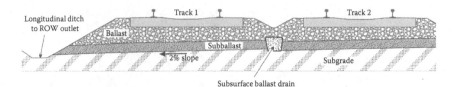

Figure 6.49 Subballast drainage for double-track right-of-ways.

development of a "slippery" surface at the top of the fine-grained subgrade. To guard against deformation of the subballast layer, the subballast should have a relatively high permeability to promote quick drainage and should not extend laterally beyond the toe of the ballast slope.

Figure 6.49 shows a recommended typical subballast drainage design for double-track right-of-ways.

6.7.3 Ditching and grading

External drainage can be provided by using conventional off-track construction equipment such as backhoes, motor-graders, excavators, gradalls, and bulldozers. External drainage can also be constructed with specialized on-track equipment such as "badger" ditchers, hi-rail gradalls, and "slot trains" with special excavators. Figures 6.50 and 6.51 show some examples of these machines.

Figure 6.50 On-track "Badger" ditching machine in action. (Courtesy of Loram Maintenance of Way, Hamel, Minnesota.)

Figure 6.51 A "slot train" excavating a ditch. (Courtesy of Georgetown Rail Equipment, Georgetown, Texas.)

Where an extensive length of track requires ditching, these production ditching machines can be very efficient. These machines require the removal of obstacles in advance that would interfere with the ditch-excavation process. Individual locations will be encountered where production equipment cannot be used. At these locations spot-digging equipment is necessary.

Chapter 7

Slopes

The design and maintenance of slopes along the railway right-of-way is a common challenge throughout the world. Slopes, whether along the side of embankment fills or through excavation cuts, are common because of the need to create this man-made topography to minimize railway grades. Hilly and mountainous terrain often necessitates alternating fills and cuts. Railways along river valleys, where grades are the most gradual, typically require numerous embankments and sidehill cut-fills (Figure 7.1).

This chapter covers new and existing fill and cut slopes, with a distinction between soil and rock material. A cursory introduction to the fundamentals of slope stability analysis is provided, and common slope instability problems are discussed along with options for stability improvement. This chapter presents the topics most relevant to railways, but is not exhaustive in its treatment of the slope topic. There is an entire body of work devoted to slope stability issues that considers problems of every conceivable type and magnitude. Of those available references that can provide more in-depth discussions, the reader is encouraged to review such references as Hoek and Bray (1981) and Turner and Schuster (1996).

7.1 NEW SLOPES

Proper design of embankment fills and excavation cuts require consideration of the type of soil and rock material that are encountered in situ or are being placed, as well as the geometric configuration of the slope (height and width). The majority of fill embankments use locally available material, which can vary from fine-grained soil to hard, competent rock. Cut slopes are made through soil as varying as loess, residual, and colluvium and through rock that can vary from weak, fractured shale to hard, monolithic granite.

7.1.1 Site characterization

Prior to designing a new slope, specific information must be gathered concerning the proposed site. Critical information for design includes the

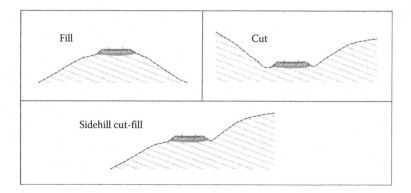

Figure 7.1 Slope nomenclature.

proposed height of the embankment or cut, type of locally available material in sufficient quantities for the embankment, the lithology through which the cut slope will be excavated and foundation conditions of the existing ground on which the embankment will be constructed.

The initial site characterization can be performed as a desk study by evaluating available topographic maps, aerial imagery, and published literature and maps on the local soil, geologic, and groundwater conditions. The desk study should be followed by field investigations and laboratory testing. The veracity of the investigation depends on the size of the project. Large fills with high embankments, and areas with foundation conditions that are uncertain, often require a robust investigation.

The field investigation should include observations of local geologic conditions and observation of the condition of existing slopes in the area by an experienced engineer or geologist. Field investigation techniques include test pits, test borings, cone penetrometer tests (CPT), and dynamic cone penetrometer tests (DCP). Geophysical techniques can also be effectively used to determine site conditions, especially for determining the depth to competent stratum within the embankment foundation. Diffraction sounding, microgravity, and ground penetrating radar (GPR) are examples of geophysical techniques that can be used to determine the thickness of compressible soil and the top of rock surface. Instrumentation, such as piezometers (groundwater monitoring wells), settlement gages, and inclinometers can be effectively used in an Observational Method approach (see Chapter 9) for large fills, where the rate of construction can be dictated by the rate of foundation settlement and corresponding dissipation of induced groundwater pressure.

7.1.2 Materials

In most instances, the material used to create new slopes is given and not chosen. The design of stable slopes, therefore, must conform to the available

Table 7.1 Desirability of soil type for new fill slopes

Material type	Erosion resistance	Embankment strength
Rockfill	Excellent	Excellent
Gravel, well to poorly graded	Excellent–Good	Good
Sand, well to poorly graded	Good–Fair	Good
Clays of low plasticity, sandy clays	Good	Good
Clays of high plasticity	Poor	Fair
Silts	Poor	Fair
Organic clay and silt	Poor	Poor

Table 7.2 Typical presumptive slope ratios for new fill slopes

Material type	Typical minimum slope ratio
Rockfill	1.5H:1V
Sand and gravel	1.8H:1V
Sand	2H:1V
Fine-grained clay and silt	3H:1V to 2.5H:1V

soil and rock types. The strength and erodability of the soil and rock will dictate slope ratios and corresponding right-of-way requirements. Table 7.1 presents a list of the desirability of various soil types for new slopes.

Based on the strength and erodability of the different soil types, Table 7.2 presents typical presumptive values of horizontal to vertical (H:V) slope ratios for these soil types. Table 7.2 assumes that the material is compacted to a high relative density.

The density of soil directly impacts the strength and, therefore, the stability of the soil mass. Most often, high densities are achieved in fill slopes by proper compaction of the material during construction. The density of the material in cut slopes is typically high due to the long period of overburden stress that strengthens the soil. However, some cut slopes may encounter low-density material (e.g., colluvium), and often the exposed material on the cut face can weather to a low density when exposed to the elements.

7.1.3 Embankment fills

The design of new slopes must consider not only the embankment fill, but also the foundation condition under the proposed embankment. Right-of-way property constraints can sometimes preclude the construction of embankment fill with conventional slope ratios. In these situations, embankment fills must be designed with steepened slopes or by incorporation of retaining structures.

7.1.3.1 Foundation

The foundation of a new slope is the existing ground on which the fill is constructed. Relatively thick and compressible soil strata, if not properly considered in design, can lead to continual long-term settlement (consolidation) of the completed fill and even shear failure of the foundation soil.

Control or prevention of settlement within the embankment and of the foundation soil can be key aspects of design. For new fill, some of the settlement is immediate and occurs during embankment construction. Time-dependent settlement, in the form of consolidation and secondary compression, may occur over time ranging from weeks to many years. Monitoring of fill and embankment foundation settlement during construction is often needed to ensure short-term stability at sites where elevated pore water pressure develops in the foundation during embankment construction. High induced pore water pressure can lead to excessive postconstruction settlement or temporary instability during construction.

Modeling, scale-model testing, and field observations have shown that stiffness of the foundation soil has a significant effect on the long-term, permanent deformations of railway embankments; that is, the stiffer the foundation material, the smaller the amount of accumulated deformation (Kalliainen and Kolisoja 2013). Therefore, considerations should be given to stiffening and strengthening the foundation soil prior to embankment construction to minimize long-term settlement of the track.

7.1.3.2 Embankment

Greater compacted density of fill material provides higher strength and lower settlement of the embankments. When constructing a fill embankment, it is important that the fill material be placed in layers and compacted near to the maximum density and close to the optimum moisture content. Figure 7.2 provides some general recommendations for new embankment construction.

Where a new fill slope is to be built adjacent to an existing embankment to widen the embankment, the interface between the old and new slopes must be well connected to prevent any slippage between the existing and new material, as shown in Figure 7.3. Ensuring proper connection of new fill to existing fill also applies to postfailure reconstruction when new fill is being added adjacent to an embankment that has failed.

When constructing new slopes, whether a full embankment or a sidehill fill, a toe bench should be constructed into the existing ground at the toe of the slope, as shown in Figures 7.2 and 7.3. The toe bench provides stability at the critical zone at the bottom of the slope where shallow surficial movement tends to originate. The toe bench ensures that the potential zone of movement (failure plane) goes through, or around, the toe bench, thus increasing the Factor of Safety (F.S.) against instability.

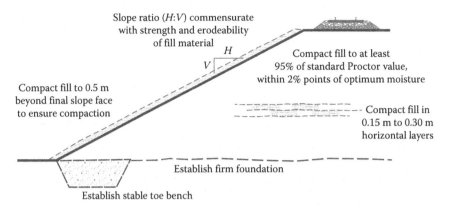

Figure 7.2 General recommendations for new fill embankment construction.

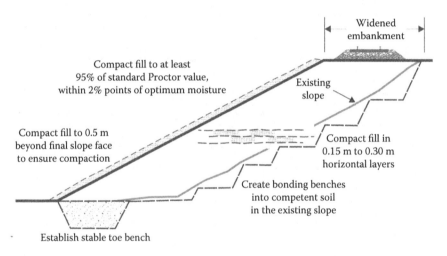

Figure 7.3 General recommendations for sidehill construction or postfailure reconstruction.

7.1.3.3 Steepened slopes

Often due to right-of-way constraints, the slope ratio of the embankment fill may need to be made steeper than what the common fill material can withstand. The slope, therefore, must be designed with a steepened slope angle or a retaining structure. An effective way to obtain greater slope ratios is to include horizontal reinforcement, as depicted in Figure 7.4.

Zoned-fill can also be used to maximize slope ratios. As the name implies, this type of slope is constructed with zones of select material, as depicted in Figure 7.5.

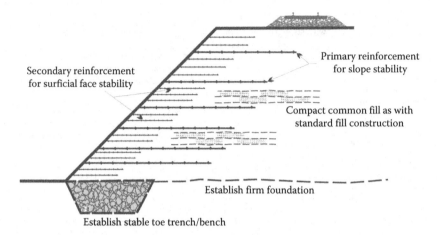

Figure 7.4 Horizontal reinforcement to obtain a steepened slope.

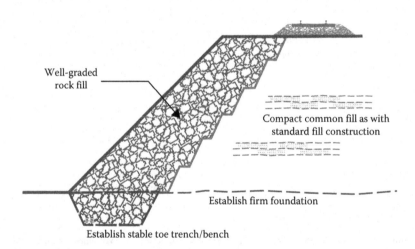

Figure 7.5 Zoned-fill embankment.

7.1.3.4 Retaining structures

Retaining structures (walls) can be used to provide for minimum right-of-way requirements, similar to steepened slopes. The walls can be constructed at the top or bottom of slopes, as shown in Figures 7.6 and 7.7, respectively. The use of a wall at the top of slope adds additional slope stability concerns from the added vertical and horizontal forces at the upper part of the

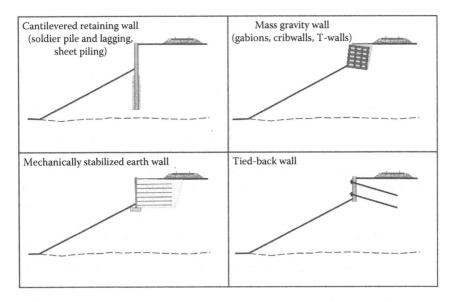

Figure 7.6 Retaining structures at top of slope.

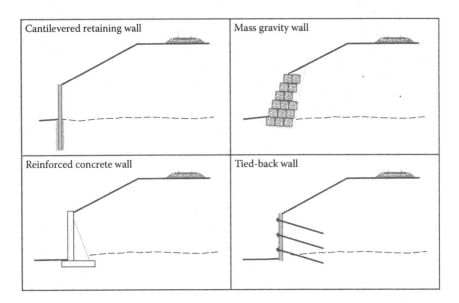

Figure 7.7 Retaining structures at bottom of slope.

slope. Whereas, retaining structures at the bottom of the slope need to be designed more substantially to account for the upper slope stresses.

7.1.4 Cut slopes

Cut slopes must be designed with slope ratios commensurate with the strength and the weathering potential of the soil and rock in the cutting. In addition, subsurface seepage and surface water runoff must be controlled to minimize erosion and strength-reducing effects on the cut slope. Design of cut slopes in soil require greater consideration than rock slopes, because soil is usually a much weaker and more erodible material than rock, and soil layers can often have zones of inherent weakness. The design of soil cut slopes is based on the same approach and method of analysis for soil fill slopes, which is discussed later in this chapter.

Design of cut slopes in rock require a good understanding of the rock lithology involved, in particular the rock's structure of joints, fractures, faults, and bedding planes. Once the structural condition and the erodability of the rock are known, then rock cut slopes can be designed to account for weathering and erosion of the slope. Rock cut slope design may need to include provisions for the protection of the track right-of-way from rock falls by providing adequate catchment areas or structural protection, such as walls, fences, or deflection sheds.

Rock cut slopes can have alternating layers that are relatively thick and well-defined, or thin and somewhat undifferentiated. Slopes with well-defined rock layers and monolithic strata can be designed with slope ratios commensurate with the strength and erodability of the strata, as shown in Figure 7.8. This approach is impractical for thinly layered rock, and therefore, cut slopes through this material are typically designed with uniform slope ratios, as shown in Figure 7.9.

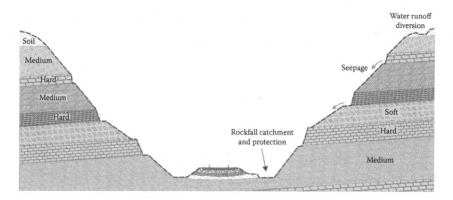

Figure 7.8 Rock cut slope of well-defined rock strata.

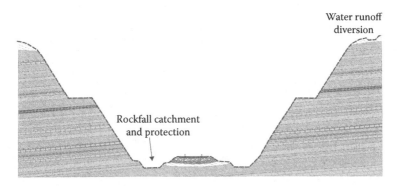

Figure 7.9 Rock cut slope of thinly bedded rock strata.

Ground water is usually present in rock cut slopes, and the flow of ground water within rock stratum is typically dictated by the dip of the bedding planes and the joint network. Control of water, whether subsurface seepage or surface runoff, is critical to minimizing erosion in rock cuts. Freeze–thaw cycles in cold weather regions produce added erosion; therefore, slope angles and talus storage may be different from one side of cut to another. South-facing slopes can be subjected to more frequent cycles of freeze–thaw.

The stability of rock cut slopes is controlled primarily by the discontinuities or joints within the rock. Rock typically has at least some degree of joint formation and these joints may be interconnected, weathered, and perhaps filled with a material that limits frictional resistance and allows movement along these joints. However, if the rock joints are sufficiently widely spaced and not highly interconnected, then the rock should be able to provide a relatively steep slope. For less-stable rock conditions, the designer can decide between using a reduced slope angle, or a right-of-way protection, or rock stabilization program. Figure 7.10 provides an illustration of some options for rock slope management, but the interested reader should consult more fully developed work, for example, by Turner and Schuster (2012).

Table 7.3 presents presumptive rock slope angles, which are based on the strength and weathering potential of the parent rock material. This table assumes the stability of the rock layers is not governed by an unfavorable system of discontinuities (joints and bedding planes).

7.2 EXISTING SOIL SLOPES

Railway embankments often exhibit problems associated with improperly prepared foundations, poorly compacted fill, and poor drainage. The stability of railway embankments is often exacerbated by relatively steep

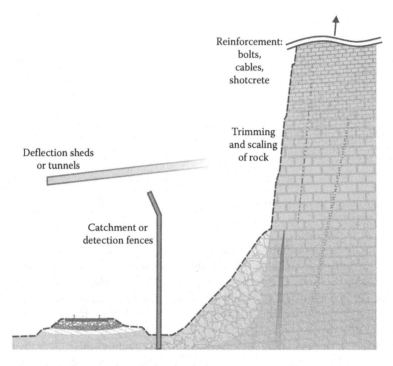

Figure 7.10 Some options for rock slope management.

Table 7.3 Design considerations for rock slopes

Rock condition	Weathering potential	Typical minimum slope ratio
Hard, competent, with no adverse discontinuities	Low	0.4H:1V to near vertical
Layered sedimentary, with low bedding dip angles	Moderate	0.5H:1V to 1H:1V
Fractured or weathered	Moderate	1H:1V
Clay shale	High	2H:1V or less

slopes due to right-of-way property limits, decades of track surfacing (adding more ballast), toe erosion, and increased train weight. Problems range from chronic track geometry roughness to sudden service disruptions.

Embankment instability is due to shear failure of the embankment fill or the underlying foundation soil. Ground water is often a major factor affecting the stability of the embankments. For railway embankments, the shear failure is precipitated by such events as build-up of ground water pressure, lateral spreading of the embankment, excessive plastic deformation under cyclic loading, vertical compression of the embankment, and hydraulic piping. The instability of railway embankments made of loose granular fill is a particularly prevalent problem along many railway corridors (Hyslip et al. 2011). The granular fill used for embankments can vary from natural sand with varying amount of fines (silt and clay) to man-made cinder and clinker mixture (bottom ash from steam locomotives).

7.2.1 Types of instability

The instability of soil slopes arises when the forces within the slope that are acting to move the material downslope (gravity, water, applied loads) overcome the strength of the material acting to keep the slope in its position. The type of instability is defined by the configuration of the movement zone, which is dependent on the interaction of the driving and resisting forces. Most types of instabilities can be categorized as rotational, translational, and earthflows, as shown in Figure 7.11.

One type of instability common and unique to railways is ballast pocket-induced instability, as shown in Figure 7.12. The development of a ballast pocket starts with load-induced depressions forming under the track from excessive plastic deformation or progressive shear failure of the embankment soil. Once the ballast pocket forms and becomes filled with water, the embankment soil can further soften from the trapped water, and in addition, the water can add considerable weight to the slope. The ballast pocket can eventually progress to a point where the water and weak soil combine to produce failure of the embankment slope.

7.2.2 Field investigation

Characterizing the mechanism causing instability of existing slopes is essential to determining the appropriate solution. Mapping and aerial photographs can provide a broad perspective of conditions outside of the railway right-of-way that may be contributing to the problem. In addition, track maintenance records and track geometry information can be used to delineate the extent and severity of the problem. Site reconnaissance is the first

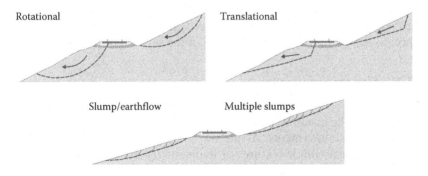

Figure 7.11 Types of slope instability.

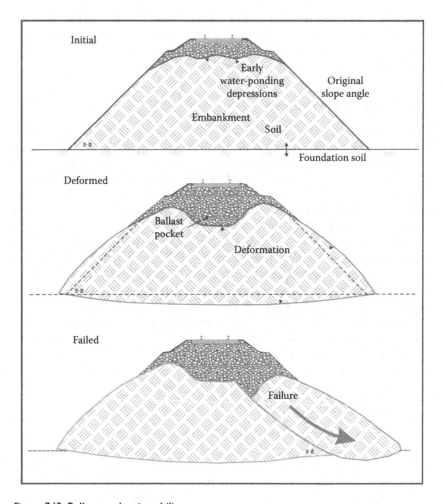

Figure 7.12 Ballast pocket instability.

step in identifying the type and extent of the problem, but often test borings and geotechnical instrumentation are needed, in addition to site reconnaissance, to define the nature and location of the subsurface movement.

7.2.2.1 Site reconnaissance

Site reconnaissance by an experienced observer can often reveal the type and extent of the instability. Many slope failures are preceded by prefailure phenomena and movements. Astute site reconnaissance and careful monitoring can reveal telltale features of impending slope movement (Leroueil 2004). Some evidence that may be apparent during site reconnaissance is shown in Figure 7.13.

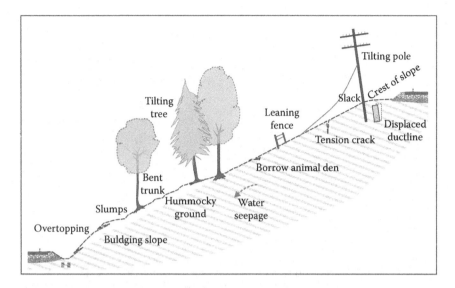

Figure 7.13 Surface features indicative of slope instability.

The extent and severity of the failure is often not evident until at least some sliding has taken place, but after such movement, key information can be gathered. Reconnaissance of postslide conditions can reveal useful information for remediation, including extent of failure, amount of material involved in the slide mass, and propensity for further movement. Figure 7.14 presents common conditions after failure and common terminology is shown.

7.2.2.2 Test borings

Test borings are needed for determining soil and ground water conditions within the slope and foundation, and for procuring samples for laboratory testing. Test boring made using standard penetration tests (SPT) is the most common method of sampling soil and of making groundwater observations. Continuous SPT sampling is important to ensure that sampling is performed through the zone of instability (failure plane). Thin-walled Shelby tubes can be used to retrieve relatively undisturbed soil. Figure 7.15 presents two examples of test boring drilled along railway tracks for slope stability investigation.

Sampling of the soil zone is often performed by conducting continuous SPT utilizing a 2-in. diameter Split-Spoon Sampler. In the SPT procedure, at each depth increment, the spoon sampler is first seated for 6 in. to penetrate any loose soil and then driven an additional 12 in. with blows from a 140-pound hammer falling from a distance of 30 in. Different types of SPT hammers can be used during the test drilling (donut, safety, and automatic) and the types of hammers used for borings have different

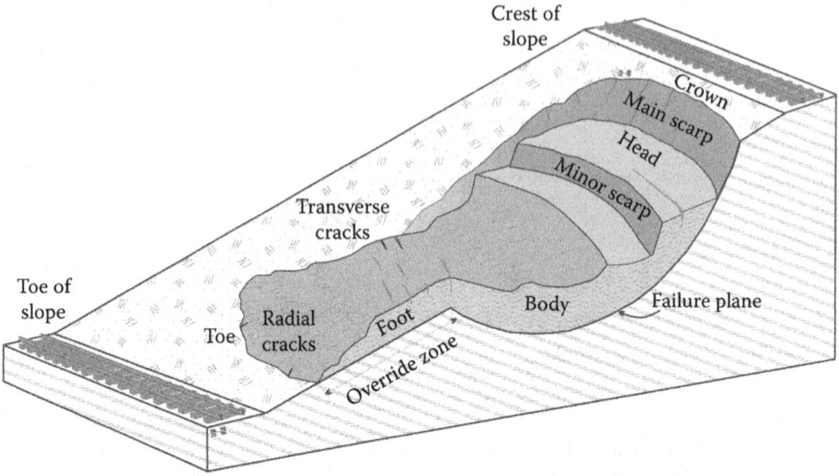

Figure 7.14 Postslide conditions and terminology. (Adapted from Cruden, D. and Varnes, D., Landslide types and processes, Chapter 3. In: *Landslides: Investigation and Mitigation*, A. Turner and R. Schuster, eds. Special Report 247, National Academy Press, Washington, DC, 1996.)

efficiency ratings. The cumulative number of blows for the second and third 6-in. intervals is designated "Standard Penetration Resistance" or "N value" and should be developed with appropriate correction factors described in the American Society for Testing and Materials (ASTM) D-1586 "Standard Test Method for Standard Penetration Test (SPT) and Split-Barrel Sampling of Soils."

The SPT *N* values are correlated to either the relative density of granular soils or the consistency of cohesive soil, based on relationships such as those shown in Table 7.4. In addition, SPT blow count data has been correlated to shear strength. Blow count data has been correlated with many design features for elements that depend on soil–structure interaction for support. Correlations should be used with caution because the variability of any statistical correlation may not be obvious and any correlation only applies to similar situations following the basic assumptions used to develop the correlation.

The information obtained from individual test borings are presented on test boring logs, which should contain information on the drilling and sampling results, and include detailed descriptions of the soil conditions encountered (e.g., soil type color, consistency/density, moisture condition) as well as the ground water level immediately after drilling and typically 24 h after completion.

DCP testing can be used to supplement test boring activities. The size and portability of the device allows for use in difficult and hard-to-reach

Figure 7.15 Drilling at crest and at midslope of a railway embankment (top). Drilling on a railway embankment using specialized drilling platform and crane (bottom). (Data from Hyslip, J. et al., Railway embankment assessment and case studies. *Proceedings of International Heavy Haul Association Specialty Conference,* Calgary, Canada, 2011.)

Table 7.4 SPT N-value correlations

Granular soils		Cohesive soils	
SPT N values	Relative density	SPT N values	Consistency
0–4	V. Loose	0–2	V. Soft
5–10	Loose	3–4	Soft
11–30	M. Dense	5–8	M. Stiff
31–50	Dense	9–15	Stiff
>50	V. Dense	16–30	V. Stiff
–	–	>30	Hard

Source: Lambe, T.W. and Whitman, R.V., *Soil Mechanics*, John Wiley & Sons, New York, 1969.

Figure 7.16 Dynamic Cone Penetrometer (DCP) on slope.

areas, such as a slope as shown in Figure 7.16. DCP information can be used to delineate the sliding mass from the underlying competent stratum.

7.2.2.3 Geotechnical instrumentation

Geotechnical instrumentation is essential to determine the nature and location of subsurface movement within embankments. This chapter serves as a brief introduction to the most commonly used instrumentation that is more fully described in the references provided. The two most important

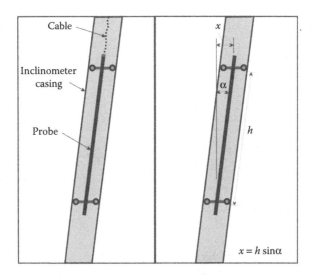

Figure 7.17 Slope inclinometer. (Adapted from Dunnicliff, J., *Geotechnical Instrumentation for Monitoring Field Performance*, John Wiley & Sons, New York, 1988.)

instruments for defining slope instability conditions are inclinometers and piezometers. Slope inclinometers are used to measure lateral movement and to identify the existence of failure or movement zones. Inclinometer information is also useful in evaluating whether the vertical settlement of the track is the result of horizontal movement within the embankment. Figure 7.17 shows a schematic of an inclinometer probe and the conversion from angle of tilt to displacement offset. Vertical extensometers can also be used to measure the vertical deformation within the embankment and thereby to evaluate the depth contributing to the settlement and to locate the zone of compressible and/or deforming soils.

Figure 7.18 shows examples of inclinometer results in the form of a cumulative displacement plot (left) and a time trend plot (right). The cumulative displacement plot shows a displacement profile, and the location of the greatest movement is apparent. An abrupt change in the cumulative displacement plot indicates shear movement in a relatively well-defined zone.

Shear tubes (Selig and Waters 1994) can also be used for monitoring lateral movement. Shear tubes are limited to approximately 10–15-ft depth. Shear tubes are considerably less expensive than inclinometers, however, shear tubes are much less precise, and the location and rate of movement of the actual instability zone (slip plane) is often difficult to determine and must be inferred. Figure 7.19 shows a typical shear tube insertion device and examples of movements that can be detected by shear tubes.

Piezometers, also referred to as ground water monitoring wells, are used to measure ground water level to determine the existence, location, and direction/magnitude of flow of the groundwater to see if these are key

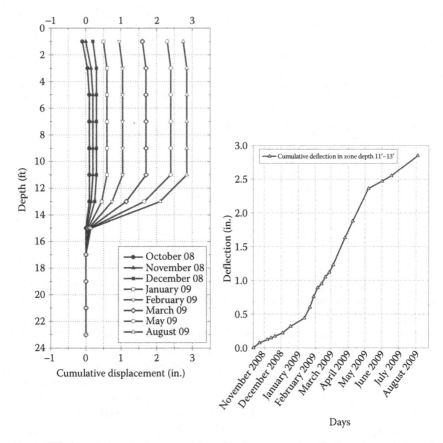

Figure 7.18 Cumulative displacement plot of inclinometer data (left), and trend plot of movement of inclinometer data (right).

factors driving the settlement condition. An open-standpipe piezometer, shown in Figure 7.20, is a simple and common type of piezometer. Water pressure sensors can be installed at the bottom of the screen pipe for continuous and remote monitoring.

7.2.3 Examples of railway embankment instability

Examples of railway embankment instability are presented here to provide instruction through actual case studies (Hyslip et al. 2011). The embankment conditions were ascertained from investigations that consisted of site reconnaissance, aerial *light detection and ranging* (LIDAR) mapping, test borings, geotechnical instrumentation (inclinometers, extensometers, and piezometers), laboratory testing, and other subsurface exploration work, including GPR and DCP testing. These tools provided information on the existing condition and behavior of the embankments, including the nature

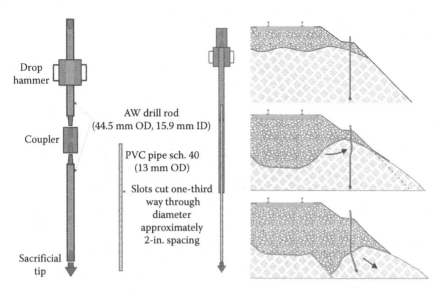

Figure 7.19 Shear tubes for detection of shallow lateral movement.

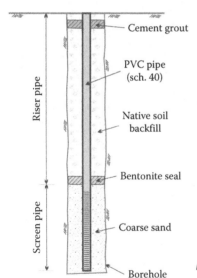

Figure 7.20 Open-standpipe ground water monitoring well.

and location of subsurface movement. Information at critical locations along the embankment was gathered from the test borings and from the instrumentation installed at the crest (outer edge of the top surface) of the embankment slopes, and at midslope and toe locations.

Figure 7.21 through 7.24 present examples of railway embankment instability (Hyslip et al. 2011). The most critical zones of instability are superimposed

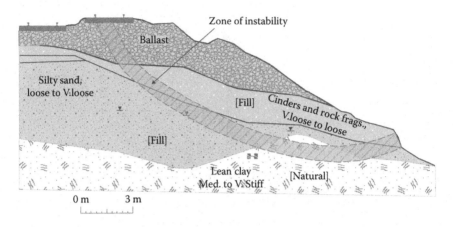

Figure 7.21 Site 1: Typical cross-section.

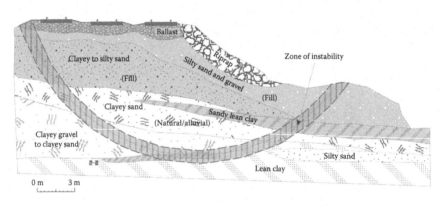

Figure 7.22 Site 2: Typical cross-section.

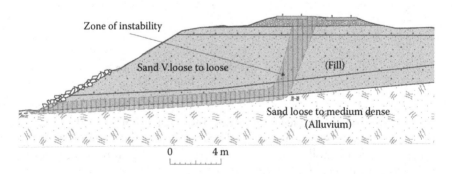

Figure 7.23 Site 3: Typical cross-section.

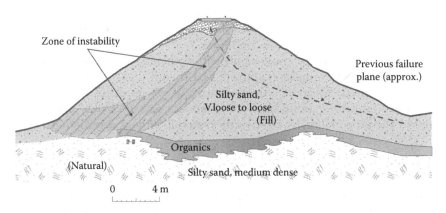

Figure 7.24 Site 4: Typical cross-section.

on the figures. The "zone of instability" is the location within the slope where the soil strength is overcome by forces trying to make the soil mass move. These driving forces are primarily from gravity and water pressure. The zone of instability is also the location where the unstable soil mass on the outer part of the slope moves away from the stable soil on the inner part of the slope.

Site 1 (Figure 7.21) is located on a moderately high (10± m) fill embankment with side slope ratios varying from 1.4H:1V to approximately 1.6H:1V. The bottom half of the slope exhibits a "bulging" that is the result of slope movement and from material that was placed on the edge as part of an earlier attempt to fix the stability problem. Site 1 exhibits the typical circular type of rotational failure.

Site 2 (Figure 7.22) is situated on a high fill embankment with a nearly 7 m slope height and side slope ratios of 1.5H:1V to 2.0H:1V. A portion of the slope is blanketed with riprap stone that was placed in response to shallow slope sliding. The embankment has a deep-seated, rotational failure zone that resides under all three mainline tracks at depths of up to 11 m. There is a large amount of soil contained within the unstable mass, and consequently, any type of remedial work needs to be sizeable in scope.

Site 3 (Figure 7.23) is located on an embankment that is approximately 6 m high and ends at the edge of a river, with side slopes of approximately 2H:1V. This site is particularly sensitive to the elevation of the ground water surface, which fluctuates by 1–1.5 m in height along the embankment due to variable river level and rainfall events. As the ground water level rises, the effective stress on the failure plane is reduced allowing movement. The instability at Site 3 is due to a translational type of failure, where a wedge of material is gradually sliding on a relatively planar interface between the granular fill and natural alluvium.

Site 4 (Figure 7.24) is a high fill embankment (~15 m) with long-term evidence of instability, including numerous curved or arching trees. The instability is due to rotational, circular failure. The problem arises from a combination of a steep embankment slope due to property boundary limits, loose fill, and poor foundation conditions.

The presence of loose granular fill within the embankment is common throughout these examples (Figures 7.21 through 7.24). The low relative density of railway fill embankments is often due to immature construction practices (built prior to understanding modern soil mechanics principles) and by the use of relatively lightweight material without compaction.

7.3 SOIL SLOPE STABILITY ANALYSIS

To explain the mechanisms that contribute to slope instability, a basic description of analytical methods to assess stability is presented. Slope stability analysis is typically a limit equilibrium analysis, in which the overall strength of the soil mass is compared with the forces, primarily gravity, at work acting to move the mass. When the driving forces equal the resisting forces, the slope is in equilibrium with a F.S. = 1.0. Analysis is based on the specific slope geometry and subsurface conditions, and the calculations assess the driving forces on the slope (weight, water, earthquake) and the overall strength of the slope.

In addition to limit equilibrium analyses, finite element modeling (FEM) has also been used to assess slope stability. FEM has desirable characteristics for application to slope stability analysis in that the results for FEM can be used to evaluate stresses, deformation and pore water pressure variations within the slope, as well as both local and global failure modes (Duncan 1996). FEM can be also used to model conditions both during construction and after construction as excess pore pressures dissipate. FEM also allows for modeling of nonlinear stress–strain conditions. However, limit equilibrium-based slope stability analysis methods appear to be more widely used because of the highly specialized nature of finite element techniques and often the need for a specialized computer program.

7.3.1 Basic method of slices

The most common limit equilibrium methods are based on the method of slices, where the soil mass is represented by discrete vertical slices that are first analyzed individually, followed by assessing their collective behavior aggregated for the entire slope. The method of slices allows for the calculation of the distribution of stresses along the failure plane, which is useful when analyzing slopes with nonuniform soil and pore water pressures. The basic mechanics of stability analysis is presented here to provide background and an initial understanding of slope stability analysis. There are various refinements to this basic method of analysis (Bishop 1955; Morgenstern and Price 1967; Spencer 1967; Janbu 1968).

In the method of slices, the slope is divided into a number of vertical segments, bound on top by the ground surface and on the bottom by the assumed or known failure plane. In assessing a slope, when the failure plane is not known, the location of the failure plane is determined in an iterative way by assuming a failure plane, performing the calculations, then repeating until the failure plane is found with the lowest F.S. Earthquake effects on slope stability analysis are often considered as an additional inertia force added to each slice.

Figure 7.25 provides an illustration of the method of slices. The left side of Figure 7.25 shows the slope divided into 18 slices. The right side of Figure 7.25 presents a free body diagram of one of the slices. The forces acting on the slice are determined based on the equilibrium of the segment, and then the equilibrium of the entire slope is determined by summation of the forces on all of the segments.

The forces shown in Figure 7.25 consist of

W is the weight of slice
Q is the surface load
u is the pore water pressure
F_n and F_t are the normal and shear forces on the failure plane on the bottom of slice
P_n and P_{n+1} are the normal forces on vertical faces
T_n and T_{n+1} are the shear forces on vertical faces
b, h are the width and height dimensions of slice
α is the angle of bottom of slice along failure plane

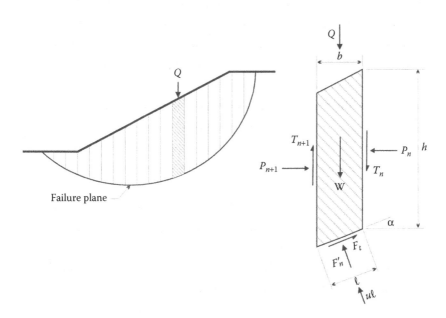

Figure 7.25 Method of slices for slope stability analysis.

The system is statically indeterminate; therefore, assumptions must be made regarding magnitudes and points of application of the P and T forces. As an approximate solution, the resultant of P_n and T_n is taken to be equal to that of P_{n+1} and T_{n+1}, and their lines of action are assumed to coincide. This simplification leaves only forces W, Q, F_n, and F_t to be resolved.

Applying the condition of equilibrium to each segment:

$$F_n = (W + Q)\cos\alpha \tag{7.1}$$

$$F_t = (W + Q)\sin\alpha \tag{7.2}$$

The unit pressure on the bottom of the slice is equal to

$$\sigma_n = \frac{(W + Q)\cos\alpha}{l} \tag{7.3}$$

$$\tau_n = \frac{(W + Q)\sin\alpha}{l} \tag{7.4}$$

The total shear force over the entire failure plane is the summation of the shear forces on individual segments, that is

$$\sum (W + Q)\sin\alpha \tag{7.5}$$

The strength of the soil (shearing resistance) is defined by

$$s = c + \sigma\tan\phi \tag{7.6}$$

The shearing resistance on the bottom of the slice is

$$s \cdot l = (c + \sigma\tan\phi) \cdot l = c \cdot l + (W + Q)\cos\alpha\tan\phi \tag{7.7}$$

The total shearing resistance over the entire failure plane is equal to

$$\sum \left[c \cdot l + (W + Q)\cos\alpha\tan\phi \right] \tag{7.8}$$

Then, for the method of slices, the F.S. is then

$$\text{F.S.} = \frac{\sum \left[c \cdot l + (W + Q)\cos\alpha\tan\phi \right]}{\sum (W + Q)\sin\alpha} \tag{7.9}$$

Considering pore water pressure and effective stresses, the F.S. becomes

$$\text{F.S.} = \frac{\sum \left\{ c'l + \left[(W + Q)\cos\alpha - ul \right]\tan\phi \right\}}{\sum (W + Q)\sin\alpha} \tag{7.10}$$

A complete stability analysis requires iteratively checking the F.S. for many presumed failure planes, and then determining the critical failure plane with the minimum F.S.

7.3.2 Example of method of slices

As an example of the method of slice calculation, Figure 7.26 presents a 10 m high embankment with a 2H:1V slope ratio and three different soil types in three layers: bottom or foundation, middle, and top of slope. The bottom layer is the mainly cohesive soil, the middle layer is more frictional than the foundation, and the top layer is mainly frictional. This example represents a relatively common slope situation.

The slope in Figure 7.26 has been divided into 10 slices, and the information on each slice is presented in Table 7.5.

Table 7.6 presents a tabulation of each slice and the summation of the driving and resisting forces.

The F.S. is determined from Equation 7.10:

$$\text{F.S.} = \frac{\sum \left\{ c'l + [W\cos\alpha - ul]\tan\varphi \right\}}{\sum W\sin\alpha} = \frac{1458 + 483}{1352} = 1.4 \tag{7.11}$$

7.3.3 Factor of safety

The stability of an embankment slope is assessed by comparing the forces available to resist movement to those that are trying to drive movement of the slope. The resisting forces are mainly functions of the strength of the soil, reinforcement, and embankment configuration. The driving forces are a function of slope height and configuration, external loads, ground water condition, and transient forces such as earthquakes. The F.S. against slope failure is defined as the ratio of the resisting forces to the driving forces, that is,

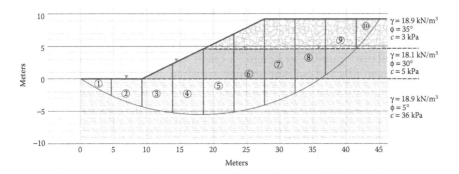

Figure 7.26 Example of method of slices approach to slope stability analysis.

Table 7.5 Information on the individual slices of Figure 7.26

Slice no.	b (m)	h (m)	γ (kN/m³)	W (kN/m)	ΣW (kN/m)
1	4.6	1.5	18.9	130	130
2	4.6	3.5	18.9	304	304
3	4.6	1.2	18.1	100	
		4.8	18.9	417	517
4	4.6	3.5	18.1	291	
		5.5	18.9	478	769
5	4.6	1.2	18.9	104	
		4.6	18.1	383	
		5.3	18.9	461	948
6	4.6	3.5	18.9	304	
		4.6	18.1	383	
		4.6	18.9	400	1087
7	4.6	4.6	18.9	400	
		4.6	18.1	383	
		3.2	18.9	278	1061
8	4.6	4.6	18.9	400	
		4.6	18.1	383	
		1.1	18.9	96	879
9	4.6	4.6	18.9	400	
		2.3	18.1	191	591
10	3.6	2.3	18.9	156	156

$$F.S. = \frac{\text{Resisting forces}}{\text{Driving forces}} = \frac{\text{Shear strength}}{\text{Shear stress}}$$

A F.S. < 1.0 indicates that the slope is not stable and that failure is imminent. A F.S. > 1.0 indicates that the slope is stable, and the amount that the F.S. exceeds 1.0 is a measure of the strength in reserve that is available to resist movement. An acceptable F.S. for slope stability is typically taken as 1.3–1.5. As a comparison, the F.S. for other geotechnical design, such as for structure foundation design, is taken as 2.5–3.0. Man-made slopes (embankment fills and excavation cuts) are often designed with slope ratios that mimic natural slopes, which have a F.S. at least slightly greater than one (F.S. > 1.0). So, with proper consideration given to design and construction (e.g., drainage, compaction, elimination of weak zones), a F.S. 1.3–1.5 is usually acceptable.

A large F.S. against failure is appropriate when the mechanisms and theory related to failure are not well understood, or if there is much uncertainty regarding the magnitudes of the driving and resisting forces, or if there is a

Table 7.6 Calculation table for determining Factor of Safety for Figure 7.26

Slice No.	ΣW (kN/m)	l (m)	α (deg.)	c (kPa)	φ (deg.)	u (kPa)	ul (kN/m)	W cos α (kN/m)	W cos α -ul (kN/m)	(W cos α -ul) tan ϕ (kN/m)	cl (kN/m)	W sin α (kN/m)
1	130	5.3	-32	36	5	15	78	110	32	3	191	-69
2	304	4.9	-22	36	5	34	168	282	114	10	176	-114
3	517	4.7	-13	36	5	59	277	504	227	20	169	-116
4	769	4.6	-4	36	5	88	406	767	361	32	166	-54
5	948	4.6	4	36	5	103	474	946	472	41	166	66
6	1087	4.7	13	36	5	98	461	1059	598	52	169	245
7	1061	4.9	22	36	5	83	409	984	575	50	179	397
8	879	5.3	32	36	5	59	312	745	433	38	191	466
9	591	6.7	43	5	30	20	131	432	301	174	34	403
10	156	5.6	55	3	35	0	0	89	89	63	17	128
									$\Sigma =$	483	1458	1352

relatively large risk associated with the slope such as with a slope support-ing a structural foundation. As slope failure is a relatively well-understood phenomenon, and the data required to derive the forces is typically obtained in practice, there is then little risk of using a relatively low F.S. of 1.3–1.5. In contrast, the larger F.S. values associated with dam foundations, for exam-ple, reflect the reality that conditions related to dam stability are often more subject to a critical geotechnical condition that may have been missed, even with detailed exploration, such as a small soil seam that will allow piping of water along it due to the large hydraulic gradient.

7.3.4 Computer models

Slope stability analysis can be performed using computer models that consider the specific slope geometry and subsurface conditions of the site. Computer programs can successively iterate the location of the assumed failure surface within selected regions, until a minimum F.S. is attained. Slope stability analysis can then also be conducted on models reflecting remedial methods. The analysis can be used to gain a crucial understanding of the sensitivity of the slope to specific remedial work and ground water lowering. Sensitivity analysis is useful to ascertain which parameters are most critical to slope stability and for ensuring that changes in input to a computer model cause sensible results, providing confidence in the calcu-lations. For instance, lowering the ground water table in the model, and thereby decreasing the pore water pressure within the embankment, would be expected to show the beneficial effects from internal drainage within the slope.

The proper use of slope stability computer programs requires an under-standing of the mechanics and analysis methods being used by the pro-gram. Designers also need a mastery of soil mechanics and soil strength, appropriate for the conditions and analysis. A designer must also be able to judge the reasonableness of the analysis results. When performing slope stability analysis, it is very important to have a good command of the com-puter program being used and the ability to evaluate the results of the anal-ysis to avoid mistakes and misuse (Duncan 1996).

7.3.5 Soil strength and stress condition

Soil shear strength within a slope can be difficult to accurately estimate. Often shear strength parameters for the soils are based on published empirical relationships correlating soil classification and standard pen-etration blow count data, known movement zones, and experience with similar types of soils. Even if careful sampling and testing is performed, the strength of the soil at the location of the test or the sample in the field will likely not represent the entire zone of interest. Additionally, the strength

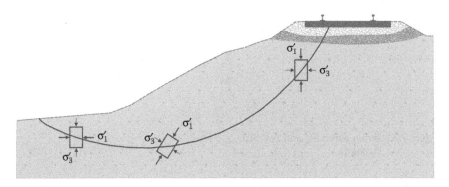

Figure 7.27 Varying stress state along a movement zone in a slope.

parameters used for the soil should also reflect the stress state and resulting failure modes shown in Figure 7.27. This variation in stress state within the slope will influence the analysis, particularly for anisotropic soil strength conditions.

For slope stability analysis, the strength of soil on the critical surface is defined by Equation 7.6 with either cohesion, frictional resistance, or a combination of both. Care must be taken when estimating soil strength parameters using either field or laboratory tests. The simplifying assumptions that form the basis for Equation 7.6 consider an average strength value, which applies across the entire movement zone, and that the strain along the entire surface is assumed to be large enough to mobilize all available shearing strength. In actuality, the shearing resistance does not develop uniformly along an incipient failure surface. The shear strain along the failure plane can vary considerably, and the resulting shear stress may be far from constant. Progressive movement often occurs within an unstable slope, where shear stresses exceed the shear strength in localized areas causing failure and redistribution of applied stress and potential failure of the entire soil mass if the localized failure zone propagates to affect a significant portion of the slope. In addition, local zones of instability could produce large local strains. Soil that has experienced significant strain generally has passed the peak strength and thereafter maintains a level of reduced "residual" strength in these highly strained portions of the failure surface (see Figure 7.28).

The stability analysis should be performed using both effective and total stress strength parameters for embankment soils. Total stress strength parameters are used for plastic fine-grained soil, and reflect the embankment condition where the soil has not consolidated and failure occurs under undrained loading conditions. Effective stress parameters apply to coarse- or fine-grained soils, especially where pore water pressure development is critical.

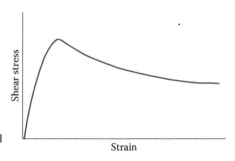

Figure 7.28 Strain softening and residual strength.

7.3.6 Effects of train load

If the tracks are close to the crest of the slope, the train load itself can have an adverse effect on slope stability. In general, the presence of track and train load should be considered in slope stability analysis. To illustrate the effect of train loading on slope stability, a series of analyses using the ordinary method of slices was performed, the results of which are presented in Figures 7.29 and 7.30. The analyses were performed for a uniform embankment and slope ratio of 2H:1V, and both 3 and 13 m high embankments (Figures 7.29 and 7.30, respectively).

Figures 7.29 and 7.30 illustrate the influence of train loading on the range of embankment strength required for a constant F.S. For example, the 3 m

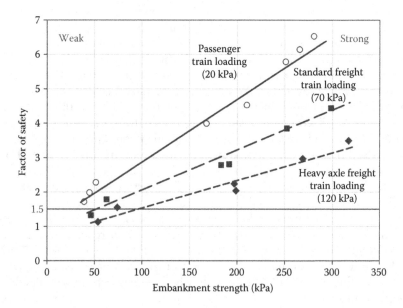

Figure 7.29 Three meter (3 m) high embankment.

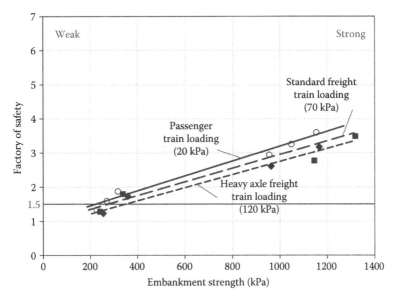

Figure 7.30 Thirteen meter (13 m) high embankment.

high embankment (Figure 7.29) and a F.S. of 2 would require the embankment soil shear strength of about 50 kPa for passenger traffic only, while soil strength of about 155 kPa would be required for heavy axle load freight traffic. In a taller embankment, the range is smaller, although the required embankment soil strength is higher. Referring to Figure 7.30, for passenger traffic, a soil strength of about 200 kPa provides a F.S. of 1.5, while heavy freight traffic would require a soil strength of about 350 kPa to attain the same F.S. against movement. Therefore, smaller embankment heights are more sensitive to train loading instabilities than are higher embankments.

7.4 ROCK SLOPES

Rock slopes are basically assemblages of intact rock blocks that are defined and characterized by discontinuities within the rock mass. The discontinuities of joints, faults, and bedding planes can be randomly oriented or can be part of a greater discontinuity set. The stability of rock slopes is determined by the orientation of the discontinuities with respect to the trend of the slope. Detailed discussion of rock slope behavior is beyond the scope of this book, and the interested reader is referred to available books on this topic (e.g., Hoek and Bray 1981; Pariseau 2011; Turner and Schuster 2012). The remainder of this section of the chapter is devoted to a basic introduction to this topic with more detailed references provided for some specific topics potentially of interest to railway personnel facing rock-related geotechnical challenges.

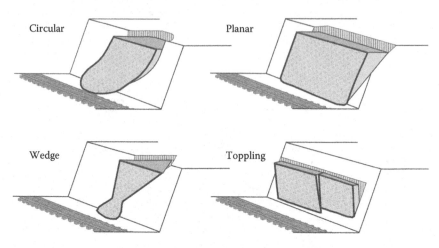

Figure 7.31 Types of rock slope instability. (Adapted from Hoek, E. and Bray, J., *Rock Slope Engineering*, The Institution of Mining and Metallurgy, London, 1981.)

7.4.1 Types of instability

Rock slope instability types are of four basic categories: rotational or circular, plane, wedge, and toppling, as shown in Figure 7.31 (Hoek and Bray 1981). The rotational failure is similar to soil failure type and occurs in heavily fractured or highly weathered soft rock. The plane- and wedge-type failures are characterized by sliding on one primary discontinuity (plane) or two intersecting discontinuities (wedge).

7.4.2 Investigation

The primary objective of rock slope investigation is to characterize the discontinuities in terms of orientation, persistence, spacing, surface properties, and infilling (Norrish and Wyllie 1996). Some of this information can be determined from geologic mapping, including the use of structural geology maps to determine the bedrock dip. However, site-specific collection of bedrock discontinuity data is essential for any rock slope investigation and should include mapping of bedrock exposures and determination of the orientation of discontinuities in the field. A geologic compass is commonly used to measure the orientation of the rock discontinuities. Figure 7.32 shows a Brunton geologic compass.

The dip of the discontinuity is between 0° and 90°, and the dip direction is defined by its azimuth (0°–360°). The strike of the discontinuity is, by definition, 90° from the dip direction, as shown in Figure 7.33.

Once the bedrock discontinuities are characterized, they are compared to the orientation of the cut slope face to determine if the potential for instability exists. A single geologic fault or a single set of discontinuities

Figure 7.32 Brunton compass.

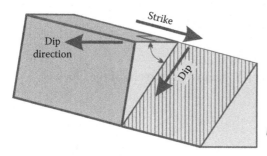

Figure 7.33 Features (nomenclature) of rock discontinuities.

can create the potential for instability, if the discontinuities intersect an unstable orientation.

7.4.3 Basic mechanics of rock slope stability

The basic mechanics of rock slope stability for sliding and toppling is relatively simple and straightforward; however, the stability analysis of actual rock slopes requires accurate information of discontinuity and slope orientation for which the basic mechanics apply.

From Figure 7.34, the unit pressure on the bottom of the block is equal to

$$\sigma_n = \frac{W \cos \alpha}{A} \tag{7.12}$$

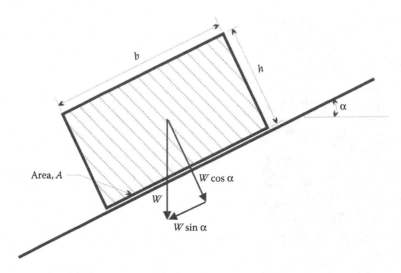

Figure 7.34 Forces acting on a sliding rock block.

The shearing resistance on the bottom of the block is given by Equation 7.6. The resisting force on the block is

$$\tau \cdot A = c \cdot A + W \cos \alpha \tan \varphi \qquad (7.13)$$

The driving force on the block is: $W \sin \alpha$.

For the limit equilibrium, the F.S. against movement of the block is

$$F.S. = \frac{c \cdot A + W \cos \alpha \tan \varphi}{W \sin \alpha} \qquad (7.14)$$

By considering pore water pressure in tension crack, the F.S. is described in Figure 7.35 and must be considered in the analysis.

The factor of safety considering pore water pressure then becomes

$$F.S. = \frac{c \cdot A + (W \cos \alpha - U) \tan \varphi}{W \sin \alpha + V} \qquad (7.15)$$

7.4.4 Rock slope stability analysis

Rock slope stability analysis is often referred to as kinematic analysis and is performed using stereographic projections that provide the ability to use three-dimensional discontinuity and slope orientation information to be represented and analyzed in two dimensions (see Figure 7.36). Knowing the orientation (spacing, etc.) one can then determine if the discontinuity system

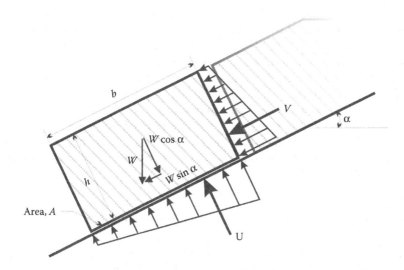

Figure 7.35 Forces acting on rock block with water in tension crack.

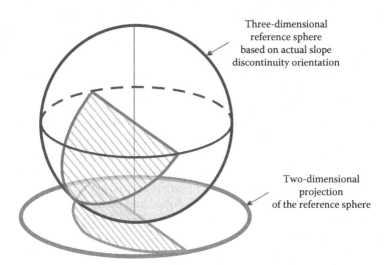

Figure 7.36 Example of stereographic projection for planar features.

is prone to sliding or toppling into the cut. Aided by computer modeling, rock mass models can be prepared that can visually portray the potential for any type of rock mass failure. Discussion and examples describing the use of stereographic analysis can be found in Hoek and Bray (1981), and Norrish and Wyllie (1996). For weak or highly fractured rock, the stability analysis approach is the same as for soil.

Once the rock mass models have been prepared, other nongeologic information can be assessed to determine the stability of the rock slope. Much of this information can be used to develop a rock cut slope rating system, which is a system to help identify critical slopes and hazards. A rock slope rating system can be used to aggregate the wide variety of information needed to quantify the risk of slope instability, and for overall slope and rock fall management. Help in determining the need for developing a rock slope management system, assessing the inventory of rock slopes to identify critical stability conditions for ranking within a rating system can be found in the variety of references on this topic, such as Bieniawski (1989) and Pierson (1990).

7.5 SLOPE MONITORING AND STABILIZATION

The major constraints placed on remedial solutions for application in the railway environment are limited work windows, remote locations with difficult access, and limited right-of-ways. These constraints often result in the need for vigilant monitoring to ensure safety while plans for maintenance or repair are developed. The selected remedial solution must minimize disturbance to the track, minimize the disruption to train operations, and cannot create any temporary instability. In electrified territory, overhead wires and catenary also present constraints.

7.5.1 Monitoring

The monitoring of existing slopes to identify problems can range from visual inspection, LIDAR scanning, and the use of failure detection systems which range from the very basic up to using extensive geotechnical instrumentation. In locations where slope failure is predicted, the use of instrumentation or sensors to detect failure is relatively common.

Some simplistic sensors have been developed and marketed over the years. One particular example is the so-called "Track Presence Detector" that consisted of a spring-loaded polyvinyl chloride (PVC) pipe with caps that are installed in track. If or when the track washed out or otherwise failed, a circuit would lose contact in the device and trigger the signal system to stop approaching traffic. However, the placement of any type of device in the track for this purpose can be problematic, and the durability of the sensor must be high to ensure reliability over time. In addition, track maintenance and operations can often trigger false alarms for these types of devices.

In rock outcropping areas where rockfalls presents a specific hazard, rockfall detection devices can provide a reliable means to detect the failure and alert approaching traffic. Often, slide fences are used, which consist of a fence connected to a detection system that triggers the signal when a slide is detected.

7.5.2 Slope stabilization

The best stabilization of slopes is accomplished when the true mechanism driving the instability is identified, allowing the most appropriate and economical remedial alternative to be selected. Remedial solutions often vary widely in construction effort and cost. Solutions may include

- Providing improved drainage of the embankment
- Use of retaining structure
- In-place reinforcement or improvement of the embankment soils
- Reconstruction of the embankment

7.5.2.1 Drainage

Providing proper drainage, through the diversion of water or by providing internal drainage, will improve stability. Periodic saturation of embankment fill from downward percolating rain water or from upward seeping ground water can weaken the soil by reducing the effective stress and reducing cohesion, as well as adding weight to the soil. Figure 7.37 shows the use of cutoff and horizontal drains to lower the ground water surface (phreatic surface). Drainage is often increasingly effective when it is improved in combination with other remedial solutions.

7.5.2.2 Structural restraint

Structural restraint using retaining walls is an approach often used by railways to address embankment instability problems. However, due to the location of the failure zone, the corresponding large amount of soil included in the sliding mass, and the potential disruption to operations, a robust and expensive retaining structure is often required. Figure 7.38 presents an example of a retaining wall for slope stability improvement.

To minimize the disruption of train traffic, retaining walls often need to be constructed away from the track (partially down the slope), and either

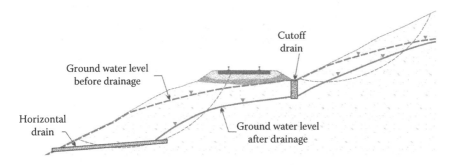

Figure 7.37 Drainage example.

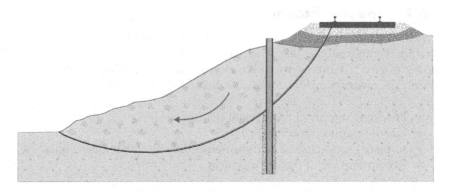

Figure 7.38 Retaining wall example.

in a top-down manner or by the use of temporary embankment support. Types of retaining walls that are found to be effective in restraining railway embankment slopes include sheet piling, soldier pile and lagging, mechanically stabilized earth, and gravity structures. All wall options must have provisions for adequate drainage of the backfill.

Continuous sheet pile walls and discrete soldier pile with lagging walls can be constructed in a top-down manner, thus eliminating the need for temporary embankment support during construction. Thus, sheet pile walls are often preferred over gravity structures. The typically high loads associated with railway embankments, combined with depth to a competent bearing stratum, often requires walls of this type to utilize tiebacks for restraint. Additional structural restraint options include anchored blocks and micropiles, as shown in Figure 7.39.

Railways have also driven rail or timber along the crest of the fill slope to help improve track stability (Figure 7.40). Although some benefit from rail/ timber driving has been reported, the benefit is likely from densification of the soil during driving of the rail/timber as cantilevered members and from some restraining effects.

The restraining effects are usually small because the rails/timbers can typically move or bend under lateral load because of flexibility and rotation about the bottom end, as depicted in Figure 7.41. In most instances, the effectiveness of these simple methods is limited because they do not directly address the mechanism causing the instability. However, in some instances, these techniques at least slow the movement to a manageable rate.

7.5.2.3 Buttress restraint

A common method employed by railways to improve the stability of loose fill embankments is to flatten the slope and/or buttress the slope

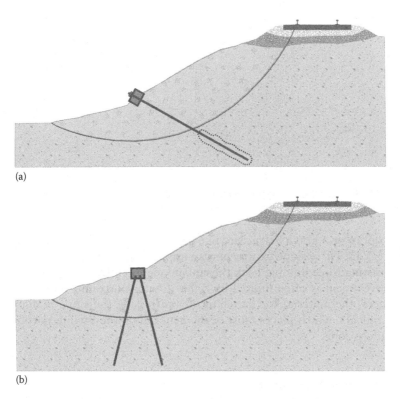

(a)

(b)

Figure 7.39 Structural restraint examples: (a) anchored block and (b) micropiles and cap beam.

Figure 7.40 Photo of driven pipe piles to improve embankment stability.

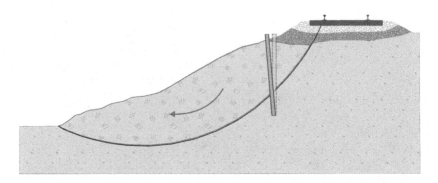

Figure 7.41 Driven piles to improve embankment stability.

with riprap or a gravity structure like gabions or cribwalls, as depicted in Figure 7.42. These slope treatments can improve the stability by increasing the resisting force holding back the embankment. However, the success of slope treatment is often limited by right-of-way constraints and, in some instances, the material for flattening or buttressing the slope is dumped from railcars, which places most of the material on the driving portion of the failure zone.

7.5.2.4 Soil improvement

Soil improvement is capable of increasing the slope stability by strengthening the weak soils while they remain in place. Soil improvement not only can strengthen the soil but also can intercept and reinforce across the failure plane (zone of instability). Many proven and cost-effective soil improvement techniques are available and can be adapted to a wide range of remedial schemes. Some techniques can be adapted to use relatively small equipment, produce little vibration, and perform outside the active track envelope or on the track using agile track equipment. However, performing the remedial work off of the track often poses both practical and safety concerns because this means performing the work on or near steep and marginally stable slopes.

Soil improvement systems are amenable to on-track production systems and can therefore be applied in remote locations with difficult off-track access and limited right-of-way. Soil improvement is also usually considered a maintenance-free option. For addressing embankment instability problems on a corridor-wide basis, soil improvement can be an effective and economical alternative to retaining structures. Figure 7.43 illustrates basic soil improvement for increasing slope stability.

Soil improvement can specifically address the root cause of the embankment instability, that is, the unstable soil, and therefore tends to be an

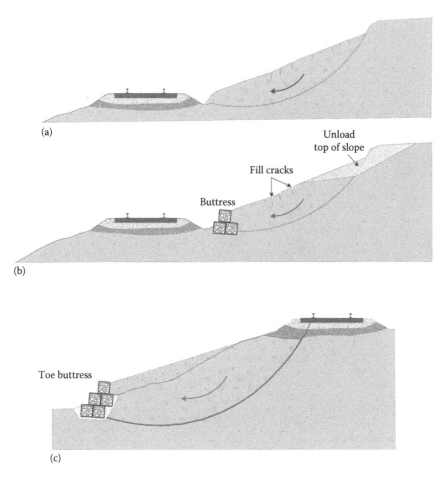

Figure 7.42 Buttress examples: (a) postslide condition; (b) remedial options; and (c) flat-
tened slope.

efficient solution. This efficiency can often translate into an economical solution when dealing with large masses of soil and at locations where depth to
competent material for anchoring of retaining structures is large. Often very
robust and expensive retaining structures are required to mitigate instability
problems due to the location of the failure zone, the great amount of soil
included in the sliding mass, and the relatively far proximity of competent
anchoring material within the embankment or foundation.

7.5.2.5 Rock slope considerations

Rock slopes present particular challenges for stabilization and protection,
and deserve the attention of specialists in these areas. As an introduction

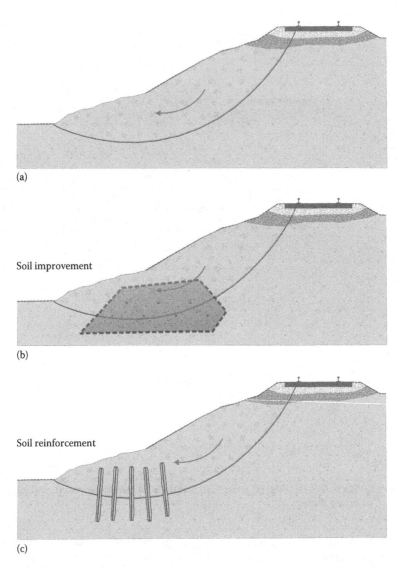

(a)

Soil improvement

(b)

Soil reinforcement

(c)

Figure 7.43 (a-c) Soil improvement concepts for slope stability.

to this topic, Figure 7.44 lists some options for both stabilization and protection from rock slope instability. It is important to distinguish between stabilization and protection measures for rock slope hazard mitigation. Developing a rock slope hazard system can be a useful way to systematically identify hazards to develop appropriate mitigation measures, as previously described in Section 7.4.3.

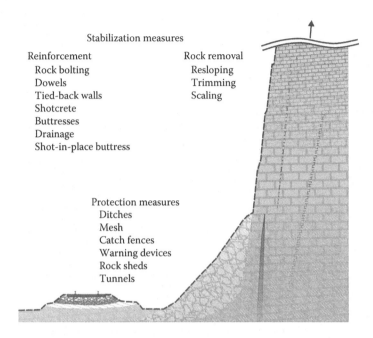

Stabilization measures

Reinforcement   Rock removal
  Rock bolting   Resloping
  Dowels   Trimming
  Tied-back walls   Scaling
  Shotcrete
  Buttresses
  Drainage
  Shot-in-place buttress

Protection measures
  Ditches
  Mesh
  Catch fences
  Warning devices
  Rock sheds
  Tunnels

Figure 7.44 Improvement techniques for stability improvement of railway rock cuts. (After Wyllie, D. and Norrish, N., Stabilization of rock slopes, Chapter 18. In: *Landslides Investigation and Mitigation*, A. Turner and R.Schuster, eds. Transportation Research Board Special Report 247, National Academy of Sciences, Washington, DC, 1996.)

Figure 7.45 Manual rock removal and reinforcement. (Courtesy of R. Kölsch.)

Often a combination of rock slope stabilization measures are employed on a rock cut slope. Manual trimming and scaling of loose rock, along with rock bolting and doweling, can be done with relatively small-sized and manually operated equipment, as shown in Figure 7.45.

Chapter 8

Measurements

The source of most track geometry deviations measured at the rails is due to settlement within or movement of the geotechnical track components, of which only the uppermost surface is available for visual examination. Therefore, reliable measurements are needed that can provide information on the substructure condition, well below this surface, to diagnose the cause of an instability and prescribe an appropriate remedy. This chapter first discusses soils investigation and methods to assess existing conditions, and then it describes the use of a remote sensing technique known as ground penetrating radar (GPR) that can probe the upper substructure layers nondestructively. Next, track geometry measurements are described and related to substructure behavior and vehicle–track interaction. The last section discusses track deflection-based measurements and what they indicate about the track substructure.

8.1 SITE INVESTIGATIONS AND CHARACTERIZATION OF TRACK SUBSTRUCTURE CONDITIONS

The first step in designing and building an adequate subgrade for new track construction or rehabilitation of an existing track is to characterize soil, water, and other environmental conditions for the proposed or existing route. To do so, a geotechnical investigation is required. In general, a geotechnical investigation includes a desk study based on the available soil, geologic, and hydrologic information in the literature along the track route, and a field and laboratory exploration and testing program.

Many references are available that explain how a geotechnical investigation should be conducted. For example, Chapter 1, Part 1, and Chapter 8, Part 22, of the American Railway Engineering and Maintenance-of-Way Association (AREMA) Manual for Railway Engineering provide general guidelines for exploration and testing of roadbed and geotechnical subsurface investigations for structures (AREMA 2012). Other references include Chapter 2 of the Soil Mechanics section and Chapter 5 of the Pavement

Design for Airfields section of the Unified Facilities Criteria (2005). Also, the US Bureau of Reclamation provides comprehensive guidelines for geotechnical investigation (Bureau of Reclamation 2004).

8.1.1 Desk study

Although subgrade conditions may vary over short distances along a railway route, before making a detailed accounting of these conditions, it is useful to first make a general assessment of soil and climatic conditions from information covering large areas. This information includes soil, geologic, and hydrologic maps, and reports available from the US Geologic Survey, the US Department of Agriculture Soil Conservation Service, as well as state and municipal highway or public works departments. Climatic and weather data for the United States are available from the National Oceanic and Atmospheric Administration and the National Weather Service.

The proper use of this information can provide identification of the origin of soil deposits (residual, alluvial, glacial, colluvial, aeolian, marine, and organic), which has direct relevance to engineering properties of subgrade soils. This information also provides the influence of environmental factors (such as precipitation and freezing index values) along the railway line.

Soil, geologic, and hydrologic maps are especially useful for estimating soil properties in large scale. For example, Figure 8.1 shows a simplified soil strength

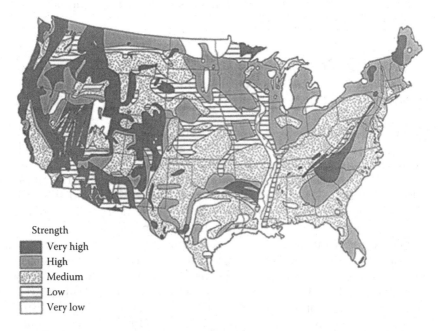

Strength
- Very high
- High
- Medium
- Low
- Very low

Figure 8.1 Soil strength map of the United States.

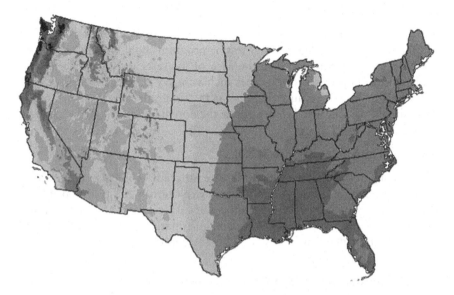

Figure 8.2 General precipitation conditions in the United States. (Data from National Oceanic and Atmospheric Administration, U.S. Department of Commerce, 2000.)

distribution map of the railroad track subgrade and was developed based on the available information in the literature (Li and Selig 1995a and 1995b).

Other maps, such as the annual precipitation map (Figure 8.2) and the freezing index contour map, can be useful to evaluate the effects of water and temperature on the subgrade. Li and Selig (1995b) provide a detailed discussion about the use of this type of information.

The investigation for an existing line should also include information on the condition and performance of the existing track, such as history of slow orders, maintenance frequency, and drainage condition changes.

A desk study does not provide all information required for the design and construction of the track subgrade. Therefore, another use of a desk study is to help obtain the additional information needed by using it to plan the extent and scope of a field geotechnical exploration and laboratory testing program. This is especially important for those locations where soil conditions are considered poor, that is, locations with presumed low strength soil or expansive soils or shallow ground water table.

8.1.2 Field reconnaissance

Site reconnaissance includes noting and photographing the right-of-way conditions. This initial field reconnaissance specifically looks for such things as evidence of ballast fouling and drainage problems in existing track, assessing surrounding topographic conditions, and the gathering of information for finalizing the field investigation plan.

8.1.2.1 Soil types and strata

Soil type has a direct effect on the properties and performance of the subgrade. The field and laboratory testing program should identify the types of soils (e.g., clay, silt, sand) in the area of interest, and their depths and distributions. In particular, soil types such as heavy clay and expansive soils should be identified, because their use as the subgrade can cause instability. Tests should be conducted to determine the potential of soil volume change.

8.1.2.2 Ground and surface water

Water is one of the main factors that cause subgrade problems. As such, one of the objectives of a field and laboratory exploratory and testing program is to identify possible sources of ground and surface water and to determine the ground water table depth. With this information, the requirements for drainage can be determined and designed (Chapter 6).

8.1.2.3 Soil strength and stiffness properties

Field in situ and/or laboratory tests should be conducted to determine soil strength and stiffness properties that are required for the design of the subgrade. A variety of in situ and laboratory testing techniques are available. Chapter 5 gives soil strength and stiffness parameters that can be obtained from various testing methods.

Boring and sampling locations must be selected to define prominent soil and rock strata and other subsurface features that will affect track performance. For investigating a proposed track corridor, a program of primary sampling points (borings) should be established and supplemented with additional sampling points whenever variations in soil conditions or unusual features are encountered. Table 8.1 gives the recommended number of primary sampling points for the soil definitions shown in Figure 8.1.

Soil samples should be obtained 10–15 ft below the proposed grade as a general rule, although deeper sampling is warranted if poor soil conditions are known or suspected at greater depths. Borings may generally be stopped when rock is encountered or after penetrating 5 ft into dense or very stiff soil.

Soil samples should be obtained from sampling locations and identified in accordance with the American Society for Testing and Materials

Table 8.1 Frequency of sampling

Soil strength zone in Figure 8.1	Longitudinal spacing of sampling points along track alignment
Very high—high	One boring every 500 ft
Medium	One boring every 200 ft
Low—very low	One boring every 100 ft

Table 8.2 ASTM standards for soil exploration

ASTM method	Procedure
D 420	Site Characterization for Engineering, Design & Construction Purposes
D 1452	Soil Investigation and Sampling by Auger Borings
D 1586	Penetration Test and Split-barrel Sampling of Soils
D 1587	Thin-wall Tube Geotechnical Sampling of Soils

(ASTM) D2488. Representative soils should be selected for strength and compressibility testing. All soils should be classified in accordance with the Unified Soil Classification System, as given in ASTM D2487.

Standard techniques for soil exploration should be followed, and ASTM standards used as guidelines. Table 8.2 lists the applicable ASTM standards for soil exploration.

Many in situ and laboratory testing techniques are available to obtain the information essential to the design and construction of the subgrade. Table 8.3 lists the applicable ASTM standards for in situ and laboratory testing.

8.1.3 Mapping

Methods for collecting and utilizing spatial data have been rapidly developing. Mapping of asset locations and the development of topographic maps (e.g., digital terrain models [DTM]) has historically been accomplished using optical ground surveying or airborne photogrammetry. Presently, technology combining laser measurement systems and highly accurate Global Positioning System (GPS) allows high-speed and high-accuracy data acquisition.

Mapping of spatial data is important for diagnosing and managing the track substructure. Spatial data is any set of data that relates to a geographically distributed set of data identified or mapped by location. Topographic data is a subset of spatial measurements of surface topography. Accurate location of track assets and right-of-way features is useful in diagnosing root cause of substructure problems as well as determining remedial solutions. Accurate location of track substructure-related assets and right-of-way assets is often used as a basis (baseline) for integrating disparate information, such as track geometry, GPR, and stiffness measurements. Accurate topographic measurement provides the necessary data to develop improved drainage designs, including drainage invert, track profile, and right-of-way cross sections.

A variety of aerial laser survey platforms are available for collection of topographic data using a combination of laser and inertial guidance systems. The accuracy of aerial surveying is dependent not only on the quality of the hardware collection system but also on the height and speed of the aerial

Table 8.3 ASTM standards for in situ and laboratory testing

ASTM method	Procedure
C 117	Material finer than No. 200 sieve in mineral aggregates by washing
D 422	Particle size analysis of soils
D 698	Laboratory compaction characteristics of soil using standard effort
D 854	Specific gravity of soils
D 1140	Amount of material in soils finer than No. 200 sieve
D 1556	Density and unit weight of soil in place by the sand-cone method
D 1557	Laboratory compaction characteristics of soil using modified effort
D 2166	Unconfined compressive strength of cohesive soil
D 2167	Density and unit weight of soil in place by the rubber-balloon method
D 2216	Laboratory determination of water (moisture) content of soils and rock
D 2434	Permeability of granular soils (constant head)
D 2435	One-dimensional consolidation properties of soils using incremental loading
D 2487	Classification of soils for engineering purposes
D 2488	Description and identification of soils (visual-manual procedure)
D 2573	Field vane shear test in cohesive soils
D 2937	Density of soil in place by the drive-cylinder method
D 3441	Deep, quasi-static, cone and friction—cone penetration tests of soils
D 4253	Maximum index density and unit weight of soils using a vibratory table
D 4254	Maximum index density and unit weight of soils and calculation of relative density
D 4318	Liquid limit, plastic limit, and plasticity index of soils
D 4546	One-dimensional swell or settlement potential of cohesive soils
D 4767	Consolidated, undrained triaxial compression test of cohesive soils
D 6938	In-place density and water content of soil and soil-aggregate by nuclear methods

survey vehicle. The most accurate aerial systems provide spatial data with an accuracy of approximately 10 cm horizontally and 5 cm vertically, but this also depends on the skill of the operator, the type of feature being extracted, and the rate at which measurements are obtained. Aerial laser survey techniques use a helicopter or fixed-wing aircraft as a survey vehicle. These aerial systems integrate inertial navigations systems, imagery collection, and scanning lasers into the survey vehicle. High-density data of up to 100 points per square meter, depending on survey speed and height above ground, can be collected of an area in a relatively short time period. Aerial systems are particularly suited for railway corridor mapping because the system can generally survey the corridor in one pass without the need for occupying the track.

Aerial data will contain coordinate measurements for all objects in the path of the laser, including brush and trees, catenary system, and other overhead objects. Processing software typically includes filters to help extract the data on the particular track asset of interest. An example of the survey data with

Figure 8.3 FLIMAP® aerial LIDAR data.

identified track assets is shown in Figure 8.3. Track cross section, which can be developed from this data, is shown both filtered and unfiltered in Figure 8.4.

Spatial data forms the basis of a good track asset management system because knowledge of track and structure locations is essential to implementing a plan for comprehensive management of the system. These applications allow graphic visualization of the various maintenance management datasets and their interrelation to a common track distance referencing system.

One example application of this data and the entire management process is undercutting in multiple track territory, where an insufficient width of the excavating bar may leave behind fouled material between tracks that inhibits drainage laterally. In this situation, it is necessary to undercut all tracks so that the undercutting widths of adjacent tracks overlap to allow water to discharge laterally through clean ballast to the outlet point. This mapping process can help with the planning of such work. Further, mapping can help with planning the optimum ditch profile referenced to the spatial track alignment in terms of ditch depth and offset at selected station intervals, which is particularly useful for on-track equipment.

Digital terrain models (DTM) are developed in these land-modeling systems using spatial (XYZ) surveyed data. A DTM is generated by forming a network of triangular elements from XYZ surveyed data points. Contours are superimposed onto the DTM's by interpolation. The spatial position of railway assets such as tracks, catenary poles, bridges, cross-drains, and other railway structures of importance can be superimposed onto the DTM to determine their effect on the design.

Spatial data is becoming more widely available for use in engineering and maintenance. The advantages of rapid surveying and data processing compared with other surveying techniques will ensure more widespread use of

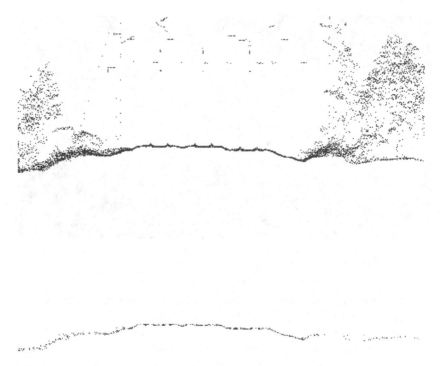

Figure 8.4 Track cross section, unfiltered (top) and filtered (bottom).

spatial data in all aspects of engineering. Another benefit is that the entire design period can be shortened because of the rapid availability of spatial data from laser and video-based surveying systems.

8.1.3.1 Ground-based LIDAR systems

Although surface mapping from airborne measurement systems have been described so far, ground-based LIDAR systems can be used to map the surface using a specially equipped road vehicle or a hyrail vehicle for railway applications. In fact, ground-based systems can be the preferred method when overhead features obscure the view from the air. Combining LIDAR surveying with high-resolution data imagery provides a highly detailed mapping capability. One advantage of LIDAR measurements made at ground level is a more detailed mapping of track assets and features than can be provided from an aerial survey. For example, ground-based LIDAR can be used to identify track locations where more ballast is needed to fill the cribs in advance of maintenance, or where the ballast shoulder may be insufficient to provide the needed lateral track stability. Although current technology does not provide sufficient accuracy for track geometry measurements, this could be a future capability for ground-based LIDAR systems.

8.1.4 Cross-trenches

Cross-trenches are shallow excavations made into the substructure perpendicular to the track centerline to observe the condition of the ballast, subballast, and subgrade and to obtain samples for testing. The samples should be described and identified in the field and the laboratory in accordance with ASTM D2488 "Standard Practice for Description and Identification of Soil (Visual-Manual Procedure)." A petrographic analysis of selected ballast samples can also be conducted to aid in determining the ballast rock type and the composition of any fouling components in the voids between ballast particles. This visual-manual procedure together with petrographic information can be used to determine the extent to which the sources of ballast fouling are from subgrade infiltrating upward, deposited from outside sources such as windblown or from traffic lading, or from the degradation of ballast itself. Choosing a remediation strategy for a heavily fouled track requires this kind of key information. A typical cross-trench with the layers documented is shown in Figure 8.5. Note that although the ballast appears to be clean on the surface, the amount of clean ballast in track is a small portion of the total, as shown in Figure 8.5. The engineering properties and mechanical behavior of the heavily fouled ballast layer will be substantially different from that of the clean ballast on the surface.

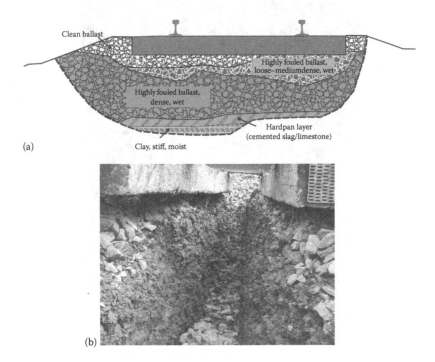

Figure 8.5 Cross-trench showing (a) substructure layers and (b) their condition.

8.1.5 Test borings

Hollow-stem auger drills, as shown in Figure 8.6, provide casing for the hole while a sampling tube is inserted and advanced further into the hole, well ahead of the end of the auger, to obtain undisturbed soil samples and to document the layers. This hollow stem auger also allows insertion of a smaller diameter casing to be left in place for instrumentation such as an inclinometer, or for an observation well. Soil layering, soil strength, and water conditions with depth are essential information that can be obtained.

Difficult drilling conditions are often encountered on the railroad right-of-way. For example, when drilling on a slope, it is sometimes necessary to construct a horizontal platform. Due to the need to minimize disruption of traffic on railways, drilling is often performed in the darkness of early morning hours, as shown in Figure 8.6. Also, any overhead wires present must be de-energized due to the proximity of the top of the drill rig to these wires.

Figure 8.6 Soil drilling in the gauge of the track using a hollow stem auger.

8.1.6 In situ tests

Many in situ test methods can be used or adopted for railway application. In particular, the Cone Penetrometer Test (CPT) has met with much success. The CPT uses an instrumented cone with a 60° apex and a diameter of 34.5 mm that is inserted at continuous speed into the track substructure while the pressure or resistance against the cone is recorded. From changes in tip pressure with depth of insertion the CPT can show the depths to subsurface layer boundaries as shown in the left-hand portion of Figure 8.7. The sleeve behind the cone is also often instrumented to measure side friction resistance. When the side friction force (f_s) is divided by the cone tip resistance to obtain the friction ratio, this ratio and the tip resistance can provide information that can be correlated to soil type classification and other soil properties, as shown in Figure 8.8 from Sharp et al. (1992). An alternative but similar soil classification based on CPT measurements is shown in Figure 8.9.

The Association of American Railroads (AAR) adapted a CPT to a high-rail truck (Chrismer and LoPresti 1996) to investigate the cause of poorly performing track substructure by determining the depth and longitudinal extent of a problem soil, its strength, and the adequacy of the overlying granular layer thickness and stiffness to produce the required stress reduction to the subgrade.

At the rate that the cone is advanced into the soil, the test simulates an undrained loading condition, and the undrained shear strength can be determined by:

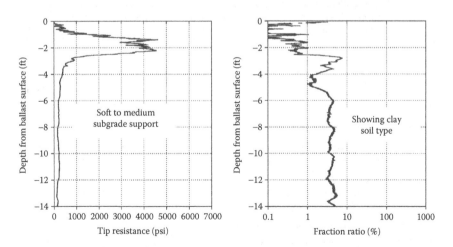

Figure 8.7 Cone Penetrometer Test results showing stiff ballast/subballast layers over soft subgrade (left), and friction ratio indicating soil types (right).

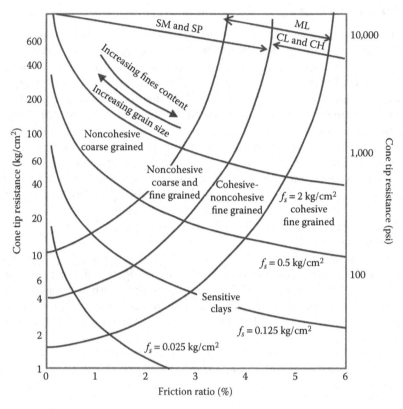

Figure 8.8 Soil classification chart for CPT. (From Sharp, M.K. et al., *Field Evaluation of the Site Characterization and Analysis Penetrometer System at Philadelphia Naval Shipyard*, Department of the Army, Waterways Experiment Station, Corps of Engineers, Vicksburg, MS, 1992.)

$$s_u = \frac{q_c - \sigma_{vo}}{N_k} \tag{8.1}$$

where:
S_u = the undrained shear strength
q_c = the cone tip resistance
σ_{vo} = the in situ total stress
N_k = an empirical factor

For cone insertion depths commonly used in railways, the correction for in situ stress, σ_{vo}, is usually fairly small and can be ignored. N_k varies in practice ranging from about 10 to 40 depending on stress history, plasticity, and stiffness, but has a mean value of about 16 (Eid and Stark 1998).

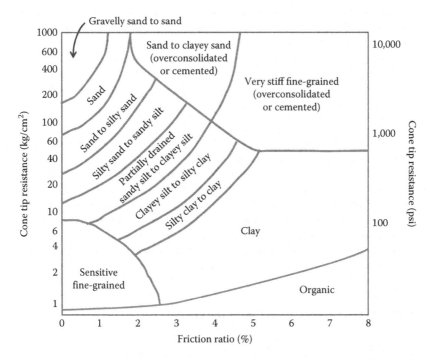

Figure 8.9 Soil classification chart for CPT. (From Robertson, P.K. and Campanella, R.G., *Can. Geotech. J.*, 20, 718–745, 1983.)

Table 8.4 CPT measurement interpretation

Cone tip resistance (psi)	Compressive strength (psi)	Modulus (psi)	Subgrade support
1,000 and over	100 and over	10,000 and over	Hard
500–1,000	50–100	5,000–10,000	Very stiff
300–500	30–50	3,000–5,000	Stiff
150–300	15–30	2,000–3,000	Medium
70–150	7–15	1,000–2,000	Soft
<70	<7	<1,000	Very soft

Table 8.4 shows the proposed correlation between the CPT tip resistance, subgrade compressive strength, track modulus, and subgrade support characteristic that has been developed over years of practice by Chrismer and Li.

Another rapid in situ test sometimes used is the flat dilatometer, which is a plate inserted vertically into the soil having an inflatable diaphragm on the face of the plate that pushes against the local soil. The measured resistance against the diaphragm can be used to determine design properties of soil such as the undrained shear strength, load/deformation behavior, and modulus.

The field vane shear in situ test can also assess the undrained shear strength of saturated cohesive soils by the insertion of a multibladed vane into the soil followed by an applied rotation while the resistance is measured.

Finally, the plate load-bearing test is sometimes used in track design to determine the deformation modulus of the track substructure before track construction, or during remediation of an existing track after the ballast, and possibly the subballast, has been removed. European practice for control of earthworks and trackbed for railways (UIC code 719, 2008) recommends that the subgrade modulus of elasticity for transmission of vertical forces (Ev) be obtained over the load range interval 30%–70% of the maximum test pressure. This maximum pressure depends on the plate diameter and is either 0.5 MN/m² for the 300 mm diameter plate or 0.25 MN/m² for the 600 mm plate. Because this test takes considerable time to perform, the number of tested locations are limited compared with a rapid assessment method such as the CPT. However, where the plate load-bearing test is performed, it can provide an excellent assessment of subgrade stiffness that can be used as the basis for design.

8.2 GROUND PENETRATING RADAR

GPR can be used to assess railway track substructure and produce quantitative information of substructure condition. GPR has the ability to map key railway track substructure conditions quickly on a continuous, top-of-rail, nondestructive basis (Sussmann et al. 2001b; Manacorda et al. 2002; Olhoeft and Selig 2002; Brough et al. 2003; Al-Nuaimy et al. 2004; Hyslip et al. 2004; Eriksen et al. 2006). In addition, analysis tools have been developed for data interpretation that can provide a variety of parameters related to track condition, including the thickness of subsurface layers and degree of ballast fouling (Olhoeft and Smith 2000; Saarenketo et al. 2003; Sussmann et al. 2003a; Clark et al. 2004; Hyslip et al. 2005; Roberts et al. 2006; Silvast et al. 2006).

8.2.1 GPR fundamentals

The principle of GPR operation is based on transmission of short electromagnetic waves from an antenna into the subsurface, the subsequent reflection, scattering, and refraction of this energy from subsurface layers, and finally, the receiving, recording, and display of this reflected energy. GPR is a pulse-echo technique using radio energy, as depicted in Figure 8.10.

As shown in Figure 8.10, reflections of the GPR pulse occur at boundaries in the subsurface where changes occur in the material properties. Only a portion of the pulsed signal is reflected and the remaining part of the pulse travels across the interface to again be reflected back to the receiver from another interface boundary, see Figure 8.10a. Figure 8.10b shows an

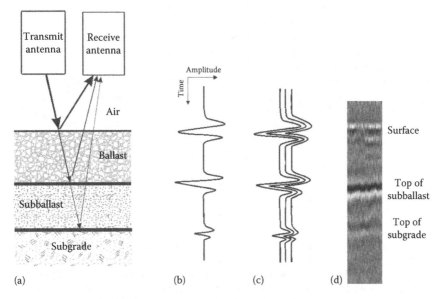

Figure 8.10 (a–d) The generation of a GPR profile. (From Hyslip, J.P. et al., Ground penetrating radar for railway substructure condition assessment. *Proceedings Railway Engineering 2004 Conference*, London, 2004.)

individual scan composed of the reflection amplitudes as a function of time, where the vertical scale is time (nanoseconds or ns) and the horizontal scale is the amplitude of the reflection. Multiple scans are generated in quick succession, Figure 8.10c and adjacent scans are combined and the magnitude of the reflection is expressed in grayscale, as shown in Figure 8.10d. In the grayscale (radargram), the reflections are displayed continuously in the horizontal direction, which corresponds to distance along the track.

The data are processed to remove unwanted background data, principally from effects of ties and rails, as well as the direct transmission of energy between antennas. One of the most common signal-processing techniques used in the processing of GPR data is the horizontal high-pass filter, which is often referred to as a background removal filter. The purpose of this filter is to clarify portions of the signal that vary with distance along the track (Sussmann 1999).

Figure 8.11c and d shows the results of processing Figure 8.11a and b using this technique. As shown in Figure 8.11d, this technique highlights the subballast layer variations and more clearly denotes a change in the signal character between the ballast–subballast layers and the subgrade below.

The key material properties of the soil that affect the GPR signal are the dielectric permittivity, the electrical conductivity, and the magnetic permeability. Electromagnetic energy is reflected and refracted at interfaces of

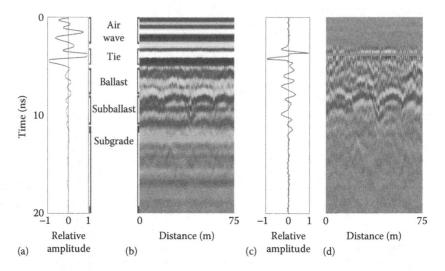

Figure 8.11 (a–d) Reflected wave and typical display. (From Sussmann, T.R., *Application of Ground Penetrating Radar to Railway Track Substructure Maintenance Management*, Doctoral dissertation, University of Massachusetts, Amherst, MA, 1999.)

material with different dielectric permittivity. The dielectric permittivity dominantly controls the velocity of electromagnetic wave propagation and is the main parameter affecting GPR wave propagation. Dielectric permittivity is a function of the density, water content, and type of material and varies from 1 for air to 81 for water. The pulse travels faster through a low-dielectric permittivity material than a material possessing a higher dielectric permittivity.

Table 8.5 presents typical values of dielectric constant for a variety of common geologic and railway track materials. Water content has the single largest influence on the dielectric permittivity of soil. Therefore, subsurface reflections are more pronounced at the interface of layers with differences in moisture content.

The velocity of the GPR wave through the material is calculated by

$$V = \frac{c}{\sqrt{\varepsilon_r}} \tag{8.2}$$

where:
    c is the speed of light in air (30 cm/ns)
    ε_r is the dielectric permittivity of the material

The electrical conductivity is the main electrical parameter controlling the depth of GPR signal penetration. Electrical conductivity is the ability

Table 8.5 Soil Dielectric Constant Values for Typical Track Substructure Materials

Material type	Dielectric permitivity	Velocity (m/ns)	Conductivity (S/m)
Air	1	0.3	0
Fresh water	80–81	0.034–0.033	0.003–0.0001
Salt water	81	0.033	4
Clayey soil (dry–wet)	4–16	0.15–0.077	$1–10^{-4}$
Sandy soil (dry–wet)	4–25	0.15–0.05	$10^{-2}–10^{-4}$
Silty soil (dry–wet)	9–23	0.1–0.063	$10^{-2}–10^{-3}$
Granite (dry–wet)	5–7	0.13–0.113	$10^{-8}–10^{-2}$
Basalt (wet)	8	0.106	10^{-2}
Other ballast			
Limestone (dry–wet)	7–8	0.11–0.106	$10^{-9}–10^{-2}$

Source: Sussmann, 1999.

of the material to conduct electrical current, which is affected by the amount of water present. The conductivity of the material dictates how quickly the pulse of radio energy decays in amplitude (attenuates) with distance and thus controls how deep the pulse will penetrate. The dielectric permittivity and electrical conductivity are mostly independent. For example, freshwater and saltwater have essentially the same dielectric permittivity (salt water is slightly lower); however, saltwater exhibits a much higher electrical conductivity than freshwater. GPR pulses travel at similar speeds through both types of water; however, in saltwater the energy is attenuated very quickly and does not penetrate deeply. The magnetic permeability also controls velocity and is commonly neglected (assumed to be the value of air), though it may be important when iron-bearing materials are present such as some slag and iron ore fouling (Sussmann 1999).

Wavelengths for GPR systems commonly applied to railway track investigation are shown in Table 8.6. The wavelength in air ranges from 15 to 75 cm, while the wavelength in ballast with a dielectric constant of 5 would be expected to range from 7 to 34 cm. As ballast particles are in the order of 4 cm or smaller in any dimension, diffractions can be expected from clean ballast when GPR operating at 2 GHz is used.

Table 8.6 GPR wavelength in air and ballast

Frequency	Wavelength in air (cm)	Wavelength in ballast (cm)
400 MHz	75	34
1 GHz	30	13
2 GHz	15	7

8.2.2 GPR for track surveys

GPR provides continuous measurement of layer thickness and configuration (lateral and longitudinal variation) of individual substructure layers, as well as moisture content and the fouling condition of ballast. GPR can be used to map the lateral and longitudinal variation of ballast and subgrade layers, and thereby provide information on failure and deformation conditions (Hyslip et al. 2005). Figure 8.12 shows a typical GPR image example along a 500 ft (150 m) section of track. Figure 8.12 also shows photos of the cross-trenches that were excavated and logged for calibration of the GPR data.

Many advantages exist for the use of GPR as a site investigation and diagnostic tool on the railroads. Principally, GPR provides information on substructure layer conditions with minimum interference to train operation. In this regard, GPR can be utilized as an effective tool for exposing the root cause of reoccurring poor geometry and unstable roadbed. GPR surveys can be collected on railroad tracks at vehicle travel speeds in excess of 100 km/h, and the use of multiple antennas can provide variations in substructure conditions

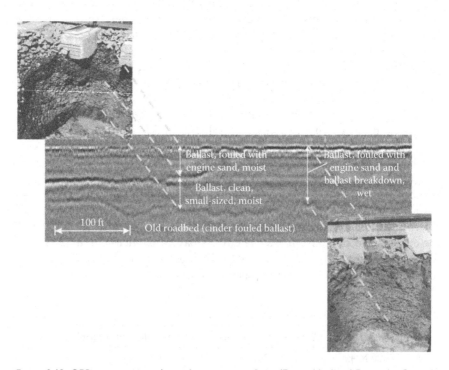

Figure 8.12 GPR image example with cross-trenches. (From Hyslip, J.P. et al., *Ground Penetrating Radar for Railroad Track Substructure Evaluation*, NTIS Report No. FRA/ORD-05/04, Federal Railroad Administration Office of Research and Development, US Department of Transportation, Washington, DC, 2005.)

across the track, which allows for this technique to be used on a network-level basis to provide indices of track substructure quality. The wide-area surveying ability also provides the means to define sections of track that have similar conditions and expected performance.

Table 8.7 presents the various substructure problems and the corresponding GPR information that can be measured for use in defining the extent and severity of the problems.

As with all subsurface investigation techniques, however, GPR does have some limitations. For instance, substructure layer boundaries may not be detectable to GPR under certain conditions, such as too little difference in electrical properties between the two adjoining layers or boundaries producing high reflections that mask radar signals from lower layers. The depth of radar penetration may be limited by such things as high moisture content or wet conditions, highly conductive (salty or clayey) soils, or highly metallic ballast components. GPR is also susceptible to interference from other radio sources, especially in urban environments, which can mask subsurface returns.

8.2.3 Equipment

A typical deployment schematic of a GPR antenna used for surveying on railroad track is shown in Figure 8.13.

The antenna is used to transmit and receive the radar pulses. The transmitting and receiving electronics are controlled by a GPR control unit,

Table 8.7 Substructure problems and corresponding GPR measurement

Substructure problem	Indicators from GPR
Poor drainage–trapped water	Intensity of GPR reflection, dispersion analyses
Poor drainage—layer depression (bathtub)	Difference in depth to subgrade surface laterally across the track, intensity of GPR reflection
Fouled ballast	• Dispersion analysis • Scattering • CMP (Common Mid-Point) • Full waveform modeling
Subgrade failure or deformation	Lateral and longitudinal variation of subgrade surface
Subgrade attrition	Lack of subballast layer in combination with fine-grained fouling
Subgrade excessive swelling and shrinking	Variation of clay subgrade surface
Transitions	Variation of the substructure layers along or across track of layer properties

Source: Hyslip, J.P. et al., *Ground Penetrating Radar for Railroad Track Substructure Evaluation*, NTIS Report No. FRA/ORD-05/04, Federal Railroad Administration Office of Research and Development, US Department of Transportation, Washington, DC, 2005.

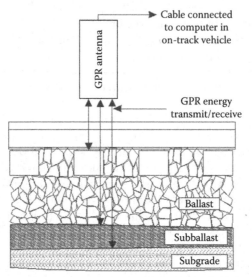

Figure 8.13 Schematic and photo of a deployed GPR antenna.

which also measures the time and amplitude of the returned signals. GPR antennas are designed to operate at various frequencies from tens of megahertz (MHz) to several gigahertz (GHz). The antenna's operating frequency must be considered for a given investigation, because it is important to address the tradeoff that exists between resolution and depth of penetration. The higher the antenna frequency, the greater the resolution it exhibits. But, higher frequency antennas have less depth of penetration compared with lower frequency antennas. Lower frequency antennas penetrate deeper than those at a higher frequency, but they have reduced resolution.

A variety of GPR systems are available and have been used for railway track investigations. Figure 8.14 through Figure 8.16 shows examples of some of the more common GPR systems.

Figure 8.14 Hyrail-based GPR system, showing a single 400 MHz antenna and 3D-radar system (left photo), and three pairs of 1-GHz horn antennas (right photo).

Figure 8.15 Push-cart application of railway GPR (left photo) and 200 MHz (left) and 400 MHz (right) GPR antennas installed on the back of hyrail vehicle (right photo).

Figure 8.16 Three 400 MHz antennas installed on hyrail vehicle (left photo). Three 2-GHz antennas on hyrail vehicle (right photo).

The three antenna-pair arrangement permits three profiles to be collected at the same time as the pairs of antennas are moved along the track. This provides three continuous parallel longitudinal images along the track, providing information on the cross-track variability.

A key part of successful use of GPR in railroads is the software that provides the ability to simultaneously view, interpret, and analyze multiple and disparate datasets, for example, GPR data from different antennas, maps, digital video, railroad databases, and condition measurements. This kind of data combination allows the user to conduct an integrated analysis of all the available datasets on a single screen.

8.2.4 Example GPR image results

The GPR grayscale image, also referred to as a radargram, provides a longitudinal view into the substructure of the track. Surveying at different longitudinal positions along the track provides information on both the longitudinal and lateral variation in the track. The following is a series of radargram images that provide examples of GPR results.

Figure 8.17 shows ballast and subballast layers interrupted by a drainage trench that was excavated into the subgrade clay and filled with ballast. The extent of the drainage trench is well-defined by the increase in ballast thickness and absence of the subballast layer.

Figure 8.18 shows an example of the subsurface conditions at a grade crossing, as detected by GPR. Trapped water immediately adjacent to the crossing is apparent by the high-amplitude reflections, that is, the pronounced white–black–white layer. The decrease in GPR reflection amplitude progressing away from the crossing in the right side of Figure 8.18

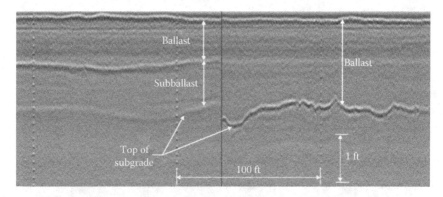

Figure 8.17 GPR profile showing interruption of ballast and subballast layer with a ballast-filled trench. (Data from Hyslip, J.P. et al., *Ground Penetrating Radar for Railroad Track Substructure Evaluation*, NTIS Report No. FRA/ORD-05/04, Federal Railroad Administration Office of Research and Development, US Department of Transportation, Washington, DC, 2005.)

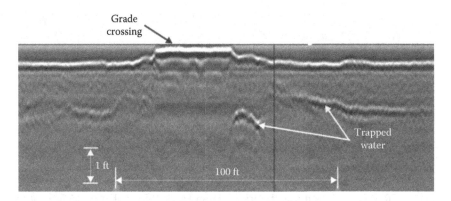

Figure 8.18 GPR profile at highway grade crossing.

correlates to decreasing water content of the subballast and subgrade. The strong reflection from the surface of the reinforced concrete grade-crossing planks momentarily reduced the signal amplitude from the subsurface layer compared with the open track.

The GPR profile in Figure 8.19 was obtained by a longitudinal survey along the access road beside the track. The variable thickness of the gravel surface layer is clearly shown. Also shown are drainage pipes from under the track installed in trenches dug into the road subgrade. The isolated reflections beneath the gravel roadbed at 10 ns are from drainage pipes at an approximate depth of 2.1 ft (0.6 m). The breaks in the otherwise continuous reflection from the bottom of the roadbed correlate with the pipes below.

The profile in Figure 8.20 shows ballast pockets that have developed in the embankment under the track. The ballast pockets are formed by failure of one side of the embankment, which bulged outward allowing the ballast to settle.

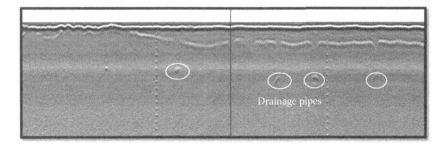

Figure 8.19 GPR image of access road along track. (Data from Hyslip, J.P. et al., *Ground Penetrating Radar for Railroad Track Substructure Evaluation*, NTIS Report No. FRA/ORD-05/04, Federal Railroad Administration Office of Research and Development, US Department of Transportation, Washington, DC, 2005.)

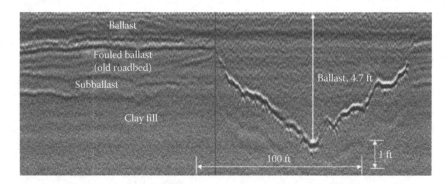

Figure 8.20 Deep v-shaped ballast pocket. (From Hyslip, J.P. et al., *Ground Penetrating Radar for Railroad Track Substructure Evaluation*, NTIS Report No. FRA/ORD-05/04, Federal Railroad Administration Office of Research and Development, US Department of Transportation, Washington, DC, 2005.)

8.2.5 Track substructure condition measurements

The main track substructure information obtained by GPR include data on depth and configuration of substructure layers, amount of ballast fouling, and relative moisture of the material.

8.2.5.1 GPR measurement of fouled ballast

Measurement of the degree of fouling in the ballast is the primary use of GPR in the railroad industry due in large part to the fact that railroads are able to react to this information with standard production-type ballast cleaning maintenance. The GPR analysis techniques to determine fouling conditions include scattering, dispersion analysis, full waveform modeling, and common midpoint.

8.2.5.1.1 Scattering

Scattering of radar waves by bouncing-off of individual particles is a function of the size of the particle and the amount of fouling material in the void space. Scattering analysis for fouling quantification is based on the degree of reduction in scattering as the void spaces are filled with fouling.

Differences in GPR signal from clean ballast edge reflections noted by Sussmann (1999) were explored by Olhoeft and Selig (2002) and used by Roberts et al. (2006) to develop a method to estimate the amount of ballast fouling. Scattering of the GPR signal results from interaction with objects of about the same size as the GPR signal wavelength. The higher the GPR wave frequency (i.e., shorter wavelength), the smaller

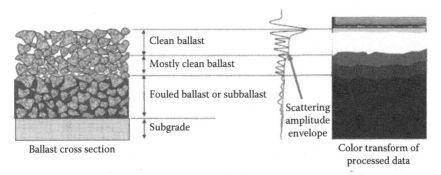

Figure 8.21 Scattering envelop development schematic. (Data from Roberts, R. et al., Advances in railroad ballast evaluation using 2-GHz horn antennas. *Proceedings of the 11th International Conferences on Ground Penetrating Radar*, Columbus, OH, 2006.)

the object needed to cause scattering, and ballast air voids can act to scatter the signal for high-frequency GPR (Al-Qadi et al. 2007). The diffracted energy from clean ballast creates a change in the energy contained within the return signal that has been shown to provide a means of quantifying ballast fouling. At high frequencies, for example, 2 GHz, the diffracted energy itself can be used to quantify the degree of ballast fouling (Roberts et al. 2006). As shown in Figure 8.21, the large amplitude reflections from the clean ballast layer that result from internal reflections between ballast particles, cause the scattering envelop to change shape in that area.

8.2.5.1.2 Dispersion

As noted in work by Sussmann (1999) and Clark et al. (2004), where changes in the reflected energy frequency spectrum were associated with ballast condition, another technique described by Silvast et al. (2006) was developed. This method uses the concept that fouled ballast will attenuate higher frequencies more than clean ballast. Developed from 400 MHz GPR data, this method is based on the frequency spectrum for the portion of the GPR signal at the depth of the ballast layer and is related to the area under the curve. As energy is attenuated, the area is reduced providing a valuable link between the GPR signal and ballast fouling (see Figure 8.22).

8.2.5.1.3 Full waveform modeling

Full waveform modeling (Olhoeft and Smith 2000; Hyslip et al. 2005) is a technique that uses the GPR signal (waveform) and a model of the soil layers to back-calculate soil properties. Through iteration, actual layer properties are determined. All other methods described to this point provide reference

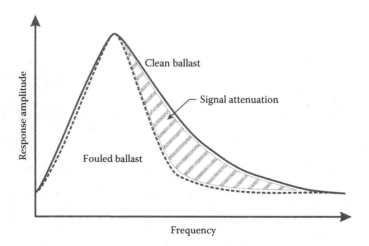

Figure 8.22 Effect of ballast fouling on GPR energy spectrum. (Idealized from Clark, M. et al., *Advanced analysis of ground penetrating radar signals on railway. Proceedings of the 7th International Conference on Railway Engineering*, London, 2004; Silvast, M. et al., NDT techniques in railway structure analysis. *Proceedings of 7th World Congress on Railway Research*, Montreal, Canada, 2006.)

or index type values that are not actual soil properties. This technique has significant merit for the interpretation of GPR data from track, but has not been fully developed.

8.2.5.1.4 Common midpoint

The common midpoint approach is based on determining the dielectric permittivity of the ballast layer, then empirically relating this dielectric permittivity to the amount of fouling material in the ballast. The common midpoint analysis uses a known geometric arrangement of antennas, where the change in reflection time to a common object is measured at antennas at increasing distances. The increasing reflection time is a function of the spacing and velocity of propagation through the material. By measuring the same layer at multiple antennas, the dielectric constant that determines that velocity in many materials can be determined.

8.2.5.2 GPR measurement of moisture

GPR is particularly sensitive to moisture levels, and GPR-derived information on the water content of the substructure indicates poor drainage and can also attest to poor subgrade. Figure 8.23 shows an example of the ability of GPR to detect moisture in the track resulting from the "wetting test" and clearly shows how the moisture is changing as the fouling remains the same.

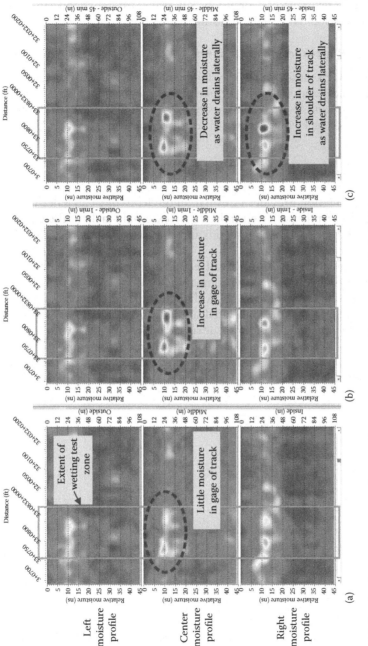

Figure 8.23 Results from wetting test showing change in moisture: (a) no water added; (b) 2 minutes after water added to center of track; and (c) 45 minutes after water added to center of track. (Courtesy HyGround Engineering and Roadscanners, Oy, Williamsburg, Massachusetts.)

In a similar manner to ballast fouling, moisture affects the GPR signal content at distinct frequencies. Using a similar process, it is possible to distinguish the track moisture condition. Full waveform modeling can also be applied to develop data on the moisture condition of track.

8.2.5.3 GPR measurement of layers

The time the GPR pulse takes to travel through the layer and back is controlled by the thickness and dielectric permittivity of the material. The travel time between upper and lower boundaries of a layer can be used to calculate the layer thickness employing a known dielectric permittivity.

$$h = c \frac{\Delta t}{2\sqrt{\varepsilon_r}} \tag{8.3}$$

Layer interpretation provides information on depth of clean ballast, depth of subgrade, depths of layers relative to external track drainage (ditch invert), and layer variations in depth and thickness both across and along the track.

A test conducted at TTC (Read et al. 2013) of six different GPR-measuring systems provided the comparative ballast layer thickness along the track, as shown in Figure 8.24 Although there are differences between measurement systems, they indicated similar ranges of ballast thickness.

The lateral variation of the subgrade surface across the track is indicative of poor drainage due to bathtub condition and subgrade failure or deformation. Figure 8.25 shows a typical track cross section with deforming subgrade. The arrow dimensions indicate the GPR antenna locations.

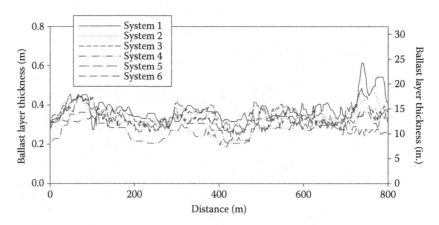

Figure 8.24 Variation of ballast layer thickness indicated by six GPR systems. (Data from Read, D. et al., *Ground Penetrating Radar Technology Evaluation on the High Tonnage Loop at the Transportation Technology Center*, DTFR53-00-C-00012, Task Order 248, Federal Railroad Administration, Washington, DC, 2013.)

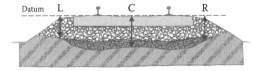

Figure 8.25 Typical track cross-section with subgrade deformation.

Figure 8.26 shows actual longitudinal scans for approximately 1,000 ft (300 m) of track. The distance from the datum to the top of the subgrade surface is indicated by the white arrows. Locations A and B are where the subgrade soil has deformed upwards, creating a situation similar to that shown in Figure 8.25.

Figure 8.27 depicts an example of a Ballast Pocket Index showing three parallel longitudinal GPR profiles indicating ballast pockets that have developed in an embankment under the influence of heavy axle load traffic.

A common, simple remedy to minimize the continued development of the ballast pocket is to drain the ballast pocket with a cross-drain (essentially a ballast-filled trench) excavated perpendicular to the track. GPR can

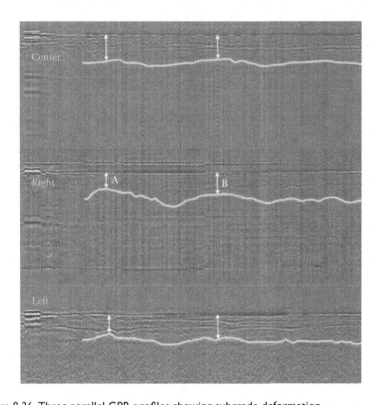

Figure 8.26 Three parallel GPR profiles showing subgrade deformation.

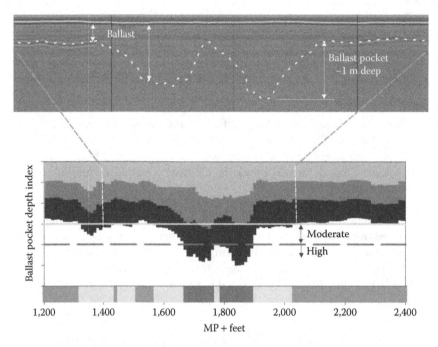

Figure 8.27 Subsurface index based on depth of ballast pocket. (From Hyslip, J.P. et al., *Ground Penetrating Radar for Railroad Track Substructure Evaluation*, NTIS Report No. FRA/ORD-05/04, Federal Railroad Administration Office of Research and Development, US Department of Transportation, Washington, DC, 2005.)

delineate the bottom of the pockets to ensure that lateral drainage is put at the most effective location (i.e., at the lowest point of the ballast pocket).

The banded image in the bottom of Figure 8.27 shows the depths of the subgrade surface added together to accentuate the areas with the ballast pocket condition. Thresholds applied to the Ballast Pocket Depth Index plot (middle of Figure 8.27) yield the color-coded bar scale plot in Figure 8.27. The bar code depicts the worst areas (deep ballast pockets) with dark banding and good areas with white.

To get the greatest benefit from the GPR, the radar information must be integrated with other information such as track geometry, track asset features, track stiffness measurements, maintenance records, and geo-environmental information. The geo-environmental information includes geology, terrain, geologic environment, clay type, drainage characteristics, and climate factors. This integration can result in a holistic Substructure Maintenance Management (Selig 1997; Hyslip 2007; Sussmann 1999) program through which correct decisions on maintenance and capital improvements can be made, as discussed in Chapter 9, Management. Several software tools have been developed to visualize and correlate the substructure characteristics derived from GPR with these other condition indicators. The management chapter shows how these geographical information system tools and linearized, track-chart based tools allow the viewing of the relationship of GPR to track condition and features, as well as the visualizing of substructure effect on geometry trends that may drive maintenance requirements.

8.3 TRACK GEOMETRY

Track geometry, as designed, is a series of classic shapes of straight (tangent) segments, spirals, and curves, as discussed in railway design textbooks. But as track settles or moves laterally under loading, deviations from that design accumulate resulting in increasing geometry error. Railway geotechnics is very important to growing track geometry deviations because they develop largely due to differential settlement or lateral movement within the geotechnical layers of ballast, subballast, and subgrade.

The goal of measuring track geometry is to quantify deviations from the ideal design with the aim of correcting this error as it approaches an established limit to maintain conditions of safety and ride comfort. As these deviations are measured continuously along the track, the locations of those that exceed the maximum allowable error based on safety limits indicate where maintenance is urgently needed. Although less urgent, those locations with deviations smaller than the safety limit, but larger than the maintenance limit, are also noted so that correction can be applied well before they become a safety risk. Geometry error safety limits are based on well-established and detailed knowledge of vehicle–track interaction and the prevention of specific kinds of events, such as rail rollover, wheel flange climb derailment, and lateral track panel shift. The size of geometry deviations corresponding to maintenance limits are typically in the range of 66%–75% of the safety limit. It is important to conduct measurements of geometry error on a sufficiently frequent basis to allow intervention before a rapidly growing error can surpass the maintenance limit and approach the safety limit.

Geometry error can be the deviation from design at a single point, an excessive rate of change of rail position over a fixed track length, or it can appear as a cyclic waveform that produces a resonant response in the vehicle. This cyclic waveform error of track profile or alignment may excite car body resonant motion and therefore poor ride quality to passengers, excessive wheel/rail forces, and possibly vertical wheel unloading. Track geometry recording cars routinely measure geometry deviations at discrete points and variations of geometry over a fixed distance, but they are not commonly configured to measure cyclic error. Therefore, a special means of processing geometry data is needed to identify cyclic error, as will be discussed.

8.3.1 Measurement fundamentals

The basic concept of geometry error measurement is the deviation from a measure of uniformity. The uniform reference is commonly a straight

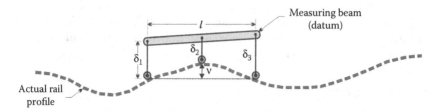

Figure 8.28 Automated MCO measurements. (Data from Hyslip, J.P., *Fractal Analysis of Geometry Data for Railway Track Condition Assessment.* PhD dissertation, Department of Civil Engineering, University of Massachusetts, Amherst, MA, 2002.)

line, or chord, where error is measured at some point along it, typically at the midpoint, as shown in Figure 8.28. The geometry deviation can be obtained by hand-held measuring devices where a string under tension held next to the rail provides the straight line reference, and the deviation from the string to the rail is measured. However, the following discussion will assume that the geometry measurements are obtained using an automated and continuous measuring system such as a track geometry car.

A midpoint error or "offset" from the chord is called a mid-chord offset (MCO), but the more general term for an offset measured anywhere along the chord is referred to as a versine. In the United States, the MCO over chord lengths of 31 ft and 62 ft are used for freight and (≤90 mph) passenger traffic, and an additional chord length of 124 ft is used for higher speed passenger track. The 62-ft chord was the original chord length chosen because of its convenient property that the MCO, in inches, equals the degree of curve. The 31-ft chord was added to control track features over a shorter distance, and the longer 124-ft chord was added with the intention of characterizing geometry error over a longer wavelength for high-speed passenger service.

The true rail profile or alignment shape is not represented by the series of MCO values. In fact, the chords tend to act as mechanical filters that provide a distorted view of the rail shape because the offset measured is affected by the chord length and because the ends of the chord are not on a level datum but are riding on rough track.

Measurements of MCO are made at regular intervals along the track, typically every foot. The safety limit applies to every MCO measurement, and so, if only one measurement exceeds the limit, this is a cause for immediate action. As chord length increases the allowable MCO value increases.

From a safety viewpoint, it is sufficient to ensure that each MCO is less than the safety limit by a significant margin, where the chosen margin corresponds to the maintenance limit, as mentioned previously. However, for ride quality in passenger operations, it is necessary to monitor and control the geometry error in the form of continuous waveforms that are present in the track geometry. This requires going beyond viewing track geometry error as discrete MCO measurements to a representation of profile and alignment deviations of the rails in the form of the space curve. The space curve of the rails, as measured by a geometry car, is derived from a baseline reference acting as a filter for the data. For its high-speed operations, the Amtrak operation in the United States uses a 400-ft filter to obtain the space curve, although freight operations may use a much smaller baseline length filter. This 400-ft reference length is sufficiently long to include error with wavelengths up to this same order of magnitude, which is needed for higher speed operation. Space curve data may be measured every foot, but these data do not relate to a MCO that is compared with a safety limit at every foot. Instead, the usefulness of the space curve is that it approximates the true waveform of the rails and as such contains various wavelengths that affect vehicle dynamic response. Filtering the space curve to reveal the wavelengths that are present, and their relative amplitude, can determine the presence of cyclic geometry error that may produce a resonant motion in the vehicle, as will be described in Section 8.3.5.

The MCO values are not a faithful representation of actual geometry error but are distorted by virtue of "mechanical filtering". As stated, actual track profile and alignment with all their inherent irregularities can be thought of as a combination of many waveforms with varied periods and amplitudes. Continuous measurement of track geometry using the MCO measurement provides a distorted geometry because MCO measurements give unequal emphasis to different wavelengths resulting in distorted MCO amplitudes. Figure 8.29 shows the relationship of the MCO response to

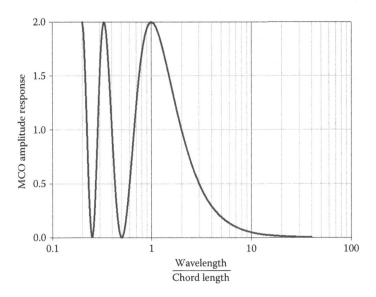

Figure 8.29 Relation of transfer functions to ratio of geometry error wavelength to measuring chord length. (Data from Hyslip, J.P., *Fractal Analysis of Geometry Data for Railway Track Condition Assessment.* PhD dissertation, Department of Civil Engineering, University of Massachusetts, Amherst, MA, 2002.)

the ratio of geometry error wavelength to MCO chord length. For example, a 62-ft MCO measurement of an error wavelength of 62 ft will have twice (two times) the actual wavelength amplitude, that is, the ratio of wavelength to chord length equals one and the corresponding amplitude response is two. As another example, measuring a 31-ft wavelength with a 62-ft MCO (wavelength to chord length ratio equals 0.5) will result in zero amplitude (flat line), that is, the 31-ft wavelength will not be "seen" by the 62-ft MCO. Additionally, a 124-ft wavelength will be faithfully represented by a 62-ft MCO, because this wavelength to chord length ratio is two and the MCO response amplitude is one (unity).

The extent to which the MCO-based measurement system can distort the actual size of geometry error is shown in Figure 8.30. The composite waveform in the upper plot of Figure 8.30 is comprised of three separate waveforms with wavelengths of 12, 6, and 3 m. If the MCO of the waveform was then measured using a chord length of 24 m, the resulting wavelength/chord length ratios are therefore 0.5, 0.25, and 0.125 ($T/l = 0.5$, 0.25, and 0.125), which, as shown in the lower plot of Figure 8.30, results in a zero MCO amplitude response for each wavelength. Despite the obvious presence of geometry error in the upper plot, the measured MCO response of zero, indicating no error, is shown in the bottom plot of Figure 8.30. Although this combination of geometry error wavelengths and chord length was deliberately and artificially selected to illustrate the case of a

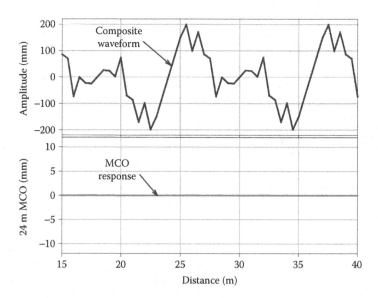

Figure 8.30 Measurement of waveform producing an erroneous zero MCO amplitude response. (Data from Hyslip, J.P., *Fractal Analysis of Geometry Data for Railway Track Condition Assessment.* PhD dissertation, Department of Civil Engineering, University of Massachusetts, Amherst, MA, 2002.)

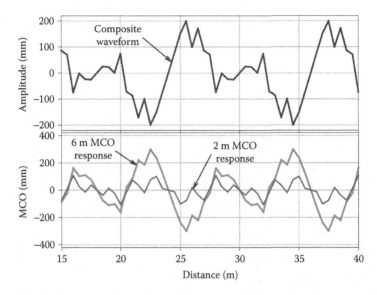

Figure 8.31 Measurement of waveform with nonzero MCO amplitude response. (Data from Hyslip, J.P., *Fractal Analysis of Geometry Data for Railway Track Condition Assessment*. PhD dissertation, Department of Civil Engineering, University of Massachusetts, Amherst, MA, 2002.)

null response, a more typical result is the nonzero but varying responses shown in Figure 8.31.

If the composite waveform was measured using chord lengths different from the 24 m, the MCO amplitude will vary accordingly, as shown in Figure 8.31, for the arbitrarily chosen chord lengths of 2 and 6 m. The MCO amplitude response is now nonzero for the composite waveform wavelengths. As shown in Figure 8.31, there is a definite MCO response and the MCO values are distorted in terms of amplitude, but the wavelengths of the responses are the same.

8.3.2 Vertical profile geometry measurement

Because the geotechnical components of track mainly provide vertical settlement, the measurement techniques of profile geometry error will be described. Track geometry recording cars basically use two different methods to measure vertical profile geometry. These are (1) three-point contact measurement using MCO and (2) noncontact measurement using acceleration relative to an inertial platform device (profilometer). Regardless of the measurement technique used, the vertical profile is typically expressed in MCO values.

The three-point measurement system can be performed by hand by the technique known as string lining, or by automation with track geometry cars. Figure 8.28 illustrates how automated MCO measurements are made. As the beam passes over the rail, the two sensors at the end of the beam define the chord and the middle sensor reading provides MCO with respect to this chord. The MCO of Figure 8.28 is calculated by

$$v = \delta_2 - \frac{\delta_1 + \delta_3}{2} \tag{8.4}$$

Inertial referencing systems based on acceleration measurement devices are used by many modern high-speed track geometry recording cars to make vertical profile measurements. Modern inertial-based track geometry recording cars use a combination of two measurement methods: (1) using the inertia method to determine the location of the beam and (2) using the laser/cameras to determine the distance between the beam and the rails. These inertial-based devices (profilometers) work by measuring the accelerations of the car body in relation to an inertial platform. The accelerations are used to calculate the track deviations by double integration. Figure 8.32 shows a profilometer schematic.

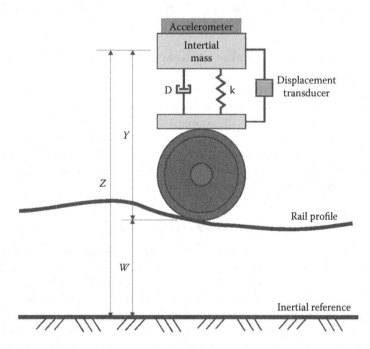

Figure 8.32 Inertial platform measuring device schematic. (Data from Hyslip, J.P., *Fractal Analysis of Geometry Data for Railway Track Condition Assessment*. PhD dissertation, Department of Civil Engineering, University of Massachusetts, Amherst, MA, 2002.)

The main benefit of inertial reference systems over direct MCO measurement methods is that they have a "flat" frequency response, unlike the distorted MCO measurements. Profilometers do have difficulty in accurately measuring very long wavelengths because the accelerometer signal becomes so small for long wavelengths that it cannot be recognized from background noise. The signal to noise ratio is also very small at slow measurement speeds; therefore, many systems are not useful at speeds below approximately 15 mph. However, newer systems that use laser gyroscopes instead of mechanical gyroscopes have been reported to provide high measuring accuracy even at low speeds (Presle 2000). Profilometer measurements provide a "relative" space curve because the measurements are derived from a continually moving (relative) datum.

8.3.3 Measures of geometry variation

8.3.3.1 Spatial correction to "line up" successive geometry data

To determine track geometry degradation rates over time with traffic loading accumulation, it is necessary to remove any shifting of the data between measurement runs to ensure that successive measurements of geometry "line up" for the trends to be apparent. Although geometry car data is usually referenced to fixed points along the track, there will be some error that will make subsequent geometry measurements longitudinally offset from each other by a variable distance that can make comparisons difficult, as shown in Figure 8.33.

This shifting of the data can result from "drift" in tachometers that indicate geometry car distance traveled, and this drift can be variable from run to run. Methods are available to produce the result shown in Figure 8.34, where the longitudinal shifting error from repeated measurements is removed.

8.3.3.2 Standard deviation over fixed segment track length

Geometry error is typically expressed as variation from a uniform mean, which can be represented statistically as standard deviation (SD). This SD

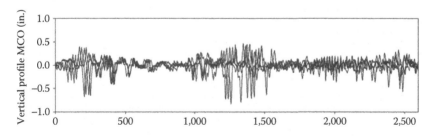

Figure 8.33 Repeated measurements of track geometry that are longitudinally shifted.

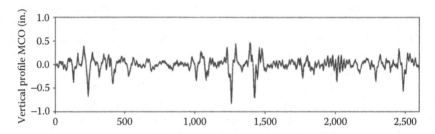

Figure 8.34 Repeated measurements of track geometry that are longitudinally corrected to remove shift error.

can be calculated from the series of MCO's measured over a fixed track length and represented as a single value indicating SD track roughness over this segment, as shown in Figure 8.35 for block segments of 220 yards (200 m).

The magnitude of MCO values measured and, therefore, the SD values that are derived from them are unique for the chord length used. For a larger chord length, the measured MCO values increase as does the corresponding SD value for the track segment. For example, the practice in the United Kingdom is to use chord lengths of 35, 70 m, and, in some cases, 140 m, and in the United States, these are 31, 62, and 124 ft, and the SD value obtained from these is always referenced to the chord length over which it was calculated. Taking this concept shown in Figure 8.35 farther, these segments can be represented as a series of values, as shown in Figure 8.36, where each track segment is one-fifth of a mile long (161 m). The values in the cells of Figure 8.36 could represent the SD error of vertical profile, lateral alignment, or other geometry parameter.

If the geometry car measures error every foot (0.3 m) along the track, the SD values shown then represent the variation about the mean of approximately 1,056 data points in each segment.

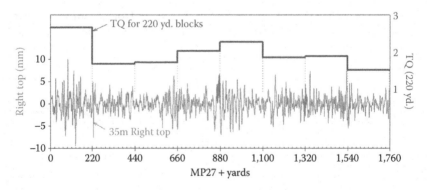

Figure 8.35 Series of MCO's and resulting standard deviation for fixed segments.

	MP100					MP101					MP102					MP103					MP104			
2014	1.4	1.9	2.5	2.7	3.6	5.9	4.1	2.4	1.3	1.6	2.5	2.5	3.1	2.0	1.7	2.1	1.9	3.4	2.6	1.5	3.4	4.3	3.7	1.5
	1.3	1.8	2.5	2.7	3.5	5.7	4.0	2.4	1.3	1.6	2.5	2.4	3.1	2.1	1.6	2.0	1.9	3.4	2.5	1.4	3.3	4.2	3.7	1.5
	1.4	1.6	2.4	2.6	3.5	5.5	4.0	2.3	1.3	1.5	2.4	2.2	3.0	2.0	1.6	1.9	1.8	3.1	2.5	1.4	3.3	4.2	3.6	1.4
	1.2	1.5	2.4	2.6	3.4	5.2	3.9	2.2	1.2	1.4	2.3	2.1	2.9	1.9	1.6	1.9	1.8	3.0	2.4	1.3	3.3	4.2	3.5	1.4
2013	1.3	1.4	2.4	2.5	3.4	4.9	3.8	2.2	1.2	1.4	2.3	2.0	3.0	1.9	1.5	1.8	1.8	2.9	2.3	1.2	3.2	4.1	3.5	1.3
	1.1	1.4	2.2	2.5	3.2	4.5	3.7	2.1	1.1	1.3	2.2	1.9	2.9	1.8	1.5	1.8	1.7	2.8	2.3	1.2	3.2	4.0	3.4	1.3
	1.0	1.3	2.2	2.5	3.0	4.2	3.4	2.1	1.0	1.3	2.1	1.7	2.8	1.7	1.4	1.7	1.7	2.8	2.3	1.1	3.2	4.0	3.4	1.3
	1.2	1.3	2.3	2.4	2.9	3.9	3.2	2.0	1.0	1.3	2.1	1.6	2.7	1.7	1.4	1.7	1.6	2.7	2.3	1.1	3.1	3.9	3.4	1.2
2012	1.0	1.2	2.3	2.4	2.8	3.2	2.9	1.9	1.0	1.3	2.1	1.6	2.5	1.7	1.4	1.8	1.5	2.5	2.2	1.2	3.1	3.9	3.3	1.2
	1.1	1.2	2.2	2.3	2.7	2.8	2.6	1.9	0.9	1.3	2.2	1.5	2.4	1.7	1.4	1.7	1.5	2.5	2.2	1.1	3.1	3.8	3.2	1.3
	1.4	1.9	2.5	2.7	3.6	6.1	4.1	2.4	1.3	1.6	2.5	2.5	3.1	2.0	1.7	2.1	1.9	5.8	2.6	1.1	3.0	3.8	3.2	1.2

Figure 8.36 SD values every one-fifth track mile from previous geometry car measurements.

Figure 8.36 also indicates historical SD values with the most recently measured on the top row, and those down the column represent values from measurements further in the past. Looking at the variation of data in a single column shows geometry degradation trend over time for that track segment. The thick black horizontal line shows when geometry correction was applied to an approximately 3.6 mile segment of track, and the amount of improvement after the correction is indicated by the difference between the SD in the cell below and above the line. The long-term durability of this correction is indicated by how quickly the SD value returns to the premaintenance value. Note that, for example, the poor geometry from approximately MP 100.6 to MP 101.2 returns after 10 more geometry measurement runs, but a much more durable geometry correction was provided at MP 103.2 to MP 103.4. This way of showing error along the track allows the planner to focus maintenance correction based on severity by location. It can also indicate where repeated applications of geometry correcting methods are effective, and where they are not.

If the trend of geometry error growth over time is to be determined as shown in Figure 8.36, it is imperative that the data from consecutive runs are aligned with minimal shift of the data along the track distance from run to run. Misalignment of data could result in the location of a rough section of track being just inside one segment for a run and then just inside the neighboring segment for the next run, resulting in SD values that fluctuate accordingly and falsely indicate that one segment is getting rougher while the other is improving.

Although arranging track roughness data as shown in Figure 8.36 has its advantages, an undesirable outcome of using these fixed length segments is that the track maintainer may apply maintenance correction only from start to end of a segment with a large SD. The potential problem here is that if a portion of rough track geometry extends somewhat into the next segment that happens to have a low SD, this portion of the segment with rough geometry will not be corrected. Another potential problem with this approach is that the chosen segment length is usually much longer than the actual rough track section within the segment that can tend to mask the presence of these shorter rough sections. Because track roughness is rarely uniformly distributed in a segment, but instead exists in short clusters within a segment, it might be better to target just those rough spots for maintenance and leave the remaining well-performing track within the segment untouched. When rough track extends over only a small portion of the segment length, a method that better identifies these locations and their severity is needed so that corrective action can be focused only where it is needed. Such a method is discussed in the next section.

8.3.3.3 Running roughness

Using SD values calculated over relatively long track sections is often not sufficiently precise to pinpoint problem locations to assign roughness values and plan for geometry correction. Track geometry car data reveals that geometry error is not distributed as evenly along the track as implied by Figure 8.36, but instead tends to exist in clusters of varying severity over much shorter track distances, as shown by the "right top" MCO data in Figure 8.35. The MCO data in Figure 8.35 displays a roughness distribution that is only approximately reflected by the SD value for that track segment.

To more precisely identify the location and quantify the geometry error of a rough section, a moving window filter could be applied to the MCO data where the error over this window length is used to assign a roughness value at the midpoint of the moving filter. By moving the window incrementally along and repeating this calculation, a series of roughness values can be plotted that is known as the Running Roughness (Ebersohn 1995) as shown by the curvilinear line over the MCO data in Figure 8.37. Note that the Running Roughness better identifies rough locations requiring geometry correction. For example, the segment between yardage 880 and 1,100 has a high SD roughness, but this misrepresents the roughness condition because correcting the geometry only over this segment would miss the substantial roughness going into the next segment to the left (between yardage 660 and 880).

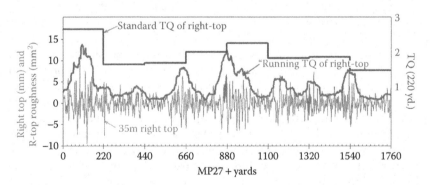

Figure 8.37 Example of running roughness derived from variation of MCO data.

The Running Roughness is a mean square statistical calculation that provides a magnitude analysis of the geometry measurements. Running Roughness, or R^2, is essentially an average of the deviations squared within a specific moving window, that is

$$R^2 = \frac{\sum_{i=1}^{n} d_i^2}{n} \qquad (8.5)$$

where:

d_i = MCO deviation
n = number of measurements in the length of track under consideration

Figure 8.38 presents another example of Running Roughness as applied to profile MCO data. The 62-ft MCO data is shown along with R^2 computed in a 200-ft moving window. It is apparent that the R^2 values well represent the location and severity of MCO deviations.

Just as the SD magnitude is dependent on the chord length used to measure the series of MCO's, the R^2 magnitude is also linked to the baseline length used to calculate it. To make R^2 more performance-related, the baseline length of this sliding filter may be chosen to correspond to the truck spacing of typical rail vehicles or to the longest wavelength of concern at the highest speed of a passenger car. For example, in the United Kingdom, the practice is to use 35, 70, and occasionally a 140 m reference length for SD calculation, which captures the medium, long, and very long wavelength error. In the United States, to improve ride quality, Amtrak has developed a method to search for geometry error up to 400 ft in wavelength (Wnek and Chrismer

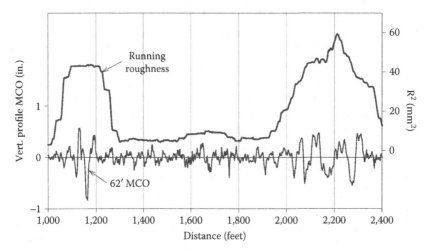

Figure 8.38 Example of running roughness derived from variation of MCO data. (Based on Ebersöhn and Selig, 1994a.)

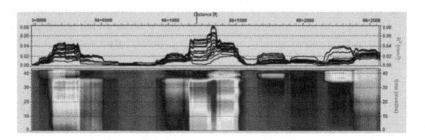

Figure 8.39 Rate of growing track geometry error shown by R^2, and by changing color/shades of data over time.

2012) by analyzing the cyclic content in the waveform of track geometry car space curve alignment and profile data obtained from a reconstructed baseline of the same length.

As track geometry degrades with time and traffic accumulation, R^2 increases accordingly at different rates along the track, as shown in the upper portion of Figure 8.39. Along with the error magnitude, the rate of error growth is also a measure of severity that must be monitored for maintenance planning purposes. However, this rate may not be readily apparent by the R^2 representation of data. A better method of visualizing this rate of increasing geometry error is to represent the error magnitude by color or shading where the change in these tones over time indicates the error growth rate, as shown in the bottom portion of Figure 8.39 (obtained using Roadscanners' Railway Data Management System).

8.3.3.4 Other measures of track roughness

Methods in addition to the SD and the R^2 measurements of track error include histogram analysis, cumulative distribution, and power spectral density (PSD) evaluation. A PSD plot of geometry error (most appropriately derived from space curve data) provides a measure of the predominant wavelengths that contain most of the error, as shown in Figure 8.40. Complex waveforms such as track geometry error over many miles of track are composed of a series of sine waves of various wavelengths and amplitudes, but some error wavelengths will have larger amplitude (more error) than others. For example, Figure 8.40 shows track profile error measured over 130 miles of track that carries predominately passenger traffic. One of the noticeable features is the peak values representing harmonics of the standard rail length of 39 ft. Although this rail is continuously welded, these recurring peaks are due to irregularities remaining at the rail welds and possibly to the substructure due to "joint memory" from the time before CWR when rail joints were located every 39 ft longitudinally along the track.

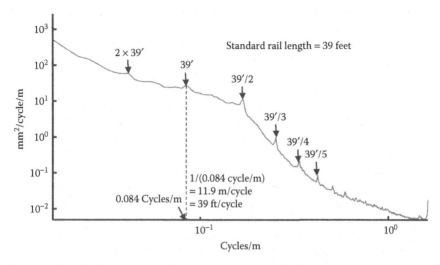

Figure 8.40 PSD of vertical profile space curve geometry data for 130 miles of track.

The trend of larger error amplitude with longer wavelength (on the left of the plot) as displayed in Figure 8.40 is the general characteristic of the PSD curve. With smaller wavelength, the error amplitude becomes smaller as shown.

Lastly, some railways use unique combinations of two or more geometry parameters in the development of their own railroad-specific track quality index or TQI. For example, in the United States Amtrak combines the track geometry car measured MCO profile error over a 124-ft chord with cross-level error by adding the two values of error together at every foot along the track, and then calculates the cumulative distribution of this series of data over the desired length of track, be it as large as an entire division or as small as a segment of open track between interlockings. This cumulative distribution is used to determine the geometry error trends and to prioritize maintenance.

8.3.4 Track alignment control for high-speed rail

High-speed passenger rail requires a high-standard for quality and durability of geometry correction. Some European countries with high-speed passenger rail operations correct alignment to an established design geometry using fixed references that are spaced regularly along the wayside. These references may be monuments established on catenary poles, posts, or other permanent structures along the right-of-way. The tampers then align the track to a set distance from these references to bring the track back to the design. This method can provide an excellent control and precise geometry corrections, assuming that the design position is known.

For cases where the design alignment position of an existing track is not known, or perhaps was never established, producing alignment geometry of sufficient quality for high speed is more of a challenge. Even if track was originally laid out with a high-quality design alignment using classic shapes of tangents, spirals, and curves, as described in track design textbooks, many years of "smoothing" maintenance correction (which does not return track to its design) may have distorted the track further and further from this design position. For such a case, the amount of distortion from the original "perfect" alignment design may have become so large that it may not be possible to return to the design alignment without lateral track throws that are prohibitively large.

However, there is a tamper control system developed to bring such a distorted alignment to high-speed quality without the need for large track throws and without costly monumenting along the right-of-way to establish a design alignment. Tamper-guided control system (TGCS) was developed by British Rail researchers, as described by Marriott (1996), as an improvement over existing control systems provided with tampers, to move from "smoothing tamping" to design tamping. Tampers in the United Kingdom have been retrofitted with TGCS, and have reportedly been producing much improved geometry corrections and ride comfort at the highest passenger speeds. Amtrak has installed TGCS on its tampers and is also finding substantial improvement in geometry error removal and ride quality improvement (Wnek and Chrismer 2012), with relatively small lateral track throws needed to produce high-quality alignment geometry.

8.3.5 Advances in analysis of track geometry data

8.3.5.1 Analyzing track geometry space curve data

A waveform might not communicate all of the information it contains from its outward appearance. Fourier's crucial insight is that even the most complex waveform is composed of a series of simple sine waves of various amplitudes and wavelengths that are superimposed, and his analysis can be used to see those sine waves that are hidden within track geometry data and that may be affecting vehicle behavior. For example, consider the outward appearance of the space curve geometry profile of two rails in Figure 8.41. A trained eye would note the y axis and x axis scales and see track that has a relatively modest amount of profile roughness. But even that trained observer is not likely to see that within this waveform exists a cyclic profile error that has the potential to produce large vertical accelerations in certain rail vehicles as they traverse this track.

Even waveforms with seemingly small amplitude error can contain cyclic components that cause a buildup of vehicle resonance and result in poor ride quality. A Fourier analysis of these profile space curves reveals the hidden cyclic profile error shown in Figure 8.42 with relatively strong cyclic

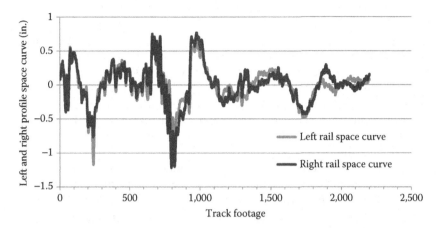

Figure 8.41 Profile space curve track geometry data.

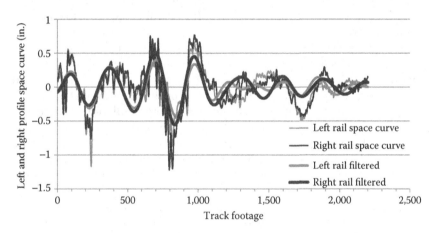

Figure 8.42 Track geometry space curve and filtered data revealing cyclic waveforms within.

amplitude with a wavelength of approximately 300 ft, starting from the origin to an approximate footage of 1,000. The analysis also reveals that the cyclic profile error of both rails is in phase over this distance and that they steadily increase in amplitude. The combination of these three features of wavelength, the two rails being in-phase, and ramping up of the amplitude add to the vehicle resonance effect.

The technique to find this cyclic error uses a band-pass filtering function to look for component sine waves that are at wavelengths known to excite the car body, in the case shown in Figure 8.42, in the vertical bounce mode. This method has been developed to identify the presence of these multiple cycles of harmonic error within track geometry measurement waveforms (Chrismer and Wnek 2011). Cyclic geometry error is usually of relatively small amplitude and so is often unobserved and left uncorrected by maintenance forces, because the associated mid-chord offset geometry error is typically far less than the safety limit. However, by using a band-pass filtering function on profile and alignment space curve data from the geometry car, the presence and magnitude of any cyclic error becomes apparent. The effect of cyclic track geometry error should be assessed on car body accelerations for purposes of ride quality, and on wheel rail forces to prevent vertical wheel unloading.

It is difficult to establish a limit for the amplitude of cyclic track geometry error with regard to ride quality, because the cyclic error amplitude that can be expected to produce excessive car body acceleration depends on rail vehicle dynamic characteristics that can vary widely. In general, however, it can be stated that because it is easier to excite a passenger car body laterally than vertically, passenger car body acceleration is more sensitive to lateral rather than vertical cyclic error, and so any limits placed on the amplitude of cyclic error should be more restrictive for lateral cyclic error for the sake of ride quality.

Although cyclic geometry error of any appreciable amplitude can reduce ride quality, it has been observed that if the peak-to-peak cyclic profile error amplitude of the filtered space curve data attains 20 mm. or greater (as in Figure 8.42), this tends to produce peak-to-peak car body vertical accelerations close to the limit allowed by some safety regulations. Therefore, Amtrak scans geometry data for cyclic error approaching or exceeding this amplitude. For example, the cyclic error shown in Figure 8.42 can produce vertical car body acceleration above the ride quality limit since the vertical space curves have cyclic profile error peak-to-peak amplitudes greater than 20 mm.

A certain pairing of cyclic error wavelength and speed causes resonant car body motion for a given vehicle type, but for a vehicle with different car body mass and vertical spring rate of the secondary suspension another wavelength/speed combination will cause resonant car motion. For example, the error wavelength that causes resonance in Amtrak's Acela Power Car is very different from that for the Acela Coach due to the large difference in the dynamic behavior of these two rail vehicles. It has been determined that the Acela coach responds in bounce mode for a wavelength of approximately 170 ft (at 125 mph) and the Power Car reacts similarly for wavelengths of 320 ft, as shown in Figure 8.43. Note that the nearly two-to-one difference between resonance-producing wavelengths in the Power Car compared with the coach is due to the differences in car body mass and vertical stiffness of the suspension. The 170-ft and 320-ft wavelengths can both be characterized as "long" wavelength error, well beyond the longest chord length (124 ft) used for mid-chord offset geometry error measurement. Ride quality improvements require geometry cars to be able to detect,

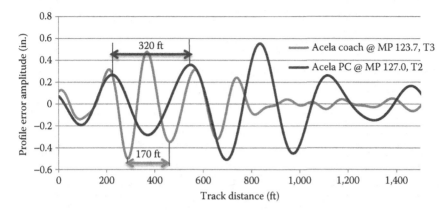

Figure 8.43 Track profile cyclic error wavelengths that drive Acela car body resonance at 125 mph.

and tampers to be able to correct, track error over a considerably longer distance than a 124-foot chord.

Although the discussion about cyclic geometry error has focused on high-speed passenger traffic, these same concepts apply to slower speed freight traffic as well, at smaller wavelengths. For freight vehicles, the concern shifts from passenger comfort to issues of wheel unloading for vertical cyclic profile error, and possibly excessive lateral wheel loads from cyclic alignment error. Due to the heavier wheel loads, freight car resonant response is more prone to provide large vertical and lateral loads that may increase track damage and rail rollover.

8.3.6 Filtered space curve cyclic characteristics and vehicle dynamics

There is still more information that can be obtained from analyzing cyclic geometry error beyond the presence of certain wavelengths with a strong amplitude. It has been observed that cyclic error produces an especially strong resonant response in passenger cars and locomotives, if three characteristics are all present:

1. Both rails being in phase with each other
2. Maximum peak-to-peak amplitude of approximately 0.8 in. (20 mm) or more
3. Pattern of "ramping up" in amplitude for two or more cycles

All three features are present in the cyclic waveform shown circled in Figure 8.44 that was observed to produce a large lateral dynamic response in the vehicle. It has also been noted that the geometry error waveforms on either side of that circled did not produce a significant car body reaction,

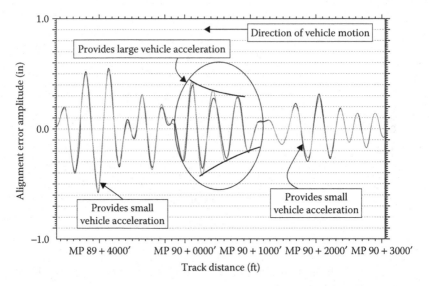

Figure 8.44 Cyclic alignment error data: band-pass filtered from 230 to 460 ft wavelength.

because they do not have all three requirements. Cyclic error waveforms should be reviewed to find track locations with these patterns so that corrective maintenance can be applied.

8.3.7 Determining effectiveness of geometry correction from waveform analysis

Waveform analysis of space curve geometry data can also be used to determine the effectiveness of geometry correction. For example, a passenger railway was concerned that its geometry correction procedures were not effective in removing long wavelength alignment and profile error, and that ride quality would continue to suffer as a result. An improved method of alignment correction (Klauder et al. 2002) was applied in a curve in the hope that it would significantly reduce longer wavelength error. To determine how effective this new maintenance method was, the amplitude of geometry error over a range of wavelengths was analyzed before and after maintenance. Because alignment error at wavelengths of 500 ft and slightly more were thought to affect lateral ride quality at the highest speeds, the cyclic error amplitude for wavelengths over a range of 500–600 ft was determined. The cyclic error over this wavelength range before maintenance was found to be approximately 0.10 in., as shown in Figure 8.45, for the entire 3,000 ft of the curve. Although this cyclic error amplitude is unlikely to produce large peak-to-peak car body accelerations, it could result in an elevated and sustained car body acceleration that reduces ride quality.

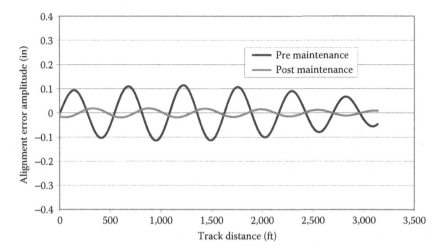

Figure 8.45 Alignment error amplitude over 500–600 ft wavelengths in curve.

This same analysis was performed on the space curve alignment error measured after maintenance and showed that the error amplitude decreased to approximately 0.015 in. Therefore, this improved alignment correction method provided an 85% error reduction from the original. This analysis was also performed over a band-pass range of 600–700 ft, showing that the improved alignment approximately 50% of the original error was removed, as shown in Figure 8.46.

The analysis showed that the error over a wavelength range of 700–800 ft was slightly increased by the maintenance, as shown in Figure 8.47, although

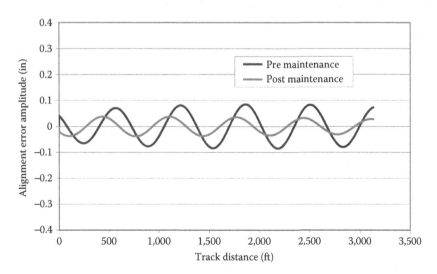

Figure 8.46 Alignment error amplitude over 600–700 ft wavelengths in curve.

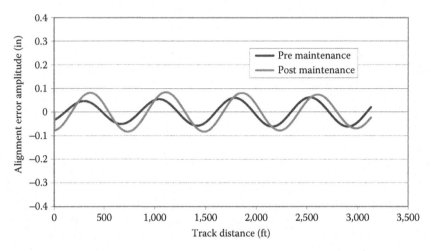

Figure 8.47 Alignment error amplitude over 700–800 ft wavelengths in curve.

an increase in the error at these very large wavelengths has no effect on ride quality even at the highest speeds.

Performing this analysis for shorter wavelengths and plotting the results in Figure 8.48 shows the effectiveness of this improved maintenance method at reducing alignment error across a broad wavelength spectrum.

The actual alignment space curve data in Figure 8.49 that was filtered to provide the before-and-after-maintenance cyclic error comparisons in Figures 8.45 through 8.47 further confirms that the correction has removed much of the alignment error, with wavelengths below approximately 700 ft. The result is track with a much improved ride quality.

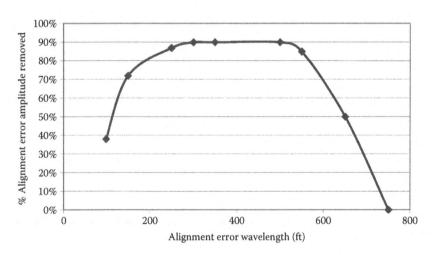

Figure 8.48 Reduction in alignment error amplitude with wavelength in curve.

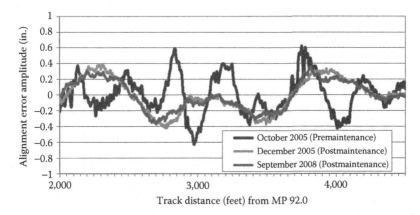

Figure 8.49 Change in alignment error with improved design and maintenance in curve.

8.3.8 Performance-based track geometry inspection

The fundamental purpose of monitoring and correcting track geometry error is to control the interaction between track and vehicle. This provides safety against derailment by controlling forces at the wheel-to-rail interface, and passenger comfort by controlling car body accelerations. Because the link between track geometry and vehicle response is so direct, it is desirable to assess track geometry based on vehicle performance. A performance-based track geometry inspection system refers to any system that can assess track geometry conditions based on dynamic vehicle response.

Examples of such a system may include the following:

- "Black Box" real-time simulation system installed on a track geometry measurement vehicle (TGMV) that can predict vehicle performance, due to the actual track geometry measured and for normal vehicle operational conditions
- Instrumented locomotive or car that measures direct vehicle response as a result of track geometry
- Postprocessing of track geometry data using vehicle/track interaction software to predict vehicle response to track conditions

A hybrid system can also be possible, incorporating some of the above examples.

8.3.8.1 Objectives and background

A major goal of performance-based track geometry systems is to detect track conditions that are at risk, but that cannot be detected by traditional measurements. Typically, these track conditions are associated with two or more track geometry conditions that are "below threshold", that is, not yet at the safety limit requiring action, but in combination, they generate derailment or ride quality risk.

Objectives of performance-based track geometry inspection are as follows:

- Locate and correct those track conditions that may produce derailment risks due to poor track geometry.
- Prioritize maintenance of track geometry exceptions identified using traditional track geometry inspection practices, based on vehicle performance.

Traditional track geometry inspection practices do not always relate to vehicle performance. Specifically, traditional inspection practices may not always identify some track geometry conditions that can cause undesirable vehicle response. For example, a past study (Li et al. 2005, 2006) showed that close to 50% of those track locations that caused poor vehicle responses may be overlooked by the traditional inspection practices.

Similarly, the same study also showed that many defects identified based on the current Federal Railroad Administration (FRA) standards did not produce poor vehicle performance for the freight vehicle types tested. This study showed that it was often the combined effect of geometry deviations, track features, operating speed, and vehicle characteristics that causes higher derailment risk.

Therefore, implementation of performance-based inspection will help railways locate and correct those track locations that may produce poor vehicle performance and potentially reduce derailment risks. In the long term, implementation of this technology may also help railways prioritize maintenance of those track geometry defects, some of which may not be as critical as others in terms of vehicle performance.

8.3.8.2 Vehicle performance measures

Vehicle performance can be defined in terms of a number of parameters that can be related to track health and derailment risks. The following vehicle performance parameters are examples used in several performance-based track geometry inspection systems:

- Wheel–rail forces: Wheel/rail forces such as vertical wheel load, lateral wheel load, and L/V ratio (lateral wheel load/vertical wheel load) are directly related to derailment risk, as well as equipment and track component degradation. However, direct measurement of wheel–rail forces requires instrumented wheelset (IWS), which can be relatively expensive as a method of regular performance-based track inspection. Therefore, wheel–rail forces are often the performance parameters predicted from a black box system installed on a TGMV or postprocessed using vehicle/track interaction software.

- Accelerations: Vertical and lateral accelerations measured on car body and truck are related to ride quality that may be associated with track conditions or truck hunting. Accelerations measured on the axle (typically near both wheels) are useful as indications of abnormal track geometry from short wavelength track issues such as rail surface defect, broken rail, and rail running surface discontinuities associated with rail joint and special trackwork. Measurement of accelerations, as compared with wheel–rail forces, is a low-cost measurement and so is often used in the instrumented locomotive or car technology.
- Other performance parameters: In addition to wheel–rail forces and accelerations, there are other measures that can be used to quantify vehicle performance. For example, vertical and lateral suspension displacements can be measured to quantify the effects of longer wavelength track geometry deviations on truck performance (such as broken spring and spring bottoming). Strain gages can be applied on various car locations to relate track conditions to specific performance issues such as top chord buckling and axle damage. Other transducers, such as those used for coupling force and brake pipe pressure, can also be installed on an instrumented freight car to monitor train operation conditions, but these performance measures are not much related to track geometry.

8.3.8.3 Types of performance-based track geometry inspection systems

There are three main types of performance-based track geometry inspection systems. One uses actual track geometry measurements and allowable speeds as input into a "black box" to predict vehicle performance. This "black box" is essentially an add-on to a TGMV, without the need for any additional instrumentation. Figure 8.50 shows how such a black box works.

An instrumented locomotive or car is another type of such technology. With this type of technology, sensors are installed on a locomotive, freight, or passenger car to measure vehicle responses directly, which are then used to assess track geometry conditions and their maintenance needs. In general, such a system is unattended and measurement results are transmitted to the office using wireless communication methods. With this technology, track inspections can be conducted frequently without affecting track occupancy. Figure 8.51 shows an example of instrumented locomotive, and Figure 8.52 shows an instrumented freight car.

A third method of performance-based track geometry inspection is postprocessing track geometry data using vehicle/track interaction simulation software. This method uses the measured track geometry to create a computer model of the track, then simulate given vehicles over the track at various speeds. Commercial simulation packages are available for this type of analysis. The advantage of this method is that it can give detailed analysis

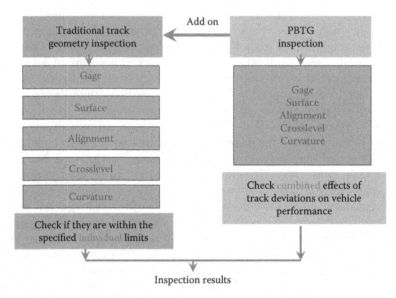

Figure 8.50 "Black Box" installed as an add-on system to traditional TGMV.

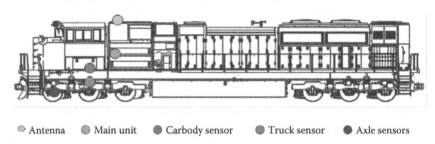

Figure 8.51 Instrumented locomotive. (Courtesy of Ensco, Springfield, Virginia.)

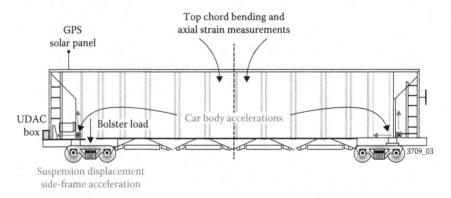

Figure 8.52 Instrumented freight car. (Courtesy of Transportation Technology Center Inc., Pueblo, Colorado.)

such as what-if scenarios, if the railroad already has access to track geometry measuring vehicle and a vehicle–track dynamic simulation software package. The drawback is that without automation it is not suited for real-time reporting. In addition, use of such a vehicle/track interaction simulation model often requires knowledge and expertise in vehicle/track dynamics.

8.3.8.4 Performance criteria

To identify whether an actual track geometry condition may generate poor vehicle performance or a vehicle response "exception" under a given operation condition, it is necessary to have vehicle performance criteria as part of the performance-based inspection system. In general, adopting or developing vehicle performance criteria is just as critical as measuring or predicting vehicle performance from actual track geometry and vehicle operation conditions.

Following are three commonly used performance criteria that can be used to reduce derailment risks in terms of wheel/rail forces. These criteria are intended to reduce derailment risks from flange climb or wheel lift.

- (L/V) max > 1.0
- V min < 10% of static wheel weight
- L max > static wheel weight

The performance criteria for instrumented locomotives and cars are typically a customized set of thresholds. Because of various types of vehicles and rail operational conditions, it is difficult to have a universal set of performance criteria for all rail operations, including freight and passenger. Therefore, appropriate thresholds are determined using statistical analysis of historical exceptions and using vehicle/track simulation software to help relate measured accelerations to other safety criteria. The ultimate goal of the thresholds is to accurately identify risk locations and have appropriate responses of the field personnel for those exceptions. Table 8.8 describes thresholds recommended by the Department of Transportation, FRA (2011) for a locomotive.

8.4 TRACK DEFLECTION UNDER LOAD

According to Hay (1982), one requirement for a stable and well-performing track is relatively uniform track stiffness that can support the applied vehicle loads with uniform track deflections of <0.25 in. The measured magnitude

Table 8.8 Example locomotive acceleration performance criteria

Measured acceleration	Performance criteria
Car body vertical acceleration	1.25 g peak-to-peak
Car body lateral acceleration	0.75 g peak-to-peak
Truck lateral acceleration	0.3 g Root-Mean Squared (RMS) over 2 s

and uniformity of lateral and vertical track deflection under load is often used as a proxy for track condition and performance, as shown by Hunt (1999) and Sussmann et al. (2001a). Track locations with large deflection/low stiffness may indicate the need for remedial maintenance to increase track strength and deformation resistance. Stationary track stiffness measurements performed at point locations can be useful to infer the strength and resistance to deformation local to the load application, while automated continuous measurements made over relatively long sections of track are useful to identify locations with soft and variable support (Sussmann et al. 2004). These two deflection measurement methods are complimentary in that continuous track deflection measurements can identify zones that should be further characterized by follow-up stationary measurements to help determine where and what kind of remedial treatments should be applied.

Vertical track deflection can be influenced by substructure conditions to a considerable depth. Therefore a large track deflection may indicate weak soil to a significant depth and the need to investigate soil conditions over a large depth to determine if a soft deforming subgrade is contributing significantly to track deflection. For a stiff subgrade, the vertical rail deflection under loading may be caused by soil strains developed over the relatively shallow subgrade depth of perhaps the top one or two meters, but this depth of influence can double or triple for soft subgrade.

Lateral track deflection and the lateral resistance of rail is a measurement of the frictional resistance at the tie–ballast interface and resistance from within the ballast immediately surrounding the ties. In general, lateral track strength is more critical to safety because it relates directly to resistance against track buckling that can occur very quickly, whereas vertical track strength is more important for long-term functional stability of track geometry rather than as a guard against impending structural failure.

8.4.1 Track stiffness/deflection measurement techniques

Techniques to measure track deflection are being implemented internationally as an indication of track stability and track condition. Measurement methods include

- Measuring the complete load-deflection behavior in a stationary load-deflection test
- Measuring changes in the slope of the load-deflection curve
- Measuring track deflection under moving load

Several testing methods have been developed for testing track stiffness, including the Track Loading Vehicle (TLV) testing method (Li et al. 2002b, 2004) by the Transportation Technology Center, Inc. (TTCI), a subsidiary of the AAR, the method by the University of Nebraska (Lu 2008; Farritor and Fateh 2013), and the method by Banverket (Berggren et al. 2006).

8.4.1.1 Track stiffness testing using track loading vehicle

TTCI has developed an automatic technique for vertical and lateral track stiffness testing using its TLV, as shown in Figure 8.53. The TLV test train includes an instrumentation coach, the TLV, and an additional empty tank car (see Figure 8.54). The tank car is required only for vertical track stiffness testing, providing a platform for measuring the unloaded vertical rail profile. In the case of lateral stiffness testing, the unloaded lateral rail profile can be measured within the length of the TLV.

Figure 8.53 TTCI's Track Loading Vehicle (TLV). (Courtesy of Transportation Technology Center Inc., Pueblo, Colorado.)

(a)   (b)

Figure 8.54 (a) Unloaded profile vs. (b) loaded profile for vertical track stiffness testing using TLV. (Courtesy of Transportation Technology Center Inc., Pueblo, Colorado.)

The TLV was built with a high-stiffness load structure (car body) and is equipped with a fifth wheelset (load bogie) mounted underneath the vehicle center. The load bogie is suspended from the car body and operated by computer controlled, servo-hydraulic actuators. As a result, this wheelset can apply the required combinations of vertical and lateral wheel forces on the rails for various modes of track stiffness testing.

The TLV can be used in both stationary test mode or in continuous measurement mode. Continuous vertical and lateral track stiffness tests are two separate tests that cannot be conducted during the same test run. Nevertheless, both testing techniques were developed based on the similar measurement methods. During an in-motion test either laterally or vertically, the TLV applies a set of test loads using the center load bogie. To determine lateral or vertical track stiffness characteristics, the TLV consist is instrumented to measure lateral or vertical track deflection responses via two sets of measurements: a loaded profile and an unloaded profile.

The loaded profile, whether lateral or vertical, is always obtained under the TLV load bogie. In the lateral track stiffness testing, the unloaded profile is also obtained within the length of the TLV. This is because the wheels from the TLV leading or trailing bogie generate little lateral wheel loads, thus, allowing the measurements of unloaded lateral profiles within the length of the TLV.

However, for the vertical stiffness testing, this unloaded profile comes from the empty tank car trailing the TLV. This is because the 138 kN (31 kip) vertical wheel loads under the TLV end bogies are sufficient to create a basin that prohibits an existing (unloaded) profile measurement from being taken within the length of the TLV. The weight of the empty tank car, however, exerts only a 62-kN (14-kip) load at the end bogies, creating a smaller basin and allowing for a no-load vertical profile measurement under the center of the tank car. The tank car is also equipped with a center load bogie, much like the TLV. However, this bogie is equipped with pneumatic actuators and is only capable of applying a nominal wheel load of 9 kN (2 kips). This load is used to assure wheel/rail contact at all times.

To quantify loaded or unloaded profiles, noncontact laser/camera sensors are installed on two independent but identical reference frames that are attached to the car body in a way allowing their movements relative to the car bodies. One reference frame is used for the loaded profile while the other is for the unloaded profile.

Three distance measurements are taken from each reference frame to the rail, allowing loaded or unloaded profile quantified by these chordal offset measurements. That is, calculated offsets are used to quantify the rail profiles. Figure 8.55 shows how an offset is calculated. The continuous offsets determined with no load yield the existing or unloaded profile, while the continuous offsets obtained directly under the test loads yield the loaded profile.

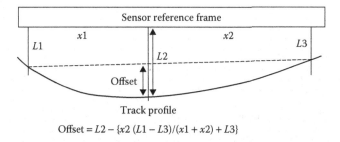

Offset = L2 − {x2 (L1 − L3)/(x1 + x2) + L3}

Figure 8.55 Offset calculation.

A synchronized subtraction between the loaded profile and the unloaded profile is thus the track deflection due only to the TLV test loads. For a given set of constant but moving test loads, it is obvious that larger deflections correspond to lower track strength or stiffness.

During an in-motion test, the TLV travels at 30 km/h to maintain the constant test loads. The accuracy of deflection measurements is 0.25 mm (0.01 in.), which is sufficient to cover the common range from 0.5 to 10 mm (0.02–0.4 in.) of lateral or vertical deflections generated by the TLV test loads. As discussed, these deflection measurements are offset-based measurements; thus, they are relative measure of track deflections under the test loads.

8.4.1.2 University of Nebraska track deflection measurement

Researchers at the University of Nebraska (Lu 2008; Farritor and Fateh 2013) developed a continuous vertical track deflection measurement from a moving railcar (Figure 8.56). The system consists of a loaded hopper car with a camera/laser sensor system to detect the vertical position of the rail relative to the wheel/rail contact point 4 ft away. This measurement system has been shown to be sensitive to most track support-related features of interest to railroad maintenance and inspection, including tie failure, ballast condition, and weaknesses due to poor rail welds that can reduce the bending stiffness of the rail and cause increased deflection and increased track maintenance if not addressed.

In Figure 8.57, the sensor head is the measurement point 4 ft from the wheel contact location.

The sensor head consists of two lasers and a video camera system. The two line lasers are installed at an oblique angle such that the lasers create two lines crossing the rail head. The camera records this image and digitizes the image to compute the distance between the lines. The distance between the lines is geometrically related to the deflection of the rail head, with the closer the spacing of the lines indicating the nearer the rail is to the camera. The sensor head is rigidly attached to the side frame of the truck such that

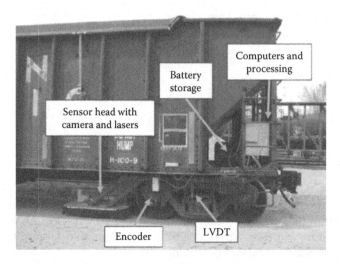

Figure 8.56 UNL track deflection system instrumentation. (Farritor, S. and Fateh, M., *Measurement of Vertical Track Deflection from a Moving Rail Car,* Report No. DOT/FRA/ORD-13/08, Federal Railroad Administration, Washington, DC, 2013.)

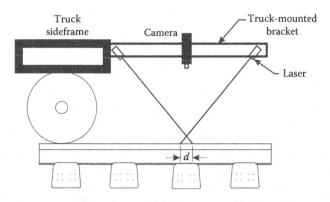

Figure 8.57 Measurement concept for UNL track deflection system. (Farritor, S. and Fateh, M., *Measurement of Vertical Track Deflection from a Moving Rail Car,* Report No. DOT/FRA/ORD-13/08, Federal Railroad Administration, Washington, DC, 2013.)

it tracks the position and inclination of the truck. In this way, the system creates a three-point chordal measurement with the measurement points of the chord being the two axles spaced 6 ft apart and the measurement projecting 4 ft past the closest wheel–rail contact. When combined with similar track geometry chordal measurement referenced to the loaded profile, this measurement system can be used to calculate vertical track deflection and to estimate track modulus.

The short end-chord offset distance provides a localized view of track support allowing individual joint, tie, and rail support anomalies to be identified. This detailed assessment shows a high degree of variability of track deflection. As track deteriorates, the continuously supported structure degrades into a discretely supported structure that will increase loads on individual components (Farritor 2013).

The estimation of track modulus from highly variable track deflection data can be accomplished by averaging the data over some length. In general, a length of 100 ft has been used to estimate track modulus. However, in highly variable track support zones, it might be necessary to account for singular events that can shift the average without compromising track response.

This single-point, single-load measurement of deflection does not provide for the separation of contact or seating deflection as some other techniques may. But, the overall measure of deflection provides suitable data available for assessment of both localized variations in track stiffness that can affect track safety and performance, while providing suitable data for assessing global variations in track support to provide an estimate of the magnitude of track modulus.

8.4.1.3 Rolling stiffness measurement vehicle—Banverket

The track stiffness technique implemented on the Banverket Rolling Stiffness Measurement Vehicle (RSMV), as shown in Figure 8.58 (Berggren et al. 2006), uses an oscillating mass to load a common railway axle. In this measurement scheme, the oscillating load is used to define the load-deflection curve of the track. For each load oscillation, a portion of the load-deflection curve is obtained. When the load oscillation is sufficiently fast relative to the vehicle speed, the load-deflection curve is obtained over a small track length. In this manner, the slope of the load-deflection curve is estimated providing a direct measure of the track stiffness on a continuous basis.

(a)   (b)

Figure 8.58 The Banverket rolling stiffness measurement vehicle (RSMV): (a) rolling stiffness measurement vehicle; (b) oscillating mass mechanism. (Courtesy of Eric Berggren of Eber Dynamics, Falun, Sweden.)

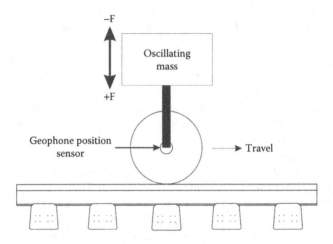

Figure 8.59 Schematic of RSMV measurement concept.

The RSMV instrumentation is partially shown in the schematic (Figure 8.59), including a load sensor and accelerometer. As the mass oscillates, the load varies and the load transmitted to the track and the deflection are measured. The accelerometer is essentially a geophone that is used to double integrate acceleration to position for the purpose of deflection measurement. One potential problem with this system is the lack of reference to the unloaded track position to obtain an estimate of the overall track deflection.

This measurement technique not only provides a static load-deflection curve for the range of loads measured but also provides spatial data that allows spectral analysis of the data. For instance, the constant driving force from the oscillating load will result in a measured delay in peak deflection that is dependent on the track damping. In this regard, the Banverket system provides a unique set of data that can provide much more insight to track support than any pure track deflection system.

8.4.1.4 Falling weight deflectometer

The falling weight deflectometer (FWD) is a method commonly used for testing highway pavement and airport runway. It is a device that applies a transient force impulse to measure the pavement surface deflection and deflection basin shape. The falling weight is intended to represent the magnitude and duration of a moving wheel load applied on the tie. The surface deflection magnitude is measured using velocity transducers located at increasing distances away from the point of load application. A back-calculation process is then used to determine the stiffness-related parameters of substructure layers to create the observed deflection magnitude and basin shape. The FWD is a stationary measurement, where one load impulse and the resulting deflection

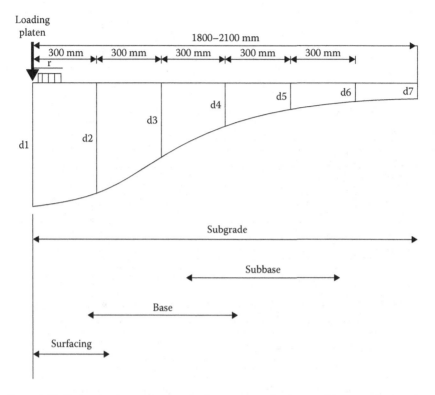

Figure 8.60 Deflection basin shape and influence zones for base, subbase, and subgrade. (Data from Christopher et al., 2006.)

measurements are measured at one location before moving the test setup down the track to the next test location. Figure 8.60 shows the deflection basin and influence zone depths.

The FWD data has also been used as a measure of the critical velocity in the subgrade. The critical velocity is often taken as the Rayleigh wave velocity of the soil, which represents a velocity over which traffic above may resonate with subsurface waves and amplify deflection of the track. The FWD can measure this response directly to evaluate the potential for excessive track vertical vibrations due to high-speed traffic approaching the Rayleigh wave velocity in soft soil.

8.4.1.5 Spectral analysis of surface waves

The spectral analysis of surface waves (SASW) technique is an emerging technology for track support measurement. SASW provides a unique measure of layer interfaces and associated stiffness (modulus) for each individual layer. The technique is based on Rayleigh wave dispersion (Figure 8.61),

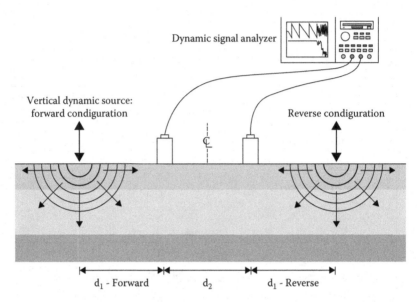

Figure 8.61 SASW test setup. (Data from Wightman, W. et al., *Application of Geophysical Methods to Highway Related Problems*, Report No. NHI-05-037, Federal Highway Administration, Washington, DC, 2003, http://www.cflhd.gov/resources/geotechnical/documents/geotechpdf.pdf.)

where a seismic wave source is needed to generate the seismic waves that are analyzed in this technique. Either transient (sledge hammer impact) or continuous source (oscillating mass) provides the necessary data, and the return signal is monitored by geophones. The recorded data are used to measure the phase change in the signal between geophones and to develop a dispersion curve. The dispersion curve is the basis for developing a shear wave velocity profile and associated layer stiffnesses (Yuan et al. 2000; Wightman et al. 2003).

This technique is emerging as a possible method to identify and quantify poor subsurface layers (Figure 8.62). Dumas et al. (2003) describes the use of advanced techniques from Europe for geotechnical aspects of roads and a railway line, indicating the use of SASW to verify field conditions.

8.4.2 Vertical load-deflection behavior

In some cases, excessive vertical track deflection under load may be due to void space between the tie bottom and ballast, or between the rail and tie, rather than a soft subgrade. For purposes of selecting the proper remediation method, it is important to determine if soft subgrade is the cause. Conceptually, it is possible to measure track deflection under load in such a way as to distinguish between voids and soft subgrade.

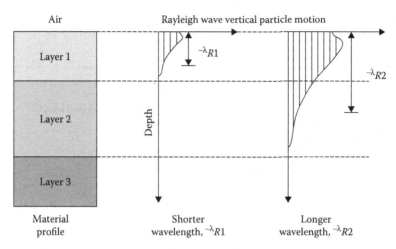

Figure 8.62 SASW wavelength relationship to material type and thickness. (Data from Wightman, W. et al., *Application of Geophysical Methods to Highway Related Problems*, Report No. NHI-05-037, Federal Highway Administration, Washington, DC, 2003, http://www.cflhd.gov/resources/geotechnical/documents/geotechpdf.pdf.)

For the idealized case of track in excellent condition, the track load-deflection behavior is nearly linear and only two measures are required to characterize the track: the test load and the total track deflection, as shown by "linear" relationship in Figure 8.63. The slope of this load-deflection line is the track stiffness, k (units of force/length).

For track that has been in place for a considerable time and developed voids between the tie bottom and ballast and between rail and tie, track deflection with load has an approximate bilinear appearance, as shown by the nonlinear curve that has been broken into two linear portions to illustrate the distinct behavior in both portions. The slack or void is taken up by a relatively small seating load that acts to compress the tie against the ballast. The contact load is the force added to the seating load that compresses the substructure layers beneath the tie. Track stiffness and modulus are calculated from the contact load and the resulting incremental track deflection is shown by the upper slope of the nonlinear curve (Selig and Li 1994; Ebersohn and Selig 1994). To ensure that the seating load is large enough to remove the voids, it should be no less than 10% of the full load, but need not be greater than 25%.

The seating load theoretically should be the lightest load necessary to eliminate all voids resulting in all track components being in contact, so that the contact load then provides compression of the layers under the tie. In practice, this requires a seating load of several thousands of pounds. The void is a measure of the nonlinearity of the load-deflection curve, and is

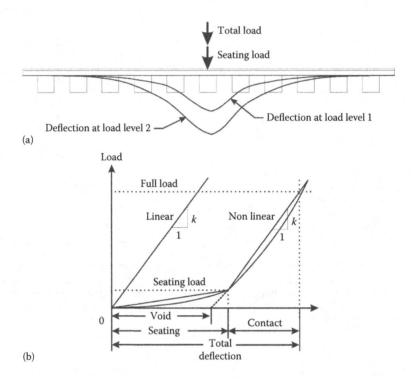

Figure 8.63 Track deflection basin and track stiffness definitions: (a) deflection basin; (b) load-deflection trends. (Data from Ebersohn and Selig, 1994b.)

measured as the deflection value at the point where the extrapolated slope of the track stiffness line intersects the deflection axis (Selig and Li 1994). In this way, the void is a measure of the track deflection that theoretically occurs with no load-carrying capacity.

As stated, although these deflection measurements can provide useful information if only a few are performed on a stationary basis, measuring while moving is better at finding locations of soft subgrade support. Figure 8.64 shows an example of this using TTCI's TLV test method where deflection profiles between two track sections show one to be strong or stiff and the other having less strength or stiffness. To validate the in-motion deflection profile results shown in Figure 8.64, a few static or stationary tests were conducted in these two sections. Figure 8.65 shows the static load-deflection results, corresponding to the stiff and weak areas indicated in Figure 8.64. As illustrated, the section with less strength showed much lower track modulus (approximately 8 MPa or 1,100 lb/in./in.) than the stiff section (65 MPa or 9,400 lb/in./in. track modulus); that is, the static track modulus test results were consistent with the in-motion track profile results.

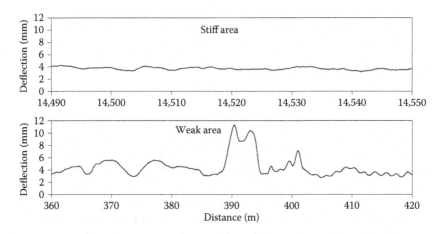

Figure 8.64 In-motion vertical track deflection test results in stiff and soft track.

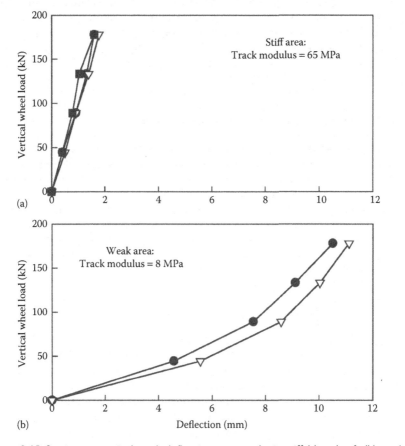

Figure 8.65 Stationary vertical track deflection test results in stiff (a) and soft (b) track.

The characteristic of the curve in the lower graph of Figure 8.65 is indicative of a soft subgrade because of the relatively large track deflection under the contact load. If the cause of the large track deflection in the weak area shown in Figure 8.64 was due to a void rather than soft subgrade, the curve in the lower graph of Figure 8.65 would have a large deflection due to the seating load but a much smaller deflection from the contact load than shown.

The strategy is then to first measure track deflection to find locations with soft support using the continuous in-motion method to obtain data, as shown in Figure 8.64, followed by stationary track stiffness measurements at soft support locations to obtain more detailed data, as shown in Figure 8.65, to distinguish between a large track deflection caused by voids or by a soft subgrade. As a follow-up to the finding of soft subgrade, the CPT could then be used at these locations to measure subgrade stiffness with depth so that remedial methods could be designed to produce a stable track structure.

Although vertical track deflection measurements can provide useful information, in practice, it can be very difficult to clearly separate deflection due to voids and soft subgrade, because voids may contribute to the deflection even at the largest loads that may give a false indication of soft subgrade. Despite the uncertainty in these measurements, there is value in identifying locations of large track deflection that require attention, be it from large voids, a soft subgrade, or both.

8.4.3 Lateral load-deflection behavior

The goal of measuring lateral track resistance may be to compare with other reference values to determine if the strength is adequate, or it may be used in analysis for determining lateral stability. Lateral track strength is usually more safety critical than vertical strength due to the very sudden nature of track buckling. While only elastic, resilient deflection is required to characterize vertical track support, track lateral resistance must account for resilient and permanent lateral deflection of a tie or a group of ties. This is because lateral track strength can only be assessed by pushing ties beyond their elastic displacement range. An example of a single tie lateral push test measurement is shown in Figure 8.66.

Note that the peak resistance in this example is 3,500 lb corresponding to a deflection of approximately 0.2 in., which is beyond the elastic limit. Rather than using the peak resistance as the lateral tie strength, the resistance at 0.10-in. displacement (3,000 lb in this example) is a more conservative choice.

As with vertical track deflection, lateral deflection measurements made on a continuous, automated basis is a more efficient means of finding those locations with weak lateral support. Unlike the stationary testing for individual ties, the automated process uses lateral loads that produce deflections that do not significantly exceed the elastic limit. If weak lateral track locations are found using the automated method, stationary single tie

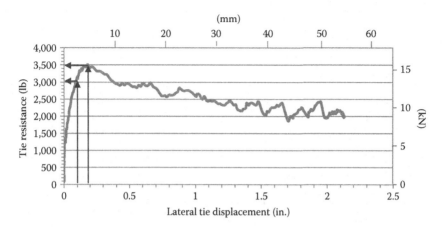

Figure 8.66 Single tie lateral push test example data.

lateral push tests could be performed as a follow-up to quantify the lateral track strength over the load range to failure.

As shown by test data in Figure 8.67, the automated technique using TTCI's TLV can easily reveal weak areas, which show much higher deflections than the rest of the test track. To confirm these in-motion test results, three stationary load-versus-deflection tests were also conducted at selected locations: one corresponding to strong track, the other two corresponding to weak spots. As shown in Figure 8.67b, the stationary test results were consistent with the in-motion test results. That is, locations D and E showed lower strength than location C.

In another revenue service testing, the effects of dynamic ballast stabilization on newly surfaced tracks and/or replaced wood ties were measured. Figure 8.68 shows an example of the lateral deflection profile over a stabilized zone (using a ballast stabilizer) and a control zone (nonstabilized) in a newly surfaced track. As shown, the mean deflection on nonstabilized track was approximately twice as much as on the stabilized track.

This increase in deflection (in response to a moving load) tends to exaggerate strength differences. Therefore, several stationary tests were performed to document the track strength differences. These showed that the stabilized strength was measured to be 104 kN (23.4 kips or approximately 85% of baseline). The average nonstabilized strength was measured to be 77 kN (17.3 kips or 63% of baseline). The baseline strength was the nominal strength of an undisturbed section of track.

Track is most resistant to lateral movement when the ballast is well compacted around the ties, a condition provided by traffic loading under normal operation. However, because tamping disturbs the ballast and may reduce lateral resistance by as much as half that existed before, ballast compaction must be at least partially restored before traffic can resume normal speed.

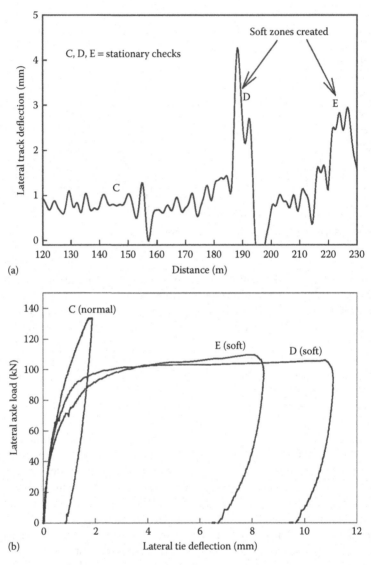

Figure 8.67 (a) Continuous lateral track stiffness test at TTC test track; (b) stationary force versus deflection at strong and weak locations.

Railways typically use mechanical ballast compaction, slow orders for resumed traffic, or both, following maintenance-induced ballast disturbance. Research (Hering 1989) has provided an indication of the contribution to ballast recompaction from traffic under a slow order, from a dynamic track stabilizer, and from traffic following both of these, as shown in Figure 8.69. The Figure 8.69 shows that the ballast compaction and lateral resistance

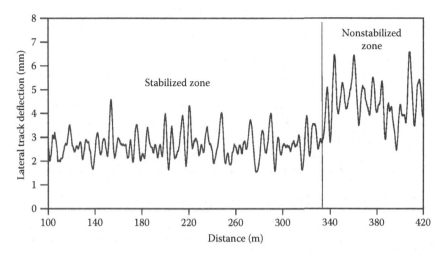

Figure 8.68 Continuous lateral deflection profile over tamped track: stabilized versus nonstabilized.

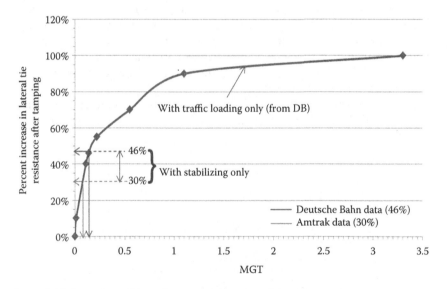

Figure 8.69 Rate of track lateral resistance increase after maintenance.

was restored to its previous level prior to maintenance after approximately 3.4 MGT. While this amount of cumulative loading may occur quickly for a heavy haul freight railroad, it may require a very long time for the lighter loading on a passenger railroad. Therefore, dynamic stabilization is very attractive for a passenger operation. The amount of increase in ballast lateral resistance and compaction from a stabilizer is shown to be between

30% (Sussmann et al. 2003b) and 46% (Hering 1989) of the value before maintenance.

The return of only 30%–46% of the ballast lateral strength that existed before maintenance disturbance may seem insufficient to allow traffic to resume normal speed. However, the initial, rapid increase in ballast compaction from dynamic stabilization, as shown in Figure 8.69, provides a large amount of lateral resistance. Additionally, there are other sources to resist track buckling and lateral track shift that contribute to resist lateral instability such as the rotational shearing of the tie to rail, and the longitudinal resistance of the tie and ballast. The group action of the ties rather than individual tie strength also provides an additional margin of safety.

8.4.4 Track buckling and track panel shift

Track strength, stiffness, and resistance measurements discussed previously are important to prevent track buckling and track panel shift. Track may buckle laterally in response to a heat-induced buildup of longitudinal compressive stress in the rails. Buckling usually occurs rapidly as it produces a sudden release of energy, resulting in a lateral track movement in the order of inches or perhaps several feet. Two examples of buckling are shown in Figures 8.70 and 8.71. Analytical models of track buckling allow the user to determine the maximum allowable rail temperature above the rail-neutral temperature (stress-free temperature) at which buckling begins to be a concern for a given set of track and loading conditions, as specified by the user.

The resistance offered by the ballast at the tie–ballast interface is crucial to determining lateral track strength and is a key input parameter in

Figure 8.70 Example of track buckle over a short distance.

Figure 8.71 Example of continuous track buckling over extended track length.

buckling analysis. The buckling model from Volpe (1996) requires the peak lateral tie–ballast resistance as discussed, the tie–ballast friction coefficient, and the track foundation modulus. If the buckling analysis is for the case of newly tamped or otherwise disturbed ballast, then these parameter values must reflect this changed condition.

Another type of lateral track instability is known as track panel shift. While buckling is a short time duration/large deformation event, track panel shift is a gradual lateral creep of the track in small increments under repeated loading, which can be partially driven or accelerated by heat-induced longitudinal compressive forces in the rail. As railways trend toward heavier axle loads and higher train speeds, vehicle-induced forces may lead to the loss of track lateral stability under unfavorable track conditions. The resulting misalignment, when combined with other adverse conditions, may also lead to track buckling.

Although the amount of lateral track movement with panel shift may be less than that occurs with buckling, panel shift is also a safety concern because this progressive lateral shift may eventually exceed the allowable lateral geometry error size under safety regulations. Volpe (2001) developed the *Track Residual Elastic Deflection Analysis* (TREDA 2001) model to assess the potential for this accumulation of incremental lateral track permanent deformation under dynamic loading.

An example of track panel shift due to the inability of the track to resist the imposed lateral loading is shown in Figure 8.72. Note that in this example heat was not required to play a role in panel shift due to the cold temperature at the time of lateral instability.

Panel shift is resisted by the same forces that resist track buckling. As with the buckling model, key inputs to TREDA are the peak lateral tie–ballast

Figure 8.72 Example of track panel shift in cold weather.

resistance, the tie–ballast friction coefficient, and the track foundation modulus, and degraded values of these must be used if the user is modeling newly tamped ballast conditions.

Figure 8.73 shows several critical lateral loads determined using various vertical axle loads, based on the tests conducted over consolidated ballast (Li and Shust 1997). As shown, the critical lateral loads are 15%–30% higher for concrete-tie track than for wood-tie track. The following equation gives an approximate relationship between the critical lateral load and vertical axle load for consolidated ballast and wood ties:

$$L_c = 6 + 0.5\ V \tag{8.6}$$

where:

L_c = critical lateral axle load (kips)
V = vertical axle load (kips)

For comparison, the critical lateral loads predicted by the Prud'homme criteria are shown as a dashed line. The Prud'homme criteria incorporates a

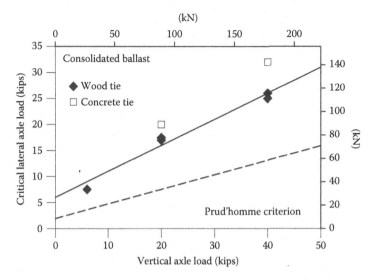

Figure 8.73 Critical panel shift loads.

safety factor of 0.85 and is written as $L_c = 0.85 \, (V/3+10)$ where all units are in kN.

Equation 8.6 indicates that the vertical axle load has a significant effect in preventing panel shift. A decrease in vertical axle load will result in a decrease in the critical lateral load.

Chapter 9

Management

It is often said that "one cannot manage what one cannot measure." This chapter discusses how a manager/engineer can be guided by measurements of track and its substructure, as described in Chapter 8, so that he or she can interpret and make decisions based on the collected data. The initial sections of this chapter discuss cost considerations in track design, track construction types, and remedial trackwork. Then management of the ballast layer and track maintenance needs as indicated by the substructure functional and structural condition are discussed. This is followed by examples of substructure maintenance and the equipment used, then examples of substructure problem management, and then methods of subballast and subgrade rehabilitation. The final section discusses how to use the observational method to manage track when there is a lack of available measured data to base engineering decisions on but when action must be taken nevertheless.

9.1 LIFE-CYCLE COST CONSIDERATIONS

In his pioneering writing on engineering economics in *The Economic Theory of the Location of Railways*, A. M. Wellington (1887) defined engineering as "the art of doing that well with one dollar what any bungler can do with two." Engineering decisions must consider economics. In particular, using life-cycle cost (LCC) analysis can ensure that the most cost-effective approach is taken when selecting among new track construction designs, track components, or track maintenance methods, by avoiding the common mistake of basing such decisions only on the purchase price of a product or on the initial, "up-front" costs of a project. Rather, the goal is to minimize cost over the life of the track component or over the life of the track as a whole. By considering the time value of money and the timing and dollar amount of required present and future expenditures for each option, LCC analysis allows the least cost option to be chosen.

For example, LCC can be used to decide between the track construction types of (1) ballast-less track, also known as concrete slab track, (2) wood or concrete tie ballasted track with a layer of asphalt under the ballast, and (3) wood or concrete tie conventional ballasted track. The first two construction types require a greater initial investment but future maintenance costs should be relatively low compared with conventional ballasted track, which can be expected to have lower initial construction costs but higher future maintenance costs. The most cost-effective choice between these types of track construction typically will not be obvious without an LCC analysis, which includes consideration of the allowable geometry error, load environment, and expected future maintenance requirements. For higher speeds and smaller allowable geometry error as with high-speed rail, the cost advantage may shift to track construction that is initially more expensive but requires less-frequent future maintenance so that the costs of service disruptions and train delay are markedly reduced.

Typically, the costs of a project accumulate as shown in Figure 9.1, where the initial efforts of planning, design, and construction accrue at an accelerating pace, and once in service, this is followed by a relatively linear increase of maintenance expenditures for most of the track life. But toward the end of this life the costs typically increase at an exponential rate as track components fatigue and require progressively more maintenance. Due to this "backend-loaded" nature of cost distribution, at least half of the overall total cost for the project may occur during this final period, which may only represent 20% or less of the total life duration. It is possible that a more cost-effective approach is provided by spending somewhat more during the initial phases if this results in a net savings from a reduction in costs toward the end of track life. An economical track design must recognize this distribution of expenditures over its life and account for it in the LCC.

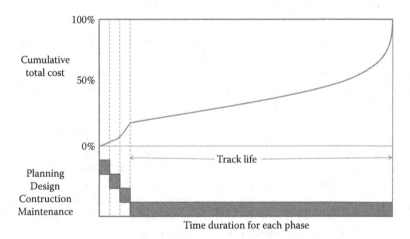

Figure 9.1 Example distribution of costs incurred during life of track. (Data from Ebersöhn, Track maintenance management philosophy, 1997.)

Very often this increased rate of required maintenance in the later stage is not recognized or accounted for by the initial LCC analysis, making the project much more expensive than first thought or planned for.

The cost of capital to the railroad is expressed by the discount rate (or interest rate), and this has a considerable effect on the calculated LCC. The discount rate available to railroads can be on the order of 10% or more, and this relatively high rate tends to make the future costs in Figure 9.1 less significant. However, for track construction projects that are in some way publicly funded or financed based on the relatively low interest rate at which the government can borrow, such as in a "public–private partnership," the distribution of these future costs would then weigh more heavily in the LCC calculation.

In any case, a proper LCC analysis requires that all track design options be considered and the costs of all initial construction activities and future maintenance associated with each design option be quantified. The following sections explore these options and considerations.

9.1.1 Designing track to minimize settlement versus allowing settlement and geometry corrections

Conventional ballasted track construction can be based on one of two basic strategies: (1) construct with substructure layers that are relatively thick and strong for durability, which not only requires a larger proportion of initial spending but also results in less settlement and so less-frequent future geometry corrections, and (2) construct with reduced layer thickness to save on the initial costs but spend more in the future to compensate for the increased amount of geometry correcting maintenance (the so-called "build now, maintain later" approach). The first strategy minimizes track settlement over the track life while the second allows more track settlement due to the subgrade being overstressed and so rely more on maintenance as needed to correct geometry. As a general rule, European track construction practice has tended to follow the first strategy, resulting in reduced need for geometry correction due to for example, removal of potentially deformable natural soil and replacement with strong and well-compacted select material. In contrast, the typical track in the United States follows the second strategy of minimizing initial construction costs for example, by using a thinner granular layer resulting in increased subgrade stress, deformation, and required geometry correcting maintenance.

Conventional ballasted track construction in the United States often uses uniform, nominal layer thicknesses of approximately 12 in. (300 mm) of ballast and 6 in. (150 mm) of subballast for multiple miles of track. But choosing these layer thicknesses because they are "industry standard" without regard to whether the subgrade will be overstressed for the loading applied cannot be referred to as a design approach; it is merely following a practice. For tracks with reasonably strong subgrade soil, the practice of using a combined ballast and subballast granular layer of approximately 18 in. (450 mm) could be adequate, but locations with soft subgrade may experience significant track settlement and geometry roughness depending on the amount of loading. By using a uniform granular layer thickness that is under-designed, the assumption is being made that it is more economical to correct track geometry later when and where required, rather than to identify and improve on these conditions before construction. This assumption should not be made without first considering its implications to LCC due to the distribution of costs over the life of track, as shown in Figure 9.1.

Whether the engineer chooses the "build now, maintain later" approach with minimal substructure layer thickness or the thicker, more durable substructure layer option, the selection of the granular layer thickness should be based on something more than merely using an industry standard. A more informed choice is to design based on measured subgrade conditions to minimize the chance of being over- or under-conservative. Based on these measured subgrade conditions, the engineer may choose to (1) use a uniform granular layer thickness based on "average" subgrade strength as measured at locations along the entire construction length or (2) use a thickness that varies over track segments based on grouping the subgrade support into "design units," as explained later in this chapter. The first option allows traffic loading to eventually identify locations of weaker than average subgrade so that the needed remediation or recurring geometry corrections can be applied locally. Although this option will result in some locations with under-designed granular layer thickness, this can be accounted and planned for in the economic analysis because the design thickness was based on a soil investigation, which is more of a controlled approach than arbitrarily choosing this thickness without regard to how much of the track will be under-designed. The second option is a more tailored-to-conditions approach of selecting granular layer thicknesses that vary with subgrade strength, as described in the next paragraph. For either option, Table 8.1 provides recommended soil sampling interval spacing along the proposed right-of-way based on soil strength.

Highway pavement design sometimes uses this "design unit" approach where an investigation characterizes soil strength distribution over long distances so that pavement thickness is proportioned accordingly. The investigation may show that the subgrade for one mile is predominately soft clay, the next mile is moderately stiff clay, and the third mile is stiff granular soil. These three segments may constitute three different design thicknesses for the pavement. By using this design unit approach for new track construction, the granular layer thickness can be proportioned corresponding to the variable subgrade strength. This variable thickness design would require a larger initial expenditure to perform the investigation and the follow-up of design work, but a LCC analysis may show this option to be favorable if it results in lower overall costs due to less need for future corrective maintenance.

The practice of choosing a granular layer thickness that is somewhat under-designed over softer soils is probably not suitable for high-speed track with its smaller allowable geometry error. However, this approach could be adequate for slower freight service with its greater tolerance of geometry error. The foundation for high-speed track is expected to require a more rigorous design approach, as discussed subsequently.

9.1.2 Track renewal with reconstruction versus limited renewal of track components

LCC analysis can help in the decision between total replacement (rails, ties, and ballast) of a degraded track with all new construction and a limited upgrading of its condition by replacing a selected track component such as only the rails, only the ties and fasteners, or only the ballast. A key element to be quantified for LCC analysis is the life of the item being assessed, and the best method to determine the life of the rails, ties, and ballast is one that uses a mechanistic approach to determine when the track component will fail and must be replaced. A mechanistic approach to quantify ballast life based on its rate of mechanical breakdown and fouling under repeated loading has been described by Chrismer and Selig (1994). These mechanistic models can be used to determine the life of the rails, ties, and ballast when these components are new or determine the remaining amount of life when they are partially degraded through use. The remaining life of a track component can be determined by measuring the current wear and knowing the cumulative amount of loading to the present, projecting out to the time at which the life would expire from continued loading. However, even if the life of the rails, ties, and ballast is known, it is difficult to perform a cost analysis for the track as a whole when various track components have lives of different duration. To resolve this problem, the LCC analysis is often determined not over track component life but over a fixed "time horizon" of perhaps 15 years or more. This approach provides the necessary consideration of the time value of money without needing to define an effective or composite life of all track components combined.

If total track reconstruction or a limited track renewal requires remediation of poor subgrade before the trackwork can begin, the cost of this must be considered in an LCC analysis as an initial "up-front" cost. It is likely that this expenditure will be somewhat compensated by a reduction in future recurring geometry correcting maintenance due to reduced subgrade deformation.

The LCC of upgrading or rehabilitating an existing track by renewing only selected components may be much less than the cost of track renewal with new construction, provided that such a limited upgrade is sufficient to bring the track to the level of service required. If the track has developed a large amount of geometry error, it might not readily be brought back to the smooth condition it had when new without a more extensive rehabilitation. For example, poor track geometry is often difficult to remove by normal maintenance procedures such as tamping when the ballast is highly degraded or when drainage is poor and the subgrade is soft. In such a case, replacing only the rails or the ties is not likely to be sufficient, and replacement of the track with new construction to the depth of the subgrade may be the best option. Again, this can be determined using LCC analysis. The sources of poor track condition must be understood when using this analysis so that the rehabilitation cost is based on the appropriate remedial method. For example, if the instability is due to a soft and deforming subgrade, then the LCC analysis must consider the cost of removing the existing track and stabilizing the subgrade over the affected track length, or other such method that addresses the nature of the problem.

The conventional ballasted track construction discussed to this point is well suited to most environments and load conditions, provided that the amount of track settlement that is characteristic of this type of track is tolerable and that the cost and service disruption of recurring cycles of geometry correcting maintenance over the life of track are acceptable to the operation. For freight and slower-speed passenger traffic, ballasted track is often the preferred choice. However, when speeds increase and the allowable geometry error decreases, an LCC analysis can show if using ballast-less track (slab track) construction provides a lower cost, as discussed in Section 9.1.3.

9.1.2.1 Track laying machines and track subgrade renewal machines

Although different terms are sometimes used to describe it, the track laying machine (TLM) will be referred to here as the equipment capable of removing and replacing the track superstructure components of rails and ties. The machine capable of renewing the track substructure layers of ballast and subballast down to the level of the subgrade will be referred to here as a track subgrade renewal machine (TSRM). (The ballast undercutter that removes and replaces only the ballast layer is discussed separately in a later section.) Both these types of machines operate in a continuous mode, as shown by the TLM in Figure 9.2.

During superstructure renewal with a TLM, the existing rails may be left in track while only renewing the ties, or new rails may be threaded in to replace the old, as shown in Figure 9.3. A shoulder cleaner/excavator can also be helpful in advance to remove shoulder ballast allowing a lateral path for, and reducing the amount of, ballast to be plowed away from the track centerline by the TLM. The TLM continuously advances along the track while removing old ties individually, which are then placed on a conveyor belt and collected by a gantry crane that shuttles along the length of the TLM. This crane places the old ties in holding cars that are in the TLM consist, but they are later separated from the consist for tie disposal. Behind the location where old ties are removed is the plow that pushes old ballast away to provide a flat excavated ballast surface for placement of new ties behind the plow. Ballast from the crib

Figure 9.2 Track laying machine in operation. (Courtesy of Amtrak, Philadelphia, Pennsylvania.)

Figure 9.3 Close-up view of TLM placing new concrete ties and new rail. (Courtesy of Amtrak, Philadelphia, Pennsylvania.)

and approximately 2 or 3 in. (51 or 76 mm) beneath the tie bottom is plowed away and later replaced with new ballast. If new rails are to be placed, the old rails are continuously lifted and pushed laterally so that new rails, placed outside the gauge before TLM operations, can be brought in under the old (see Figure 9.3) and later fastened to the new ties behind the TLM.

Behind the TLM, new ballast is placed in the cribs and subsequent tamping brings the track to its desired elevation. Because plowing removes only the upper few inches of ballast below the tie bottom, a large portion of the existing ballast layer thickness will remain, and if fouled, will not provide sufficient drainage or support to the track. Further, the fouling material in the old lower ballast may mix into the new ballast in a short time, making the life of the new ballast placed far less than would occur with full-depth undercutting, which typically provides a new ballast layer of approximately 12 in. (300 mm) under the tie. Therefore, although the TLM can improve ballast drainage conditions by getting the ties out of fouled ballast and placing them into a thin layer of clean ballast at least for some time, the TLM is not a substitute for ballast undercutting.

If a limited track renewal using the TLM is being considered where the options are to replace (1) only the rails, (2) only the ties, or (3) both rails and ties, an LCC analysis provides a way to choose between these options. If ballast renewal is also needed, then the LCC analysis should consider the case of using a TLM followed by an undercutter.

Rather than considering the economics of renewing the individual components of rails, ties, and ballast separately, European practice has been increasingly to use the approach of total track renewal to lengthen the service life of the entire track as a whole. The TSRM is used to renew the substructure down to the subgrade level by using two, and sometimes three, excavating chains: the upper chain removes the ballast layer as a typical undercutter would, followed by the lower chain, which excavates into the subballast layer. Some machines have a third chain that allows for excavation and renewal at the upper part of the subgrade. A uniform new subballast layer is then placed and compacted followed by the new ballast layer. By renewing the entire granular layer (ballast plus subballast) over the subgrade, a strong and uniform track support is provided that minimizes subsequent track settlement and geometry error, resulting in much less geometry correction and longer track life. A ballast undercutter provides these benefits for the ballast layer alone, which may be sufficient. But ballast undercutting does not address the situation of poor subballast drainage conditions and instability at the subgrade layer interface. Where these poor conditions exist, use of the TSRM should be considered to address these deeper-seated sources of track instability.

9.1.3 Cost considerations of ballast-less track (slab track)

Slab track, or ballast-less track (see Chapter 5), consists of rails supported by steel-reinforced concrete slab that rests on a stabilized base, which is placed over a prepared subgrade. This type of track construction has a relatively high initial cost, not only because of more expensive materials and skilled labor and the increased time and effort required but also because of the need for an extensive subgrade investigation to find weak soils and to remove or increase their strength in situ where these are found.

To compensate for the high initial cost, future maintenance costs over the life of the slab must be relatively small. Because repeated geometry corrections that are unexpected and unplanned for during the design phase could greatly increase the LCC of slab track, an investigation is required along the planned route to find soft soils that must be replaced or stabilized.

Certain slab track designs are more tolerant of settlement than others. Of those designs that allow vertical geometry correction, the adjustments are typically on the order of 1 in. (25 mm), and a lesser amount for lateral correction. More problematic are those designs that do not allow for an adjustment. If the needed adjustment is outside of the allowable range, other correctional methods must be used such as injecting grout or other leveling material between the subgrade and slab, or improving the subgrade soil itself after the slab track is in operation, which are usually costly procedures.

A track settlement analysis by Chrismer and Li (2000; also see Section 5.3.6), which predicted the amount of subgrade deformation under various traffic loading cases, has found, for the assumed conditions, that subgrade with a resilient modulus as low as 2,000 psi is permissible under slab track for light rail vehicle loading and for a somewhat heavier commuter car where the limiting subgrade deformation is 1 in. over the life of the slab track. For 100-ton freight car traffic on slab track, the subgrade modulus required was 5,000 psi. This shows the sensitivity of subgrade strength with loading under a slab track with such a small allowance for subgrade deformation.

There are some assumptions that are often made regarding the use of slab track that may not be justifiable. One such assumption is that slab track is better suited for high-speed passenger rail service due to its improved ability to maintain smaller geometry error compared with ballasted track construction and the elimination of ballast's contribution to track settlement and geometry roughness. Although these are advantages of slab track, they are by no means decisive factors that compel its use for high-speed service. Indeed, some high-speed rail authorities find that ballasted track provides good and durable geometry performance at their highest speed and that the required geometry correction of the ballast layer is minimal. Another assumption that should be dispelled is that because slab track is relatively rigid, it will, as a result, "bridge over" a soft subgrade. If the unsupported length of slab track becomes excessive due to subgrade deforming underneath, the result could be not just a functional failure due to excessive vertical deformation but a sudden structural failure of the slab as well. As stated, the existence and extent of any soft subgrade conditions must be identified and remedied before slab track construction.

9.1.3.1 Top-down versus bottom-up slab track construction

Top-down track construction is typically associated with slab track where the rail positions are first established to a small error or tolerance and the rail supporting structure is then built so that the rail position is locked in. Alignment fixtures provide support to the rails to maintain their position, and a survey team checks their position as concrete pouring proceeds to ensure that the finished track meets the specified tolerances. Careful construction procedures must be followed to control moisture loss in the Portland cement concrete and to provide constant curing temperatures.

Bottom-up construction for slab track follows the typical procedure for conventional track of constructing rail support from the prepared subgrade upward. With this method, the final position of the rails is established once the concrete has cured by making adjustments at the rail supports.

9.1.4 Vertical rail deflection to locate soft track support of new track

Once the track has been constructed and is ready for traffic, a vehicle capable of measuring vertical rail deflection continuously along the track under a constant applied load can serve as a check on the quality of construction and the assumption of uniformity of track support made in the LCC analysis.

Any locations with larger than nominal track deflection (lower track stiffness) can be identified and considered for possible remediation as needed before traffic is introduced. This practice is similar to the concept of proof-rolling where the deflection of a compactor roller is used to indicate soft conditions that may require further reinforcement or placement of an increased granular layer over the soft soil. This method is ideally suited to new track that typically has uniform conditions of track structure, ballast, and subballast, which allows increased measurement sensitivity to, and better isolation of, zones with larger rail deflection due to soft subgrade. Also, the voids between rail and tie and tie and ballast are likely to be small and uniform when track is new, which also helps to isolate the contribution of soft subgrade to vertical track deflection under load. For this new track condition, the rail deflection under a typical fully loaded freight car or a locomotive should not exceed 3 or 4 mm. Deflections of 5 mm or more under relatively heavy loading indicate soft support and a track that could be expected to require an excessive amount of maintenance in the future. (Refer to Chapter 8 section 8.4 on this topic.)

9.2 BALLAST LIFE

As with all track components, ballast has a finite life. Ballast life expires when the voids between the particles become filled with fines to the extent that permeability is basically lost and the ability to maintain a track lift after tamping is greatly diminished.

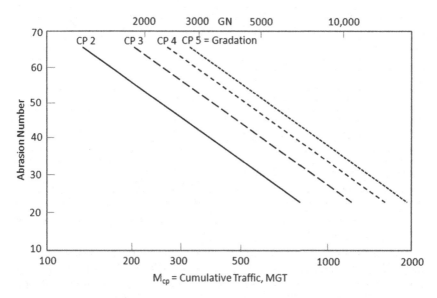

Figure 9.4 Ballast life (in MGT) based on ballast gradation and Abrasion Number. (Data from Chrismer, 1994.)

A comprehensive study that aimed to quantify ballast life was performed by Klassen et al. (1987) for Canadian Pacific Rail, where several sites on different geographic sections of the railroad were selected and ballast material was sampled from under and around the tie. Gradation analyses were performed on the samples, and the remaining life of the ballast was estimated. From these data, the ballast life chart in Figure 9.4 was derived. The chart uses the Abrasion Number (the percent breakdown in the Los Angeles Abrasion Number test plus five times the percent breakdown in the Mill Abrasion test) as a measure of material quality and an index to judge the rate at which fouling fines will be produced.

The various lines in Figure 9.4 show the amount of ballast life for new ballast of material quality designated by Abrasion Number and with gradation (when new) varying between CP designations 2 through 5. As expected, a lower Abrasion Number provides longer ballast life, and the increased life for gradations results from the increased void volume to hold the fines that are provided by the more narrowly graded but large particle sizes. Because gradation CP 4 is very close to the AREMA 4 gradation, which is commonly used on North American railroads, it will be used as a reference. Ballast with a broader gradation (a wider distribution of particle sizes, such as gradations CP 2 and CP 3) has less void volume to hold the fines and, therefore, has diminished life. Conversely, gradations with a narrower range of particle sizes than CP 4 (such as gradation CP 5) have proportionally more ballast life. The ballast life as calculated from the CP method for CP 4 gradation can be written as

$$M_{CP} = 10^a \tag{9.1}$$

Where M_{CP} is the cumulative traffic (MGT) and

$$a = \left[(-0.017)(A_n - 60)\right] + 2.5 \tag{9.2}$$

Where A_n is the Abrasion Number.

This CP ballast life concept was used in deriving the AAR Ballast Renewal Model by Chrismer and Selig (1994), although with a modified approach to obtaining the measure of the voids in various ballast gradations because the CP method was found to overestimate the influence of voids on life. To obtain an improved estimate of void ratio (e, the ratio of volume of the voids to volume of solids) for the range of typical gradations, a narrow gradation (AREMA 4) was used, and an increasing amount of fines was added so that the void ratio could be measured for broader gradations. The gradation was broadened by adding progressively more particle sizes smaller than the maximum size to provide the coefficient of uniformity ($C_U = D_{60}/D_{10}$) values shown in Figure 9.5 by the points designated by Chrismer (1994). The C_U is a measure of this broadness of gradation,

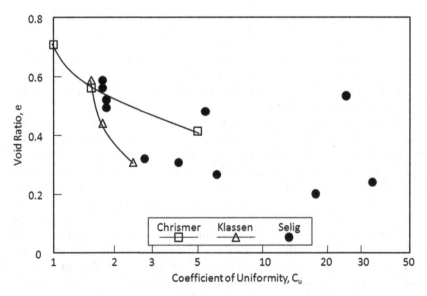

Figure 9.5 Void ratio of ballast versus gradation broadness. (Data from Chrismer, 1994.)

where D_{60} and D_{10} are, respectively, the size of the particles for which 60% and 10% of material is finer.

The void ratio is calculated by

$$e = \frac{V_V}{V_S} = G\left(\frac{\gamma_w}{\gamma_t}\right) \tag{9.3}$$

where:

γ_t (the total unit weight) was obtained by measuring the weight and volume of the ballast

G is the specific gravity

γ_w is the unit weight of water

Figure 9.5 shows the void ratio resulting from the changing C_U.

Also shown in Figure 9.5 for comparison with the data from Chrismer (1994) are the void ratio values from the CP study, designated by Klassen (1987), and the data from Selig (1989), which was obtained by field measurements of ballast unit weight. In general, the data from Selig agree with the data from Chrismer. Judging by the field data, there seems to be an overestimation of the changes in void ratio with gradation in the CP method and therefore an overestimation of the effect of gradation on ballast life. The relationship between gradation and void ratio found by Chrismer (1994), as shown by the solid line in Figure 9.5, is used in the AAR Ballast Life Model.

Because AREMA 4 gradation is commonly used, it was decided to relate the life of gradations with different void ratios to the life of gradation AREMA 4 with its own void ratio. This relationship is

$$M_e = \frac{e_x}{0.567}\left(M_{CP}\right) = \frac{e_x}{0.567}10^a \tag{9.4}$$

where:

M_e is the gradation corrected ballast life

e_x is the void ratio for the gradation used

The e_x is normalized by 0.567, which is the void ratio calculated for gradation AREMA 4, for a C_U of 1.5. The term 10^a is the ballast life for a given A_n with an AREMA 4 gradation. Although there are an infinite number of gradations that could be used, AREMA gradations 3, 4, and 5 are the typical choices. AREMA 3 has more and AREMA 5 has less voids than AREMA 4. Therefore, the model allows the user to select from these gradations, and then the model chooses the e_x for AREMA 3, 4, or 5, which are 0.605, 0.567, and 0.539, respectively. The model assumes that the rate of ballast breakdown is linear with traffic.

9.2.1 Influence of ballast depth on ballast life

The CP method considers the nominal ballast depth to be 7 in. under tie. For ballast depth, h_B, other than 7 in., the AAR model (Chrismer and Selig 1994) linearly adjusts the ballast life by multiplying by the ratio $(h_B/7)$. This makes the simplifying assumption that a deeper ballast section gives proportionally more void space or storage volume for the fines. The depth-corrected ballast life, M_{Depth}, then becomes

$$M_{\text{Depth}} = M_e \left(\frac{h_B}{7} \right)$$

(9.5)

9.2.2 Influence of tamping degradation on ballast life

The amount of ballast damage due to tamping was quantified by Chrismer et al. (1994) by placing ballast into an open-topped steel box with dimensions that slightly exceeded the effective tamping area occupied by tamping tines as they are inserted into the ballast. Two ballast materials were tested: a granite with a Los Angeles Abrasion (LAA) loss of 24% and a limestone with a LAA loss of 38%. Both ballast materials were tamped 10 times in one set of boxes and 20 times in the other. The ballast from all boxes was removed, and a gradation analysis was performed to determine the amount of ballast particle degradation. The ballast degradation due to tamping is shown in Figure 9.6 expressed as the amount of material passing the ¼-in. sieve.

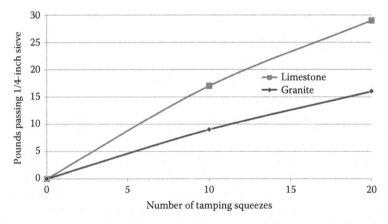

Figure 9.6 Ballast degradation due to tamping damage. (Data from Chrismer, S.M., *Track Surfacing with Conventional Tamping and Stone Injection,* Report No. R-719, Association of American Railroads, Chicago, IL, 1990.)

If the measured degradation is expressed in terms of an increase in the Fouling Index (FI), the case of limestone ballast tamped 20 times produced an FI of approximately 10%. If the ballast life is considered to be expired at an FI of 30%, this amount of tamping has reduced the life of this ballast by 33%. Based on this research, the ballast life decrease due to tamping is calculated by Chrismer and Selig (1994) as an increase in FI, according to the equation:

$$D_T = 0.5 \left(\frac{A_n}{45} \right) \left(\frac{12}{h_B} \right) \tag{9.6}$$

Where D_T is the increase in FI per tamping squeeze, A_n is the Abrasion Number that is normalized to the nominal value of 45, and the effect of ballast depth is accounted for by the correction factor of $12/h_B$, which is required because the other terms in this equation were derived based on research on tamper ballast damage using a 12- in. thick ballast layer. For every tamper tine insertion, the tines are typically squeezed either once or twice. If double insertion tamping is used, and the tines are squeezed twice both times, the increase in FI would be four times that in Equation 9.6.

9.3 MAINTENANCE NEEDS ASSESSMENT

9.3.1 Functional condition

Track condition can generally be divided into two main categories: functional and structural conditions. Categorizing indicators of track condition as either functional or structural helps support correct interpretation of data. Functional indicators refer to track serviceability generally related to the track conditions required to safely convey passing traffic, which are often determined using track geometry and ride quality. Functional condition represents the present condition of the track and is directly indicated by track geometry data obtained by track geometry recording cars. The difference between smooth track design geometry and its actual geometry indicates track roughness, which is a measure used to infer safety of train operations and ride quality experienced by the passengers. Functional measures typically reveal little about future condition trends or even the expected performance of the track under load; however, a review of historic functional condition measurements can provide a clear indication of track degradation rates and may provide insight into the root cause of track condition deterioration.

Figure 8.36 shows an example of trending geometry error over time that indicates changing historical functional condition. Shown in this manner, changes in the functional condition due to degrading condition or applied maintenance provide an indication of past behavior that can be used to determine the effectiveness of maintenance. However, past trends of functional condition are often a poor indicator of future performance due to changing conditions and nonlinear stress–strain behavior of the track substructure.

Although functional condition measurements of track such as geometry are useful as a "snapshot" in time, they do not indicate the immediate support condition of track. For example, the geometry condition may appear normal, and yet the track could be on the verge of failing due to a sudden embankment failure. Therefore, structural condition track measurements are needed.

9.3.2 Structural condition

The structural condition of track relates to its ability to support the imposed loading without excessive stress, deformation, or fatigue that would limit track life. This typically involves evaluation of the structural properties of the track such as the strength and stiffness of the superstructure and substructure but can also include surrogate measures of condition such as soil density from a nuclear probe or dielectric constant from ground-penetrating radar (GPR). These measures are often made with the intention of evaluating the strength of track relative to a specific failure mode or the remaining life of the track based on a set of specific strength or stiffness measurements.

The most common indication of track structural condition is the amount of vertical deflection under load (i.e., track stiffness), where it is generally assumed that relatively large deflection is indicative of weak track support that will settle excessively under repeated loading, while small deflections show track to be resistant to this settlement. Although this may usually be true, it can sometimes be misleading, such as where the ballast is placed directly on a stiff but fine-grained subgrade such as a hard clay or perhaps shale bedrock. This stiff subgrade may deflect little but is erodible by sharp-edged ballast particles in contact with it in the presence of water (subgrade attrition) and so produce substantial settlement. Therefore, track deflection under load may be only an approximate indicator of structural condition, potential long-term settlement, and track life, but the relative ease of measuring deflection makes it an attractive choice.

9.3.3 Vertical rail deflection to locate soft support of existing track

One measure of track structural condition is its vertical deflection measured under load continuously along the rail. Although large track deflections (>5 mm under the wheel of a typical locomotive or loaded freight car) may indicate locations of soft support from poor subgrade, deflection data are difficult to interpret for older, worn track due to the presence of voids between rail and tie, and tie and ballast, that may produce large deflections that can wrongly be interpreted as soft soil. While track deflection from voids may of themselves be worthy of measurement because they could have an effect on track performance, soft subgrade support is expected to have the greater effect on track functional condition over time, and so it is the most important indicator of track performance and maintenance needs. Despite this, however, methods to isolate the contribution of the subgrade alone using track deflection measurements have had mixed success due to the difficulty in accounting for the amount of voids in the track versus the amount of rail deflection due solely to soft soil.

To the extent that the source of measured vertical track deflection can be assigned to either void or soft subgrade, Table 9.1 shows the relation between test parameters, track problem diagnosis, and maintenance or remedial methods. A large track deflection, which is not due to voids, indicates low track stiffness primarily due to a soft subgrade condition because subgrade properties have the most influence on track stiffness. Maintenance for this condition can involve design and reconstruction of the track substructure layers (ballast and subballast) possibly in conjunction with installing a reinforcing layer over the subgrade to reduce the stresses transmitted to it or improving the subgrade's ability to support the applied loading by using such techniques as compaction grouting or micropiles (see later section in this chapter).

Variable track stiffness is taken to indicate a condition where variation in track support occurs. Conditions where track stiffness can vary significantly include track stiffness transition locations such as bridges, tunnels, special trackwork, and concrete to wood tie transitions. Other locations of variable track stiffness may include changes in local geology, track construction procedures, or any change in the properties or condition of the support provided to the track by the substructure. The variation in track support results in distorting an initially smooth rail profile due to differential elastic track deflection. This difference in elastic deflection will provide a small increase in dynamic loading, which in turn may progressively increase track deterioration and settlement. Methods to improve this track condition include design of rail seat pads,

Table 9.1 Track stiffness parameters and potential maintenance needs

Parameter	Problem	Maintenance/rehabilitation
Low track stiffness	Poor or weak subgrade or fouled muddy ballast	Adequate granular layers, improve drainage, stabilize subgrade
Variable track stiffness	Variable track support (stiffness or modulus)	Match stiffness with railseat pad or ballast mat, strengthen weaker subgrade, improve drainage
Void	Fouled ballast, local settlement, or poor fastener condition	Replace broken fasteners, tamp/ surface, stone blow, undercut

under-tie pads, or ballast mats that slightly increase the elastic deflection on the "stiff" side of the transition and modifications to increase strength and limit settlement on the "soft" side of the transition (see Section 5.4). If the variation occurs at locations other than track stiffness transitions, techniques to improve the uniformity of support and inherent track shape should be utilized.

The presence of voids can be measured with a relatively small load and indicates slack in the track structure resulting from hanging ties, loose rail seat fasteners, and possibly fouled ballast. Methods to reduce the void deflection include inspection and repair of broken fasteners, tamping and surfacing, stone blowing, and undercutting where the void is due to fouled ballast that deforms easily under load.

9.3.4 Assessment of track life and maintenance needs

As discussed in Section 9.1.2, methods have been developed to determine the LCC of the track components of rails, ties, and ballast individually based on their calculated timing of needed maintenance and ultimately replacement life for each. There has also been work to assess the life and maintenance needs and economics of track for all its components combined. For example, ECOTRACK (ECOnomical TRACK) was developed by Jovanovic (2006) as funded by the European Rail Research Institute to provide a tool for making decisions of railway permanent way maintenance and renewal management based on LCC. This program allows the decision maker to determine when track components should be renewed and how best to combine their renewal in case these components are found to require this close to the same time. Decisions are aided by an "electronic track chart" information database showing track components and their exact location, date of installation, cumulative tonnage, and many other details. The rate and trends of track geometry deterioration are forecast into the future as these track components degrade. ECOTRACK can forecast the time at which maintenance costs will begin to escalate toward the end of track life, as shown in Figure 9.1, due to its components becoming unable to maintain the required geometry.

The influence that track substructure conditions can have on the life and maintenance needs of other track components such as rail is shown in Figure 9.7, where the statistical distribution of rail bending stress from thousands of wheel loads is seen to be highly dependent on ballast fouling and moisture condition. The lack of tie support and formation of voids under the ties in muddy ballast produces increased rail bending deflection and stress compared with the nearby location where ballast was not fouled. This poor substructure condition will also damage other track and vehicle components due to increased dynamic wheel loads.

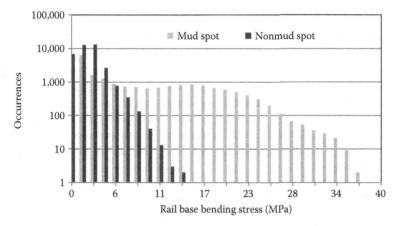

Figure 9.7 Increased stress in rail due to fouled and wet ballast.

9.4 MAINTENANCE METHODS AND EQUIPMENT

9.4.1 Tamping

Tamping of track is performed to adjust the vertical profile and lateral alignment geometry of the rails. A tamping machine first raises the track to the desired level to create a void between the tie bottom and the ballast and then inserts tines into the ballast on either side of a tie, and these tines are squeezed together so that the ballast from the crib area is moved underneath the tie filling the void. The insertion depth of the tine tips below the tie bottom is adjustable on some tampers, but a typical depth is approximately 6 in. below the bottom of the tie. Research has found that tamping actually disturbs ballast over the depth of insertion and reduces the ballast unit weight or bulk density. Measurements have shown tamping to cause typical decreases in ballast bulk density from 1760 to 1600 kg/m^3, or approximately a 10% decrease, although decreases as high as 30% were measured by Selig (1989). The ballast will regain its pre-maintenance bulk density after traffic resumes, which accounts for the rapid initial settlement, or "ballast memory," that often returns the track to its pre-maintenance profile following tamping.

Although tamping is often used as a "cure-all" because it is relatively easy to perform, it cannot be expected to provide a durable geometry correction where it does not address the root cause of track movement such as a failing subgrade. However, tamping may provide a reasonably long-lasting correction if the rate of subgrade deformation is relatively slow.

Due to certain limitations of conventional tamping, improved methods have been sought. One problem with conventional tamping is that it often smoothes geometry error rather than returning the track to its original designed shape of alignment and profile. Track that was originally laid out using classic shapes of tangents, spirals, and curves, as described in track design textbooks, often becomes distorted over many decades of traffic and maintenance. Very often the amount of distortion from the original alignment design is so large that it may not be possible to return to it without lateral track throws that are prohibitively large when the tamper is operated in conventional smoothing mode. In the case of profile, the lift provided by conventional tamping is often too small to be durable, and it quickly settles under traffic, returning to its rough surface. The following discusses improved methods that provide better quality and more durable geometry corrections.

9.4.2 Improved methods to correct track geometry error

9.4.2.1 Design over-lift tamping

Developed by British Railways, design over-lift (DOL) tamping provides a track lift in excess of that needed to bring a dipped track to a level position. As the name implies, the amount of over-lift is designed to compensate for the subsequent rapid ballast settlement with resumed traffic so that the track settles into a design smoothed profile. Further track settlement will occur very slowly so that the track stays at this smoothed profile much longer than provided by conventional tamping with no over-lift. The situation is as shown in Figure 9.8 (Chrismer et al. 1994). As illustrated, after conventional tamping and resumed traffic loading, the track quickly settles and returns to its prior rough shape, while DOL is more successful in removing this "ballast memory" producing a more durable geometry correction.

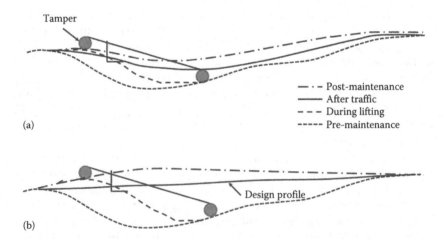

Figure 9.8 Illustrated concept of (a) conventional smoothing tamping and (b) design over-lift tamping.

The amount of over-lift to be applied to each tie is based on the measured amount of dip at each tie location. To obtain the pre-maintenance rough track profile, the tamper first performs a measurement run over the track length to be surfaced. Then the solution is obtained using the on-board computer, and the tamper applies the determined lifts (and lateral throws if required) for each tie during the return trip back to the starting point. Another measurement run may then be performed to obtain the resulting post-maintenance track profile.

Measurements of geometry correction durability provided by DOL have shown this method to last typically three times longer than conventional tamping with no over-lift (Chrismer et al. 1994). This amount of improvement has been observed for tests with traffic loading ranging from heavy haul freight to passenger trains.

The success of the DOL process depends on there being a predictable relationship between the applied and residual lifts as the track settles under loading. Applied lift is the change in rail elevation provided by the tamper. Residual lift is the remaining amount of track raise after a certain amount of accumulated traffic loading. The exact relationship between applied and residual lift to use for DOL tamping should be established from lift-settlement tests in the track over a range of track lifts from less than 1 in. to slightly more than 2 in. The resulting plot of track lift versus settlement data should appear similar to that in Figure 9.9 after a significant amount of traffic tonnage and track settlement has accumulated. Note that there is a "knee" in the distribution at approximately 20 mm of track raise, below which the track has lost most of its applied lift and above

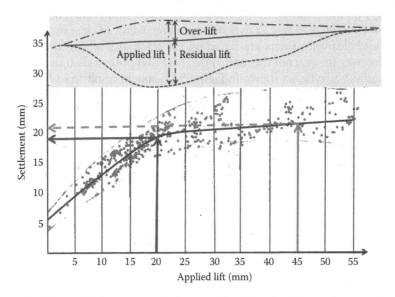

Figure 9.9 Relationship between applied and residual lift as basis for design over-lift.

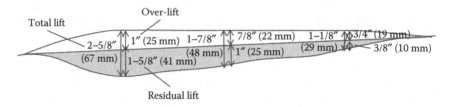

Figure 9.10 Typical amounts of applied lift, over-lift, and residual lift.

which a substantial amount of track raise still remains. Below this 20 mm, the track raise only dilates the ballast particles without allowing them to rearrange, and so they come back to their previous point contacts with settlement. A track raise greater than 1 in. (25 mm) is sufficient to allow the ballast particles to rearrange and form new point contacts in a more stable arrangement that better resists settlement, as shown by the lift of 45 mm in Figure 9.9. The relative stability of track raises above approximately 1 in. is the basis for DOL.

Figure 9.10 shows the relationships between applied lift, over-lift, and residual lift. The applied lift is that needed to raise the track from its initial "low" position to an elevation above that into which the track will eventually settle. Therefore, the applied lift includes the needed over-lift. The amount of over-lift is designed to account for the initial settlement with traffic so that the track will settle into the relatively stable position shown as the remaining residual lift. Figure 9.10 shows some typical values of over-lift for the pre-maintenance rough track profile shown. Note that even the largest dips do not require an over-lift of much more than 1 in. This fact helps to ease concern over track raises infringing on overhead clearances; because overhead clearance limits are based on the track at level position, it is the smaller over-lift and not the total applied lift that poses a clearance concern.

A durable profile, as shown in Figure 9.11, should result from DOL. Note that the initial rate of settlement decreases quickly and the track settles into a more durable smoothed profile. Often the immediately post-DOL profile roughness will not be as smooth as occurs just after conventional tamping, but the DOL profile roughness improves with settlement, reaching an optimum smoothness some time later.

It has been found that over-lifting without benefit of the design approach does not yield optimum results based on instances where a tamper operator was instructed to adjust the lifts based on his or her best guess of the amount of over-lift required along the track. Even an experienced operator with good tamping equipment will often have difficulty obtaining a smoothed profile close to that achievable by using the design procedure.

A lift/residual lift relationship established for one set of track conditions should only be used for another location if it is known to have reasonably similar annual tonnage, axle loads, subgrade type and modulus, and track

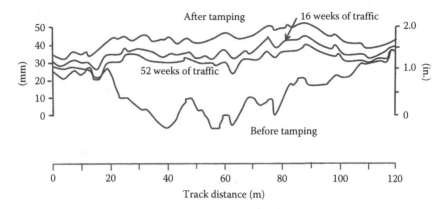

Figure 9.11 Design over-lift removing effect of "ballast memory."

structure. A reliable lift/residual lift relationship derived from these parameters will allow an engineer to specify lifts with assurance that the rail profile will remain close to the design for a considerable time.

DOL software is commercially available and can be installed on certain types of tampers. In addition to the benefits of DOL mentioned, this software was developed to overcome the limitation of tamper "blind spots" to certain wavelength faults and the three- or four-point reference system of conventional smoothing tamping that does not view the track as being continuous in space. These limitations are eliminated because the lift is determined not with reference to a three- or four-point chord but by the existing and desired continuous rail profiles.

9.4.2.2 Improved tamper control systems

Ballast tamping machines are usually supplied with the hardware needed to measure existing track geometry error before tamping and with the software to record this information and provide the required geometry corrections at each tie location. However, one commercially available alternative to user-supplied control systems (the only one known to the authors) is known as Track Geometry Control System (TGCS) (Marriott 1996). This was developed at the former British Rail Research facilities when it was noted that tampers equipped with user-supplied systems were inadequate at removing geometry error with long wavelengths as required for passenger comfort at higher speeds. When supplier-provided tamper control systems were investigated, it was found that several aspects could be improved on such as the ability to measure longer wavelength geometry error to provide better quality geometry corrections and increased durability of these corrections. These subsequent improvements led to the development of TGCS, which has been installed on almost half of Network Rail's tamping fleet to date and has become a commercial product that Amtrak has purchased for its tampers. A comparison (Wnek and Chrismer 2012) between track tamped with TGCS and that tamped with the tamper manufacturer-supplied control system has confirmed what British Rail researchers have found: that TGCS improves on the ability of the tamper to remove geometry error and especially longer wavelength error, which is essential to maintain passenger comfort.

9.4.2.3 Stone blowing

Stone blowing, like DOL tamping, was developed by British Railways. This maintenance method was adapted from the older "shovel packing" technique where jacks were used to lift the rail and ties to create a void between tie bottom and ballast so that a measured amount of ballast or smaller stone could be placed by shovel under the tie but on top of the existing ballast. Stone blowing uses this same concept to correct track profile error, except that instead of being placed by shovel, the stone is blown into the void between tie bottom and ballast bed using pneumatic injector tubes that are driven into the ballast along the side of the raised tie. The amount of stone blown in is chosen based on a design to provide the desired rail elevation. The tubes are driven into the ballast to a depth that provides the stones with a flow path and access beneath the tie; then compressed air is applied to the tubes, and a measured amount of stone is blown under the raised tie. The stone blowing tubes are then removed and placed along the side of the next raised tie, and the process is repeated. Because the stone blown in is placed on top of compacted ballast that is only minimally disturbed by the tubes driven into the ballast, the ballast and stone blown on top of it is stable, limiting the settlement post maintenance.

Stone blowing can be accomplished with handheld injector tubes and a portable air compressor or with the mechanized version. The stone blower machine's track lift mechanism is shown in Figure 9.12a, with a close-up view of this arrangement in Figure 9.12b. There are a total of eight blowing tubes that are driven next to a tie, which are arranged at approximately the same positions along a tie as the tines of a tamper.

Only high-quality stone should be used for stone blowing because the granular material directly under the tie is highly stressed from traffic loading. The injected stone must be small enough to flow into the void without blocking up in the tube and to disperse properly when injected, but large enough to interlock and support the tie and not fall into the voids of the existing ballast. The stone size that appears to work best is between ¾- and ½-in. (17 and 12.5 mm).

Stone blowing complements tamping, where the former is best suited for correcting dips on the order of 1 in. and less that typically extend over relatively short wavelength geometry faults, while the latter is mainly to address track dips larger than 1 in. that are usually associated with greater wavelength geometry faults. As stated, tamping with lifts less than 1 in. typically does not provide a durable geometry correction for the reasons

(a)

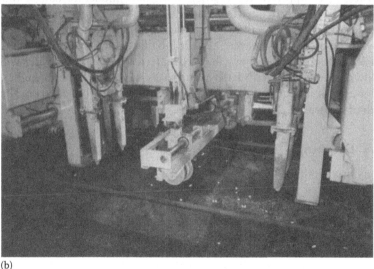

(b)

Figure 9.12 (a) Stone blower with track lift mechanism and stone blowing tubes.
(b) Close-up view of stone blowing tubes.

mentioned, and therefore stone blowing these smaller dips would provide a
longer-term correction.

Although stone blowing is mainly being used in the United Kingdom, tests
were conducted with stone blowing in the United States on track with heavy
haul freight loading. These US tests have shown that stone blowing can provide a two- to three-fold increase in geometry correction durability compared with conventional tamping (Chrismer 1990) similar to the amount of

improvement DOL tamping has over conventional tamping. Another potential application for stone blowing may be at recurring dips at track transitions such as bridge approaches. Even though a track over-lift at the transition may not be possible if the rail elevation is fixed by the bridge, a track raised only so far as to remove the dip so that stone can be blown under the affected ties might still provide a much more durable geometry correction than tamping. The inadequacy of conventional tamping to remove these dips at transitions is shown in Section 9.5.3.

9.4.3 Ballast shoulder cleaning

Shoulder ballast lies beyond the end of the ties and extends down to the bottom of the ballast layer. If fouled, it can prevent drainage of ballast from the crib area. Mechanized shoulder cleaning excavates shoulder ballast so that it can be replaced with new, clean ballast. Research has found (Parsons 1990) that cleaning one or both ballast shoulders can provide improved drainage from the crib area, provided the crib ballast FI is not highly fouled (a FI less than approximately 30%). With only moderate ballast fouling, shoulder cleaning provides an escape for any water being retained in the crib area, and this may allow some of the fouling material to be washed away in the process. However, if the crib ballast is highly fouled (FI greater than 30%), shoulder cleaning will not be effective because the water will not readily drain from the crib ballast due to its much lower permeability. For such a case, it is required to remove the highly fouled ballast from the full width of the track.

9.4.4 Ditching

Just as it is vitally important to maintain good internal track drainage by replacing fouled ballast with clean ballast so that water can escape from within the track, water draining from the track must then have a place to go. Ditches are required to provide the external drainage to rapidly carry the water first laterally away from the track and then longitudinally to a location of low elevation so it can be removed from the railroad property. The ditches must have adequate lateral and longitudinal slope and an invert elevation of sufficient depth below the subgrade under track so that water is not able to return to the track. Ditching machines can provide these ditch slopes and contours. Although external track drainage is of fundamental importance, ditches are often difficult to establish and maintain due to the presence of utilities, structures, or other features along the track. But despite these difficulties, every effort should be made to overcome these restrictions so that water does not remain long in track.

Although the cost of providing good drainage through ditching and other means is typically not large, the investment is often not made because the benefits are not seen for a considerable time after they are implemented.

But the cost of the lack of drainage will almost certainly become apparent eventually at which time the cost of remediation can be quite large. It can be very difficult to return the track to a good drainage condition once drainage has been lost.

Management of external drainage can be greatly aided by light detection and ranging (LIDAR) ground surface elevation data, which can be used to determine the adequacy of existing ditches over many continuous miles of track by reviewing the elevation along the ditch line and finding the locations of blockages. LIDAR can also show how far ditches should be extended longitudinally so they can drain into low-lying areas adjacent to the track so that water can flow by gravity.

9.4.5 Ballast undercutting/cleaning

Ballast, like all track components, has a finite life and must eventually be replaced. The end of ballast life usually coincides with the voids being filled with fouling material, which reduces the permeability to the point that the drainage function is lost. The ballast can be removed by using an undercutter that is equipped with a continuous chain that excavates the fouled ballast under tie as it advances along the track. An undercutter that does not attempt to reclaim any of the larger-sized ballast particles existing in the old ballast either will transport the entire amount of excavated ballast by conveyor belt to waiting rail cars or will dump this material to the track side as it moves along, forming a spoil pile. An undercutter/cleaner is an undercutter machine that also has one or more sieves that shake the fouled ballast to separate the large stone from the fine material. The larger ballast particles that are retained on the sieve are then conveyed back to the track while the fouling material is wasted. Because the amount of ballast returned to the track after sieving is not sufficient to provide a full-depth ballast section, new ballast is usually required to make up the difference.

Tests of the quantity of wasted and reclaimed ballast during undercutting/cleaning operations have allowed the efficiency of these operations to be estimated by examining the gradations of material in the waste pile and in the returned ballast that was placed in track. One undercutter/cleaner investigation (Chrismer et al. 1992) used a ¾-in. sieve to separate the fines from the larger particles. Figure 9.13 shows the measured gradation of the wasted ballast where no attempt was made to try to reclaim a portion of the excavated ballast.

Note that in the material that was wasted, only approximately 30% of it was smaller than the ¾-in. sieve, which means that about 70% of it was larger than this size and therefore potentially recoverable if this had been attempted. Despite this, the operator made a decision to sacrifice the recoverable ballast to increase machine speed or perhaps he or she made a judgment that there were too many fines in the fouled ballast that would have bulked up the sieve had he or she attempted to reclaim some of it. In any case, it is

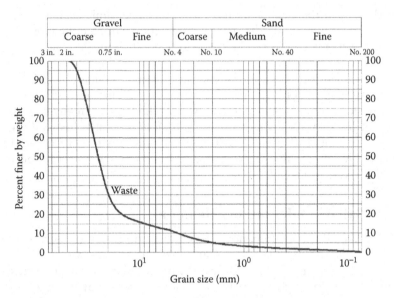

Figure 9.13 Undercutting operations resulting in low efficiency of wasting ballast.

clear that a significant amount of potentially recoverable ballast could have been available at a significant cost savings. The decision to waste potentailly recoverable ballast should not be made without full consideration of the economics of doing so.

If the reclaiming and wasting processes were 100% efficient for this operation, no material smaller than ¾-in. would be reclaimed and no particle larger than this would be found in the waste pile. Figure 9.14 shows an example that approaches this ideal condition. Because the sand-sized and finer particles were only 2% of the total sample returned to track, it appears that the undercutting/cleaning process was very efficient as it did remove most of the fines from the reclaimed ballast. The wasting process was similarly efficient because only a few percent of material wasted was larger than ¾ in.

Figure 9.15 shows a case where an attempt was made to reclaim the larger-sized ballast, but this was not very effective. The actual material that was reclaimed was very clean, but note that 40% of the potentially recoverable ballast was able to be reclaimed. The reason for the lack of reclaiming efficiency may have been due to an operator decision such as to divert most of the excavated ballast directly to the waste pile to increase production rather than attempt to sieve it. Nevertheless, as a result, more new ballast must be ordered to make up for the loss of the larger ballast particles to the waste pile.

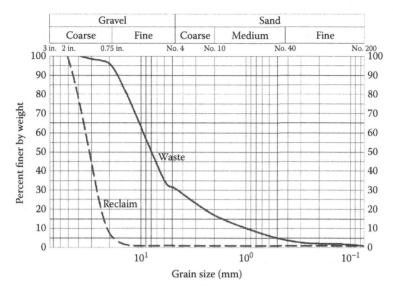

Figure 9.14 Undercutting operations resulting in high efficiency of reclaiming and wasting ballast.

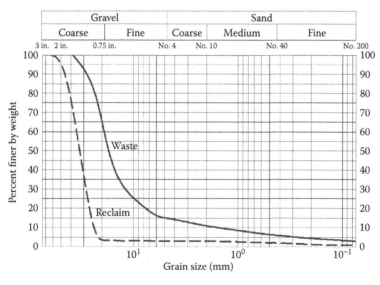

Figure 9.15 Undercutting operations resulting in high efficiency of reclaiming and low efficiency of wasting ballast.

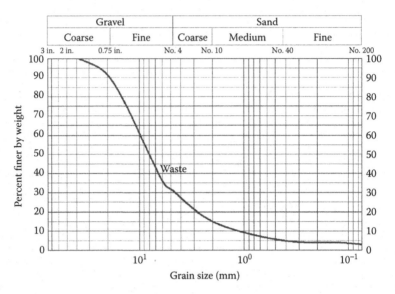

Figure 9.16 Undercutting operations resulting in high efficiency of wasting ballast.

The gradation plot in Figure 9.16 shows the result of what appears to be a correct decision to waste the entire excavated ballast material due to the lack of larger ballast particles present. There would be little economic advantage of trying to sieve this material given the measured gradation from the waste pile.

The decision of whether or not to attempt to reclaim at least some of the fouled ballast during undercutting/cleaning depends on a number of factors. A large economic incentive exists to reclaiming as much old ballast as possible due to the considerable cost to procure and transport new ballast. Although it is difficult to estimate the amount of potentially recoverable, larger-sized ballast that is present and may be recovered from the undercut ballast, generally the amount of old ballast large enough to be returned to track is less than half of the amount needed. Measurements of fouled ballast gradations have been made in advance of the undercutter/cleaner at several locations around the United States in a variety of conditions (Chrismer et al. 1992). Figure 9.17 shows the results of an analysis based on these gradations indicating how much recoverable ballast was found as a function of the depth of undercutting.

Because the gradation usually becomes finer grained with less of the larger-sized ballast particles with depth, the general trend is diminishing returns with depth. A typical depth of undercutting is 0.3 m (12 in.) under the tie bottom, which is approximately 0.53 m (21 in.) in Figure 9.17 when accounting for the 0.23 m (9 in.) height of wood ties. As shown, the average gradation provides about 36% of the needed ballast on average for a 0.3 m (12 in.) deep ballast section, with a range from 27% to 45%. If this

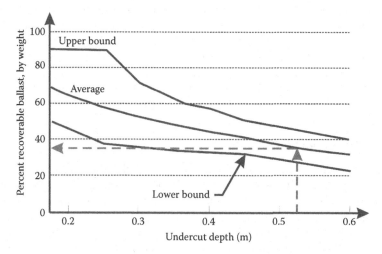

Figure 9.17 Amount of potentially recoverable ballast (>19 mm) from undercutting.

amount of ballast return is possible, a considerable savings will result from the reduced need for new ballast.

Worn ballast particles typically do not become highly rounded under mechanical wear if the ballast is reasonably hard and resistant to abrasion. When ballast is new and has sharp edges, the shapes generally have round-ness values of 0.1 and 0.2, as shown in Figure 9.18 (Krumbein 1941), which is common after quarry operation produces fractured faces. But with wear,

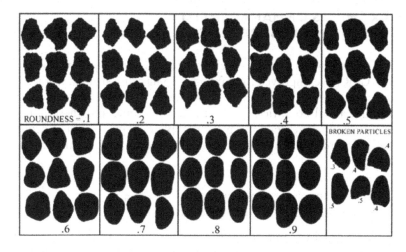

Figure 4. Chart for rapid determination of roundness (Krumbein, 1941)

Figure 9.18 Ballast particle shape. (Data from Krumbein, W.C., *J. Sediment. Petrol.*, 11, 64–72, 1941.)

the edges become slightly worn with abrasion to become the shape shown by the roundness designation of 0.3 in Figure 9.18. This slight rounding at the edges will not significantly change the mechanical behaviour of the ballast from the new. However, for softer ballast materials, there is the possibility that such particles could become somewhat more rounded with mechanical wear. Therefore, a visual inspection of the worn ballast particles should be performed to determine if they are worthy of reuse as ballast.

9.4.6 Track vacuum

A strong vacuum can be used to remove fouled ballast around the ties. Track vacuum machines have been developed with large hoses that can be positioned to excavate fouled ballast from track. The vacuum is provided by large motors, and the excavated material is deposited in a holding tank for later disposal. Some track vacuum machines have a rotating bit on the end of the hose that breaks up the fouled dense ballast so it can more easily be excavated from under the tie and further down into the ballast section than would be possible otherwise.

Because this method of vacuum excavation is slow, it is commonly limited to short lengths of track such as for making an opening for the undercutter chain or to remove fouled ballast that is limited to only around a rail joint, for example. Track vacuuming is also useful in 3rd rail electrified track territory where the power rail and appurtenances complicate the ballast cleaning process. For removal of fouled ballast over a track length of more than 10 ft, another faster method such as a switch undercutter may be preferable.

9.5 EXAMPLES OF MANAGEMENT BASED ON MEASUREMENT

Advancements in recent years in remote sensing, nondestructive techniques such as GPR have improved the capability to measure structural track conditions. Findings on the structural condition can then be better related to the functional condition of the track over time, and these data can then be brought together for a more complete diagnosis of the problem. In this way, improved measurements allow better management of track substructure.

9.5.1 Using track geometry data to determine tamping needs

All railways need to tamp track as a rapid response to the sudden development of geometry error at isolated locations that is or soon threatens to be safety-critical. Additionally, tamping programs are needed that are planned in advance and that address error over many miles. Although a common practice among railways is to tamp track on periodic basis without determining the actual need based on geometry error, scarce resources and limited budgets and manpower make it difficult to justify tamping track that does not need it. Therefore, it is more reasonable to target those sections for tamping based on their functional condition as determined by geometry measurements. When geometry data are studied over time, it becomes apparent that some track sections require monthly or even weekly geometry correction, some need correction on perhaps a yearly basis, and others that almost never do. Tamping all track regardless of its condition is not cost effective. Instead, there is a great incentive to use available functional condition data such as geometry records to distinguish between track sections requiring different amounts and timing of maintenance.

For example, in the U.S., Amtrak determines locations that require tamping based on the running roughness, as described in Chapter 8, where the R^2 value is obtained not from one measure of geometry error but from the combination of all three mid-chord deviations measured over 31-, 62-, and 124-ft chords for horizontal alignment and all three of the same for the vertical profile. In other words, all six channels of data are combined into an overall roughness, and those locations with large values are targeted for tamping. This targets locations with track geometry roughness over the lengths corresponding to these chord lengths. But error over a chord length as long as 124 ft is still not long enough for a passenger railway with high speeds. Therefore, Amtrak additionally processes the deviations in the profile and alignment space curves that are based on a reconstructed base line length of 400 ft. These deviations reflect geometry error of wavelengths as large as 400 ft, which is on the order required for Amtrak's speeds up to 150 mph. So in addition to the locations identified for tamping based on the error over chord lengths up to 124 ft, those locations with error over larger wavelengths, up to approximately 400 ft, are identified. These locations are put into a list and prioritized according to severity and given to the tamper operators. In this way, Amtrak assures that track is tamped only in response to measured performance, allowing better management of scarce resources.

Tamping decisions can also be aided by the indications of track structural condition provided by GPR measurements. Ballast that is revealed to be heavily fouled and wet by GPR measurements would not provide durable geometry corrections because of its degraded condition. Heavily fouled ballast might also indicate that it is in a very dense condition and may be highly resistant to penetration to the tamper tools. In this way, GPR can provide indications where tamping is relatively ineffective.

9.5.2 Poor ride quality: Due to track geometry error or the vehicle?

The assumption that poor ride quality is due to track geometry error is not always well founded. If passenger coaches are instrumented with accelerometers and these are monitored each time they traverse a track location

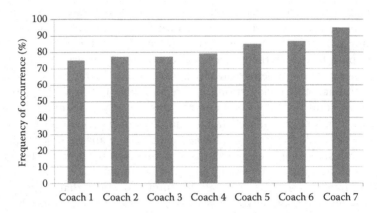

Figure 9.19 Frequency of occurrence at which coaches experienced poor ride quality by repeatedly traversing location with large track geometry error.

with a geometry error that excites the carbody, after multiple passes it can be determined how frequently the acceleration of each carbody exceeded an established limit or ride quality exception. Figure 9.19 shows a case where all instrumented coaches were more likely than not to register a ride quality exception with each pass, although certain coaches performed somewhat worse than others. This is the generally assumed condition where a relatively large track geometry error affects all coaches more or less equally. There is little doubt that track error is fundamentally to blame for these carbody reactions.

However, other track locations with a lesser amount of geometry error can show large differences in carbody reaction between rail vehicles, such as in Figure 9.20. This figure shows consistent differences between coaches

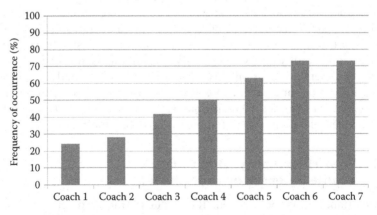

Figure 9.20 Frequency of occurrence at which coaches experienced poor ride quality at track geometry error that preferentially excites poor car suspension.

where those on the left of the graph are significantly better at negotiating these geometry errors than those on the right. It appears that where track geometry error is relatively large, this error will be the factor controlling vehicle/track interaction and differences between vehicles will be minor as in Figure 9.19, but geometry error of a less-severe magnitude can allow differences in dynamic behavior between vehicles to be more apparent. These examples shown in Figures 9.19 and 9.20 can serve as filters to expose those vehicles that are perhaps in need of maintenance. Although differences in carbody dynamic reaction between vehicles of the same type can be due to causes other than the suspension, the condition of the secondary dampers is the leading suspect in such cases. Using the 50% frequency as the basis for requiring a vehicle inspection, the secondary dampers on Coaches 4, 5, 6, and 7 should be inspected as they appear to be in need of replacement.

9.5.3 Managing the dip at track transitions

Vertically dipped rail is commonly found in track transitions such as bridge approaches and road crossings, as discussed in Chapter 3. To diagnose the cause of track substructure instability, it is useful to combine structural and functional condition information, as shown in Figure 9.21.

The upper two bands of gray wavy lines are the GPR radargram from Track 2 and Track 3, respectively, which show track structural condition below the ballast surface. GPR data clearly indicate the presence of open deck bridges in Bridges 1 through 5 by the relatively uniform horizontal

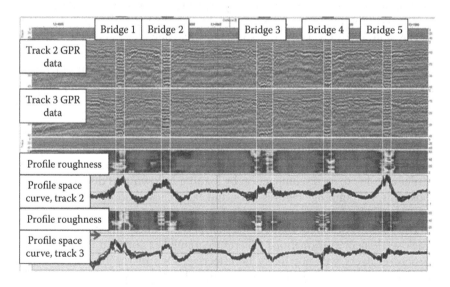

Figure 9.21 GPR and track geometry combined to aid in remediation at track transitions.

lines, and at the ends of these bridges are the transitions to ballasted track. Processing of these GPR reflections shows deformed substructure layers in some transitions, and in others GPR indicates the presence of water.

The functional conditions in these transitions are indicated by the lower bands of data showing historic profile roughness and more recent profile space curves for the two tracks. The dips in rail profile space curve are seen to vary in track length and dip amplitude ranging from the abrupt dip to the left side of Bridge 4 on Track 3 to the more gradual dip between Bridges 1 and 2 on Track 2. These dip lengths and amplitudes produce varying dynamic wheel/rail loads and possibly poor ride quality, indicating a variable degree of degraded functional condition at these five bridges. Profile roughness data indicate the time history of geometry error with the most current condition on the top of the band. A uniform vertical streak at the transitions indicates that the geometry error has been constant over time with the same magnitude. Geometry correction will appear as a horizontal interruption in the streak, and if the dip is permanently removed, the streak will appear to abruptly stop. But these profile roughness data trends indicate that geometry correcting maintenance tends not to be effective, and the dip quickly reforms, sometimes just days after the maintenance.

Bringing these disparate data together in the manner shown in Figure 9.21 allows this common problem to be better managed. The need for more effective corrective action becomes apparent from viewing the data in this way, and the effectiveness of various alternative remedies can readily be assessed.

A recent study (Tutumluer et al. 2012) of behavior at transitions used multi-depth deflectometers (MDDs) to determine the amount of settlement in substructure layers next to bridge transitions. The investigation has found that the dip grows mainly because of settlement in the ballast layer following tamping. As shown in Figure 9.22, MDDs were installed at the boundary layers of the soil types as they were encountered to a maximum depth of approximately 10 ft down from the ballast surface and anchored at the bottom in the fairly stiff lower layer. Measurements of permanent deformation within the layers were obtained by finding the difference between the settlements indicated by the MDDs at the interfaces of the soil layers.

Figure 9.23 shows the resulting deformations of each layer in the days following MDD installation. The ballast was tamped after MDD placement followed by traffic loading. The ballast layer provides almost all the track settlement due to its re-compaction by loading following its disturbance and loosening by tamping. Figures 9.24 and 9.25 show another location with the same findings.

The findings indicate that the ballast is most likely the main source of track settlement at transitions. Therefore, remedies aimed at the lower substructure layers will not likely be effective. One of the best methods to address settlement in the ballast layer may be stone blowing, as described in Section 9.4.2.3.

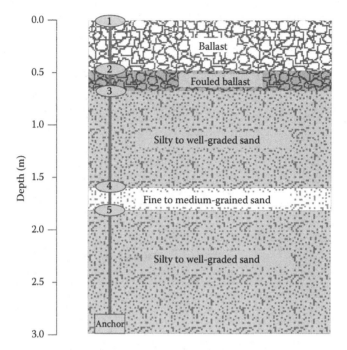

Figure 9.22 Track substructure layers and MDD instrumentation placement at bridge abutment A. (Permission of E. Tutumluer.)

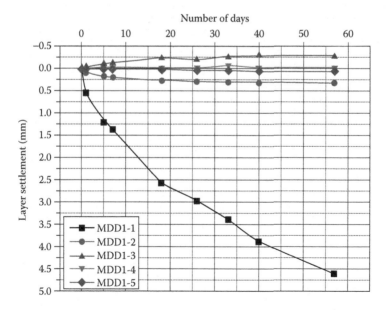

Figure 9.23 MDD cumulative deflection in layers with traffic loading at abutment A. (Courtesy of E. Tutumluer.)

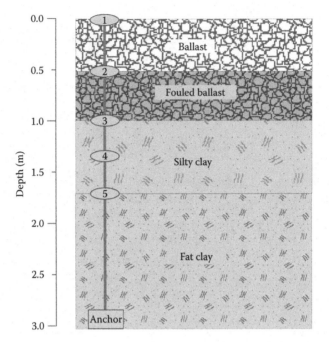

Figure 9.24 Track substructure layers and MDD instrumentation placement at bridge abutment B. (Permission of E. Tutumluer.)

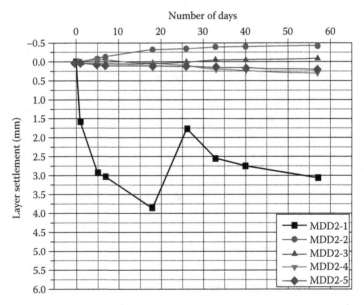

Figure 9.25 MDD cumulative deflection in layers with traffic loading at abutment B. (Courtesy of E. Tutumluer.)

9.5.4 Managing undercutting: Selecting locations, track segment length, and their priority

The GPR measurements, as described in Chapter 8, can provide much more than just the gray wavy lines that imply layer structure, as shown earlier. Further post-processing of the GPR data can reveal much more detailed information about the track substructure. For example, Figure 9.26 shows the amount of ballast fouling as measured by GPR from the view of looking down from above the track, where most fouled ballast is indicated by the darker shades and the least fouled is indicated by the lighter shades.

Ballast fouling from GPR measurements can also be represented as a numerical index. The second band shown in Figure 9.27 is also an indication of ballast fouling but is derived from the information in the band above it by quantifying the level of fouling rather than using shades or color. The third graph from the top shows the amount of moisture in the track substructure with lighter shading meaning more water. This represents water that should be released by improving track internal and/or external drainage.

The last band shown in Figure 9.27 indicates the track locations where the geometry error has historically been rough by the lighter-shaded vertical bands.

All data are combined in this view to present a composite picture of the need for undercutting. It is apparent from this figure that for this location undercutting should proceed, for example, from MP 91.6 to MP 92.2, from the beginning to the end of the fill section. The length of track to undercut is based on the length for which excess water exists that must be removed and the length of the fill section that also appears to be inhibiting external drainage. Last, the data in Figure 9.27 are used to confirm that the track length to be undercut is also sufficient to address the section associated with rough track geometry. In this way, measurement data from various sources are combined to provide a rational means to determine the locations and priority of undercutting, allowing more efficient management of their remediation.

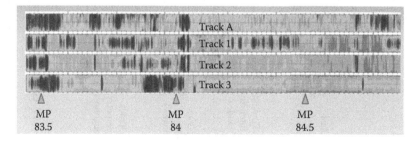

Figure 9.26 Ballast fouling intensity from GPR.

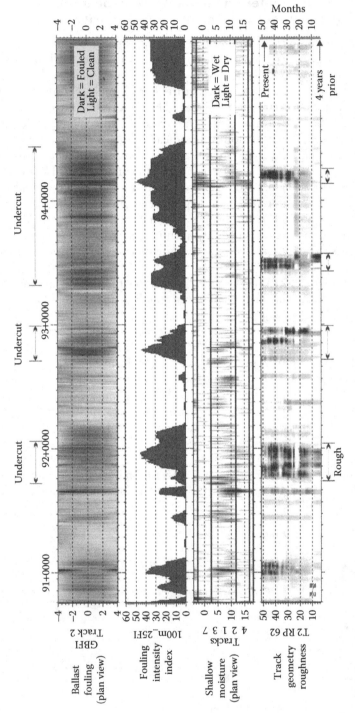

Figure 9.27 GPR and track geometry data combined to prioritize need for undercutting.

9.6 SUBBALLAST AND SUBGRADE IMPROVEMENT AND REHABILITATION

9.6.1 Subballast improvement methods

Subballast does not typically fail or require maintenance during the life of the track if it was installed with adequate thickness and gradation. Instability within the subballast has been observed on occasion when water is not readily drained from it and loading induces liquefaction of its fine sand and silt components. However, if the subballast layer is well drained, liquefaction does not occur. The improvements that can be made to the subballast are typically aimed at improving its ability to reduce the stresses from loading passed onto the subgrade and its ability to divert water away from the subgrade. Unfortunately, there are no known methods for improving the subballast in place without gaining access to it by removing the overlying material.

For example, a soft subgrade that is deforming due to being overstressed from the imposed loading can be stabilized by stiffening the subballast layer with placement of a geosynthetic material, such as a geocell or geogrid. A stiffer overlying layer acts to reduce the vertical stress on the subgrade surface. In the case of geogrid, its tensile strength can help reduce the stress transmitted to the subgrade. For the geocell, the composite action of subballast being laterally confined by a geocell can result in a much stiffer layer than with the subballast alone.

A layer of asphalt concrete placed over or within the subballast layer or a cement stabilized layer mixed in with the subballast can also increase subballast stiffness and reduce subgrade stress.

However, if the stiffer subballast layer in the form of asphalt or cement stabilized layer is not strong in bending fatigue, excess vertical deflections of a soft subgrade under loading can produce cracking of the asphalt or cement stabilized subballast layer, resulting in large track deformations and possible intrusion by the subgrade. This fatigue cracking of a cement stabilized subballast layer occurred at the Kansas City Test Track failure, as described in Section 3.4.4. Cracks in this stiffened subballast allowed the soft subgrade clay layer to infiltrate upward and into the ballast, producing large track settlement. This type of fatigue failure has also been observed in asphalt subballast layers installed in track over soft subgrade. A design procedure should be used that considers the subgrade-induced bending stresses and deflections in the stiff subballast layer so that its allowable tensile strain is not exceeded.

Another potential subballast improvement is the addition of an impermeable layer on its top surface to act as a barrier to water flowing in from above so that it is intercepted and diverted laterally to a drainage ditch. This can help to stabilize a subgrade that is moisture-sensitive. The impermeable layer could be a sprayed-on bitumen coating of emulsified asphalt or a rubberized asphalt cement poured on top of the subballast, which should allow bonding with the ballast placed on top. A layer of sand on top of the impermeable layer could assure that the ballast will not puncture this layer. A rubber membrane geosynthetic may also be used, provided it has sufficient surface friction to preclude a lateral slip plane from forming.

9.6.2 Subgrade stabilization methods

If track is to have an acceptable LCC, the design assumption must be that the subgrade remains stable with acceptable differential settlement during the entire track's life. Subgrade that performs poorly can be remediated by various means, the choice of which is dictated by the nature of the soil, the type of improvement desired, and the cost. In addition, the remedial work often needs to accommodate short track-occupancy times.

The most common subgrade improvements sought are better soil gradation, reduction of plasticity characteristics or swelling potential, increases in durability and strength, and better working platform for construction operations.

The following are brief descriptions of various subgrade improvement methods that are applicable to railways. Of those presented next, for most situations the most effective and economical subgrade treatment techniques are over-excavation and replacement, admixture stabilization, and horizontal reinforcement.

9.6.2.1 Over-excavation and replacement

Remediation of inadequate subgrade through the removal of unsuitable and deleterious soil, with the subsequent replacement and compaction of good material, is often the most cost-effective alternative, provided that the poor soil to be removed does not extend to a large depth (<5 ft), ground water is not present, and economical disposal of the excavated material is possible. Excavation and replacement is all the more a viable alternative when the soil type and conditions do not lend themselves to in-place remediation methods.

9.6.2.2 Admixture stabilization

Admixture stabilization of the subgrade is the process in which additives such as bitumen, lime, or commercial chemical products are blended and mixed into soil to improve its engineering characteristics of strength, texture, workability, and plasticity. Admixture stabilization is most effectively performed by heavy-equipment tillers, known as pulverizors (see Figure 10.7). Agricultural-type disc harrows can also be used, but these tend to be less effective and slower than tillers. Factors to consider when selecting the type of additive to apply are the type of soil to be stabilized, the purpose for which the stabilized layer will be used, the type of soil improvement that is desired, the required strength and durability of the stabilized layer, and the cost and environmental conditions.

When using lime or cement admixture, consideration must be given to the issue of "sulfate attack." This is a phenomenon in which lime-treated soils with moderate to high levels of sulfate content can swell excessively over an extended period. Prior to applying lime admixture stabilization, laboratory testing on subgrade soil samples is required to determine that the sulfate SO_4 and SO_3 content is <0.04% by weight to avoid the detrimental effects of sulfate attack.

9.6.2.3 Horizontal reinforcement

Horizontal reinforcement of track substructure layers using geogrid or geoweb can provide a good bearing medium through the mechanisms of interlocking, tensile reinforcement, and lateral confinement. A layer of reinforcement placed in a granular layer can resist tensile strain and will act to increase the horizontal confinement stress and correspondingly increase the overall stiffness properties of the soil or aggregate layer around the reinforcement. The increased stiffness results in less vertical deformation. The increased stiffness of the reinforced layer also results in lower vertical stresses transmitted below the reinforced layer. The reinforced layer also takes up some of the load-induced shear stresses that are not then transferred through this layer to the layer below.

9.6.2.4 Rammed-aggregate piers

This ground improvement method requires drilled holes followed by placement of stone, which is then compacted to create stiff piers of unbound aggregate. This method works best in fine-grained soils that can stand open after drilling, whereupon stone is then placed and compacted. The surrounding soil is improved by its composite interaction with the stone, and vertical track loads can be transmitted through the aggregate piers to a competent bearing stratum.

9.6.2.5 Compaction grouting

Compaction grouting is used to form a grout bulb that displaces the soil around it and hence increases its strength. Compaction grout is a mixture of cement, sand, clay, and water and is pumped in a stiff, low-slump state. In addition to densifying the soil, compaction grout columns can transfer load to a stronger layer.

9.6.2.6 Penetration grouting

Penetration grouting, also referred to as permeation grouting, is injection of Portland cement or chemical grout into the soil to increase soil strength and often decrease permeability. As the name implies, the grout medium is intended to penetrate into the void matrix of the soil and thereby decrease soil permeability by sealing off the voids and increase the strength of the soil it bonds with. This technique is suitable for granular soil due to the relatively large voids and high permeability. However, some low viscosity chemicals, such as certain urethanes, can be effective in penetrating fine-grained soil.

9.6.2.7 Soil mixing

Soil mixing is an improvement method that mechanically mixes the soil with cementitious binding material. This method is used typically for soft soil with high moisture content. Soil mixing is unique in that a high ground water table is actually a benefit rather than a barrier to the use of this method. The soil mixing paddles, or augers, are first advanced through the soil, and then the binder material is injected so that it mixes with the soil as the drill is retracted. This soil–cement columns increase the bearing capacity.

9.6.2.8 Slurry injection

Slurry injection, a form of penetration grouting, has been used on railways for many years and with mixed results. In this technique, lime or cement slurry is injected into the subgrade by pumping the material through injection rods or pipes as they are inserted. Typically, three injection rods are used (Figure 9.28), and injection generally proceeds from the ground surface, stopping intermittently to allow slurry to enter the soil matrix. The typical maximum depth of penetration is approximately 12.2 m (40 ft). Injection is

Figure 9.28 Slurry injection nozzle and rod system.

continued until (1) slurry emanates from the track surface, ditches, or other injection holes, (2) the rods will no longer penetrate, or (3) slurry will not flow into the soil. Spacing of injection locations along a track varies with site conditions but typically corresponds to every third tie crib area. Injection pressures of 21.8 kPa (150 psi) are commonly used.

For this method to be effective, there must be a sufficient network of voids and cracks that exist in the soil mass, or are opened up under the high-pressure injection, to allow the slurry to penetrate. However, it appears that this slurry penetration often does not occur. Karol (1983) reported that clay soils are generally not able to accept the slurry because they do not possess the necessary voids to promote flow of grout. To treat clay soils, the method of fracture grouting is generally required. Fracture grouting involves injection of slurry under adequate pressure to cause hydraulic fracture of the soil, which opens seams, fissures, and bedding planes. Although slurry injection pressures are adequate to produce such fracturing, the procedure does not lend itself to high-quality dispersion of slurry, where a network of seams would open up to allow slurry penetration.

Investigations of slurry injection sites have revealed a lack of dispersion of the lime slurry into the fine-grained subgrade. However, instead of the grout reaching out into the soil, the lime slurry has been shown to fill and solidify the thick, relatively clean, ballast in ballast pockets. Although not the theorized improvement method, these ballast pocket sites did seem to be improved by the slurry injection.

Although improvements from using this method have been reported, it should not be chosen without first conducting a soil exploration to determine if the slurry can be effectively distributed throughout the soil mass.

9.6.3 Control of expansive soil

Expansive soil is typically best treated by stabilizing using lime or Portland cement admixture. If admixture stabilization is not an option, then the problems associated with expansive soil can be mitigated by densification of the expansive soil by compaction at 1%–3% wet of the optimum moisture content and then kept from drying out before the subbase is constructed.

In fill situations, potentially expansive soil should be kept below 3 ft from final grade and non-expansive soil placed within 3 ft of final grade. The non-expansive cover will also have a surcharging effect on the expansive soil below. The cover soil should be of low permeability to limit the moisture change in the soil below.

9.7 OBSERVATIONAL METHOD

Although this chapter has stressed the need to use measurements to manage and arrive at an engineered solution, situations must sometimes be managed and decisions and engineering judgments made with little or no information before hand. The Observational Method (Peck 1969) is a useful approach that an experienced engineer can use to manage these situations.

When faced with a design challenge, the engineer has several options. The design engineer may choose to adopt a very large factor of safety to compensate for the lack of detailed knowledge; however, the over-conservatism this can produce can be very inefficient and costly. Another approach can be to make assumptions based on general, average experience; however, this can provide an un-conservative design if the actual conditions happen to be worse than this average assumed condition. A third option is to use the observational method approach that allows the design to be altered during construction based on the findings made from observation, assuming that such design changes can be implemented during construction.

An outline of the general approach to the observational method (Peck 1969) is the following:

1. Gather information.
2. Evaluate conditions and define both the most probable and the most unfavorable conditions.
3. Design based on the most probable condition.
4. Select parameters to be observed and calculate the values that are expected for the observation. Also, define the most probable and the most unfavorable observations.
5. Determine action for any possible condition encountered.
6. Measure and evaluate quantities during construction.
7. Take action/modify design.

The essential ingredient in the observational method is the visualization of all possible eventualities and the preparation in advance of courses of action to meet whatever situation develops.

The utility of the observational method increases with the experience of the engineer. To use the method with confidence, the engineer must be aware of the information needed for its success, focus on what is to be observed and measured, and select realistic courses of action based on the observations. It is imperative that data be reviewed immediately as they are collected. It may be necessary to slow down the construction in order to make continuous changes to designs and procedures as information is obtained.

The merit of the observational method is that it encourages the engineer to use a systematic procedure for each new job and to see the job without preconceptions, to see it as unique, before making interpretations. This approach allows for a large amount of uncertainty and a lack of detailed knowledge at the outset of a project, as solutions are being formulated, which is a benefit when much information is unknown. In short, the observational method recognizes the necessity of using empiricism in design.

Railways undertaking construction or remedial projects, or contracting such work out to others, should consider using or advocating this approach where it is warranted, given the large potential savings in time and money. The authors encourage use of the observational method because of the large potential for cost and time savings for typical railway geotechnical projects.

9.7.1 Examples of use

Geotechnical engineer Ralph Peck developed and championed the use of the observational method (Peck 1969) and provided a number of applications of the method. The needed amount of lateral bracing support for an excavation is one of the great unknowns even when geotechnical conditions have been fairly well defined, due to the uncertainty of the stress distribution that results from excavation. However, this uncertainty can be well managed by using the observational method, as illustrated in the following method, as described by Peck (1969).

The contractor had extensive data from excavations nearby but chose to base his bracing design not on the most conservative case but on the average conditions encountered by others—an approach that, by itself, would present considerable risk. Accordingly, the contractor proposed to design the struts at a relatively low factor of safety by using a minimum amount of bracing initially, but he continuously measured the strut loads in the bracing as excavation progressed and at all stages of construction. He was thus able to quickly place additional bracing if the need arose. Ultimately, only three additional braces beyond the 39 originally planned for in the design had to be placed using this method, the cost of which was small compared with the savings realized using this approach, even considering the costs of the observation and measurements. Even if the contractor had encountered very unfavorable conditions, the cost and time penalty would have been minimal.

Another example mentioned by Peck, also with railway implications, involved tunneling for the Bay Area Rapid Transit, where buildings with foundations in the zone of influence of the tunnels could have been severely affected. A conventional approach would have been to provide underpinnings before tunneling; however, underpinning was not possible, given the restricted access. Injection of cement grout or of chemical grouts was also not an option because the soil permeability was too low. As the normal solutions were not suitable for this case, the option considered was to construct a protective structural wall between the tunnels and the buildings to shield the buildings from movement during excavation. The cost of this protective structural wall was extremely high so the observational method was used to determine if this support was actually needed as tunneling progressed. As tunneling started far from the location of the buildings, it gave a chance to observe the amount of settlement leading up to the site.

As the settlements were minimal, it became apparent that the protective work was not needed. But this amount of time was needed to make an adequate assessment.

These are just two examples where the observational method was appropriate and yielded substantial savings in time and money.

9.7.2 Potential disadvantages of the observational method

The mistake with the most serious implications when using the method is failure to select in advance the appropriate courses of action for all foreseeable deviations of the actual conditions. The implications of such a mistake become apparent as observations begin to indicate an unplanned for deviation during the design phase. If the observations show the project to be experiencing a problem for which no defense has been devised, crucial decisions must be made under the pressure of the moment. In fact, there may be no adequate solution to be found. Therefore, foresight is needed in considering all possibilities of unfavorable conditions for any chosen design so as to avoid embarking on a design that is vulnerable to these conditions and for which there may be no adequate solution.

The required assumption in using the observational method that the design can be altered during construction can provide some contractual difficulties. Developing contracts for bidding purposes can be difficult to do. In some cases, the probability that the most unfavorable conditions are realized may be relatively high such that the corresponding remedy may not be worth the cost.

If the engineer cannot devise solutions to problems that could arise under the most unfavorable conditions, even those with very low probability of occurrence, the design must then be based on those least favorable conditions. For this case, the observational method would not apply.

Chapter 10

Case studies

Example projects are presented in this chapter to illustrate investigation, design, remediation, and construction elements of several track substructure–related projects. The first case study examines the diagnosis of the root cause of a chronic subgrade problem and remedial work that fixed the problem. The second and third case studies deal with substructure aspects of track design: one considers new track design within an existing right-of-way as part of a double-tracking project and the other considers new track on a completely new corridor. The fourth case study looks at embankment instability issues, including investigation, analysis, design, and remediation. Although a stand-alone case study is not included, track drainage is addressed as an essential element of every case study presented in this chapter.

10.1 DIAGNOSIS AND REMEDIATION OF A CHRONIC SUBGRADE PROBLEM

Implementing a proper investigation of the track substructure together with a proper engineering evaluation is critical to diagnosing the root cause of track substructure–related problems. In most cases, it is not until the underlying cause of a problem is known that a lasting and economical solution can be found. The following sections present a case study of the correction of a chronic track substructure problem (Hyslip and McCarthy 2000), the solution of which was determined after effectively diagnosing the root cause of the problem.

10.1.1 Background and investigation (diagnosis)

A heavy axle load track with approximately 140 MGT of annual traffic of primarily unit coal trains began to severely deteriorate within approximately 4 years of initial track construction. The track began to experience profound substructure-related problems resulting in frequent loss of surface and alignment and requiring frequent maintenance intervention. In the 4 years of deteriorating track surface, various treatments were applied, including surfacing, undercutting, and lime injection, as well as attempts to strengthen the track shoulder by installing shear keys, burying ties, and creating a large shoulder berm. These treatments did not work because they did not address the root cause of the problem.

The track is located in a region in which published geologic and geotechnical information revealed that the majority of the track subgrade consists of residual soils derived from non-marine claystone and silty shale. These residual soils are typically highly plastic clays with high shrink-swell characteristics. In addition, the clays were of a dispersed nature.

The field investigation began with a detailed site reconnaissance that included an assessment of the drainage condition and local topography. In addition, cross-trenches were dug along the problem section to determine the ballast, subballast, and upper subgrade conditions and to procure samples for testing. Within the cross-trenches, the substructure layers and zones were accurately measured, recorded, and sketched. Field strength testing was also performed in the cross-trenches on the fine-grained subgrade soils using handheld penetrometer and shear-vane devices. Ground-penetrating radar (GPR) was also used to determine layer configurations and moisture conditions.

Laboratory soil testing for material identification and classification was performed, which included grain size distribution of the ballast and subballast material, and liquid and plastic limits tests and moisture content determinations of the fine-grained subgrade soils. These tests confirmed the data from the background investigation that the soils were highly plastic, had a high swelling potential, and had a great affinity for water that caused softening.

Additional liquid and plastic limits tests were performed on samples of the subgrade clay to determine the effect of hydrated lime on the behavior of the subgrade soil. The results of one series of these lime tests are presented in Figure 10.1. This figure indicates that the lime effectively reduced both the liquid limit and the plasticity index and thereby made the clay stronger and less susceptible to water softening and swelling. This testing determined that at least 5% lime ($CaOH$) by weight of soil was needed for the desired effect in the field. Other tests performed on the soil included swell tests on the natural- and lime-treated soil, as well as unconfined compression tests of lime-treated samples.

The subgrade soil was also tested to determine its sulfate content to access the "sulfate attack" potential of the soil. Moderate to high levels of sulfate in lime-treated soils can result in the development of the mineral ettringite within the soil. Ettringite has an enormous affinity for water, and the swelling, upon imbibing water, can produce as much as a 500% increase in volume. This development of ettringite mineral and the consequential swelling is referred to as sulfate attack. The sulfate content of the soil (both SO_3 and SO_4) was determined to be <0.04% by weight, which indicates that the soil was not prone to sulfate attack.

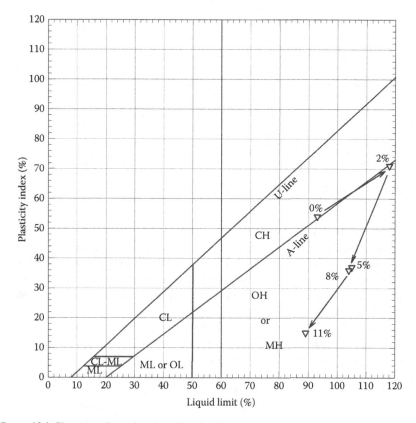

Figure 10.1 Plasticity chart showing effects of lime.

10.1.2 Substructure conditions

The ballast layer was clean to moderately fouled and varied in thickness from 9 to 20 in., which was the result of localized addition of ballast during spot surfacing work. The thickness of the ballast not only varied longitudinally along the problem section but also perpendicular (transverse) to the track. Subballast, consisting of crushed dolomite gravel, was present between the ballast and the underlying subgrade. The thickness of the subballast layer varied considerably, and it was evident that the subballast had been displaced since initial placement. Figure 10.2 shows a typical cross-section of the track.

The subgrade was residual olive gray, highly plastic clay (fat clay) derived from the in-place weathering and decomposition of the underlying bedrock. Being residual in nature, the clay was generally highly over-consolidated and therefore relatively incompressible and strong. However, these highly plastic soils have a great affinity for water and tend to soften considerably

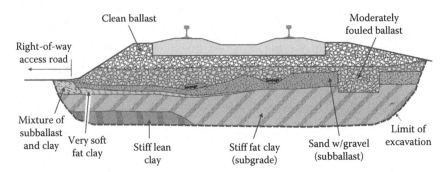

Figure 10.2 Typical cross section of chronic problem site.

when in prolonged contact with water. In addition, the soil was dispersed (not flocculated). The clay subgrade was moist with a medium stiff to very stiff consistency below the uppermost 1–3 in. of the stratum. The upper 1–3 in. of the subgrade clay, at the interface between the bottom of the subballast and the top of the subgrade, was soft to very soft and was found intermixed with the subballast.

The drainage of the track substructure at the problem section was typically poor. The subballast, which is a key component to proper substructure drainage, did not provide adequate drainage because the bottom of the subballast layer was blocked by the right-of-way access road. This initial poor drainage condition resulted in further drainage problems as depressions formed in the subgrade, resulting in ponding of subsurface water. The poor drainage conditions at the site caused the water falling in the track to remain in the substructure for a relatively long period of time, and consequently the water was able to saturate and soften the upper portion of the highly plastic subgrade soil.

It was found that the mechanism of failure was a combination of

- Prolonged contact of water with the highly plastic subgrade, combined with the repeated train loading, resulted in a loss of strength of the thin upper layer of the subgrade.
- The subgrade soil softened to the extent that a thin layer of soil remolded under repetitive heavy axle loads from the train traffic and then squeezed to the side of the track in the direction of least resistance.
- The soft soil layer acted as a lubricated surface that resulted in lateral spreading and deformation of the unbound granular subballast and ballast.
- The process continued and the problem worsened as the soft layer became thicker (deeper) and as the layers became further deformed and drainage was further impeded.

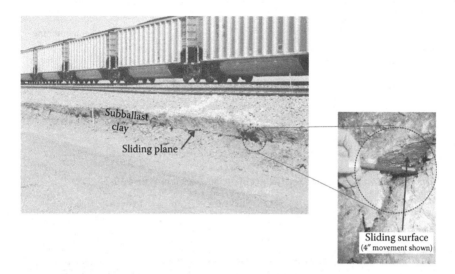

Figure 10.3 Photo of subballast sliding on weakened subgrade surface.

The lateral movement of the subballast due to the weakened subgrade surface could be sudden, as the lateral forces increased due to confinement and then released once the confinement forces in the ballast shoulder were exceeded. Evidence of this lateral displacement and the sudden nature of the movement was apparent once construction began and the right-of-way access road was removed adjacent to the track, which effectively removed the confinement of the subballast. Once the lateral confinement was removed, the track was prone to shifting laterally, as shown in Figure 10.3.

10.1.3 Solution and implementation

Once the root cause of the problem was determined, remedial options were considered, and the most appropriate option was chosen based on effectiveness and long-term costs. The investigation and analysis revealed that the solution to the substructure-related problems must address the poor internal drainage of the substructure as well as the moisture-sensitive subgrade soil. In general, the correction of the problem entailed the following:

- Removal of track, ballast, and subballast to expose the soft clay layer.
- Removal of soft, thin clay layer, leaving the underlying stiff clay.
- Reducing the clay subgrade's affinity for water by mixing dry lime with the top 6 in. of clay and then compacting the clay.
- Sealing of the clay with geomembrane to prevent surface water from reaching the clay.
- Reconstructing substructure with good internal drainage.

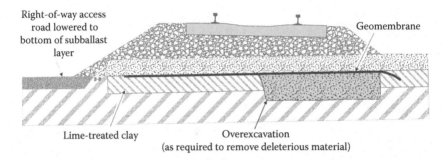

Right-of-way access
road lowered to
bottom of subballast
layer

Geomembrane

Lime-treated clay   Overexcavation
(as required to remove deleterious material)

Figure 10.4 Solution chosen for the problem illustrated in Figure 10.2.

Figure 10.5 Removing track panels with front-end loaders.

Figure 10.4 presents a typical cross section of the solution.

The remedial work began by removing the track in approximately 120 m (400 ft) panel lengths using front-end loaders, as shown in Figure 10.5.

Once the track was out of the way, the ballast and subballast layers were excavated. The adjacent right-of-way access road was lowered using heavy earthmoving equipment to provide a drainage outlet for the subballast. The soft, upper portion of the clay subgrade was removed, and the subgrade surface was then "proof-rolled" with heavy compaction equipment to densify the near surface subgrade soil and to locate any excessively soft or wet zones, which were subsequently over excavated. Figure 10.6 shows subgrade compaction equipment and a location that required over excavation.

Lime admixture stabilization was performed on the clay subgrade to a depth of 6-in. using a high-speed, production-type tiller, as shown in Figure 10.7. The lime admixture treatment of the clay soil was very effective in improving the properties of the clay for placement, compaction, and long-term stability.

After the subgrade was treated with lime and compacted, a geomembrane was placed on the prepared subgrade to prevent water from coming

Figure 10.6 Over excavation and compaction.

Figure 10.7 Lime admixture stabilization with high-speed mixer/polarizer.

into contact with the soil (Figure 10.8). The subballast and ballast layers were then placed, and the track was reinstalled.

The remedial work effectively corrected the problem. Figure 10.9 shows before and after photographs of the problem site.

10.1.4 Monitoring

This substructure problem remediation was done incrementally by first reconstructing two of the most severe locations and then monitoring the performance for a period of time. Different treatments were attempted at two of the most severe problem sections of track. One treatment used lime

Figure 10.8 Geomembrane placement (left), and geomembrane after being in service for 200 MGT (right).

Figure 10.9 Before (top) and after (bottom) the remediation.

admixture stabilization to reduce the softening potential of the highly plastic clay subgrade. The other treatment used emulsified asphalt (bitumen) spray to seal water away from the clay, thus precluding water softening. Based on the monitoring of track performance, the lime admixture stabilization was chosen as the preferred alternative.

During the excavation of the track substructure, evidence was observed of past unsuccessful attempts at stabilization using cross-drains, shear keys, and buried ties at numerous locations. In most cases, these treatments were ineffective because they resulted in trapping of water and softening of the surrounding clay subgrade.

After more than a decade and more than 1 billion gross tons of traffic, the site remains smooth and stable. The lime admixture treatment was very successful in improving the soft clay soil to the point where the upper subgrade no longer causes a problem. The lime-treated soil layer between the subballast and natural clay subgrade was found to be hard and impermeable and provides an excellent bearing and water shedding medium.

10.2 TRACK DESIGN ON OLD ROADBED

Designing new track is more of a challenge when it is being placed on an old roadbed that has pre-existing problems such as locations of soft, deformed subgrade, and hidden drainage problems. Where new track is being constructed next to an existing track, the designer typically does not have the luxury of building over newly placed and well-compacted subgrade made of select materials but must instead build over an old roadbed. If the poor conditions of the roadbed are not identified and improved before construction, they will impose themselves on the new track, resulting in much more maintenance than might have been planned (Staplin et al. 2013).

10.2.1 Investigation

10.2.1.1 Use of GPR to assess old roadbed conditions

GPR can be used to rapidly characterize poor roadbed conditions. Figure 10.10 (left) shows the common arrangement of GPR antennae during measurement where they are attached to a hi-rail vehicle. Figure 10.10 (right) shows an alternative arrangement used in these investigations, with the antennas being dragged behind the vehicle along the maintenance-of-way access road to measure the subsurface conditions of roadbed over which the track is to be built.

As described in Chapter 8, when the GPR signals are transmitted downward and subsequently reflected back to the receivers, they encounter substructure layer boundaries such as ballast, subballast, and subgrade. These signals can be processed to show depth to the subgrade and the variation of this depth along and across the track. In this way, the presence of a soft and deforming

Figure 10.10 400 MHz GPR antenna installed on hi-rail vehicle (left photo), and GPR setup to investigate right-of-way access road for new track (right photo).

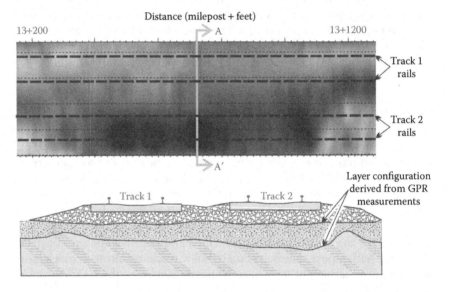

Figure 10.11 Plan view of subgrade surface elevation for two-track right-of-way (upper). Cross section of track showing deformation of subgrade surface (lower).

soil condition in need of remediation can become apparent from the resulting distorted and varying subgrade profile, as shown in Figure 10.11.

GPR measurements also provide information on the thickness of the existing granular roadbed layer over the subgrade soil so that locations where this thickness is insufficient can be identified. Last, GPR can determine the relative amount of water present that needs to be drained.

10.2.1.2 Use of track geometry car data

Track geometry car data from the existing track that is adjacent to the track to be constructed can be used to determine locations that have a history of repeated profile and cross-level geometry error that may be attributed to soft

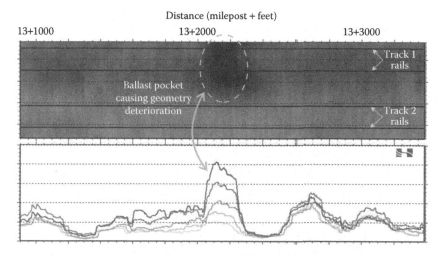

Figure 10.12 Plan view of GPR-indicated ballast pocket (top) and corresponding progressing surface roughness from repeated geometry car measurements (bottom).

subgrade. In this way, track geometry data complement the GPR measurements to indicate a deforming subgrade in need of remediation, as shown in Figure 10.12, where the new track is to be constructed next to Track 1.

10.2.1.3 Use of mapping

As explained in Chapter 8, mapping using LIDAR technology can be a valuable tool to determine ground surface features for use in design of external track drainage. Figure 10.13 shows an example of a digital terrain model (DTM) that was developed from aerial LIDAR scanning. This mapping provides a detailed rendering of the surface and shows local drainage features.

10.2.2 Granular layer thickness design

For the new track to be constructed on the old roadbed, the required granular layer thickness to control subgrade deformation was determined by using the design procedure described in Chapter 5. The following is the granular layer thickness design that was performed for the new track.

A soil exploration showed that the subgrade is clayey silt with low plasticity, an ML soil type using the Unified Soil Classification system, and had the following parameters: unconfined compressive strength, $\sigma_s = 16$ psi; subgrade resilient modulus, $E_s = 8,000$ psi; ballast resilient modulus, $E_b = 20,000$ psi. The mixed traffic of passenger and freight loading was found to produce an annual traffic tonnage of 4.25 MGT/year. The track design life is 30 years.

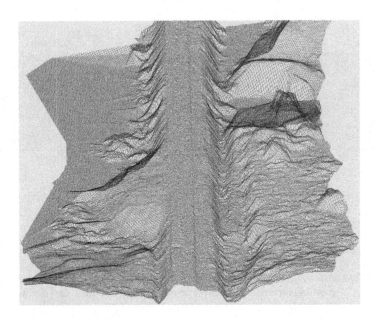

Figure 10.13 Example of LIDAR image in track cut and fill sections.

10.2.2.1 Design method I

From Equation 2.8, with adjustment for English units, the dynamic wheel load, $P_d = [1 + (0.33 \times V/D)] \times P_s = 1.55 \times P_s = 56$ kips (28 tons), using $P_s = 36$ kips (18 tons), $D = 36$ in., $V = 60$ mph.

For design period of 30 years:

$$T_i = 30 \text{ years} \times 4.25 \text{ MGT/year} = 128 \text{ MGT}$$

From Equation 5.15, the number of load cycles:

$$N = \frac{T_i}{8P_{si}} = \frac{(128,000,000 \text{ tons})}{8(18 \text{ tons})} = 890,000 \text{ cycles}$$

Assuming ML for the clayey silt, use coefficients $a = 0.64, b = 0.10, m = 1.7$, and assuming an allowable plastic strain, ε_p, of 1.5%, find the allowable deviator stress, σ_{da}, at subgrade surface, as found from Figure 10.14.

From Figure 10.14, $\beta = (\sigma_d/\sigma_s) = 0.74$.

Using $\sigma_s = 16$ psi for clayey silt of low plasticity (ML), then $\sigma_{da} = \beta \times \sigma_s = (0.74)(16) = 11.8$ psi.

The strain influence factor, I_ε was determined by the following equation using an area factor, $A = 1,000$ in.[2] (constant for English units) by:

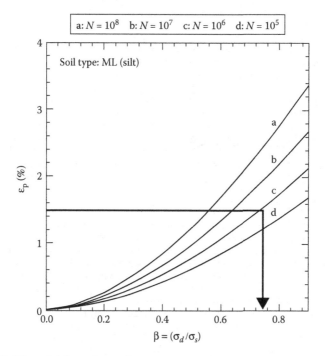

Figure 10.14 Chart A.1 from Appendix.

$$I_\varepsilon = \frac{(\sigma_{da}A)}{P_d} = \frac{(11.8)(1,000)}{56,000} = 0.21$$

From Figure 10.15, with $I_\varepsilon = 0.21$ and $E_s = 8,000$ psi and $E_b = 20,000$ psi then $H/L = 3.4$, and using an English length factor of 6 in. ($L = 6$ in.), then $H = 3.4 \times 6 = 20$ in.

Therefore, for Method 1, granular layer thickness, $H = 20$ in.

10.2.2.2 Design method 2

Using an allowable plastic deformation, $\rho_a = 1$ in., and considering a design period of 30 years, the influence factor, I_ρ, was calculated from

$$I_\rho = \frac{\rho_a/L}{a(P_d/\sigma_s A)^m N^b} \times 100$$

$$I_\rho = \frac{1/6}{0.64\left[56,000/(16)(1000)\right]^{1.7} 890,000^{0.1}} \times 100 = 0.79$$

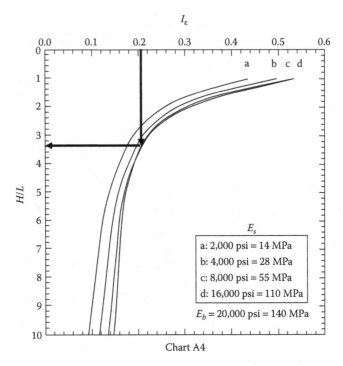

Figure 10.15 Chart A.2 from Appendix.

With a deposit of clayey silt of 13ft thickness (156 in.), $T/L = 26$ was considered, using an English length factor of 6 in.

From Figure 10.16 with $I_\rho = 0.79$, $H/L = 2.8$, where $H = 2.8 \times 6 = 17$ in.

Therefore, Method 1 governs, and the design granular layer thickness was determined to be 20 in.

Use of this design method showed that for this project the minimum required granular layer thickness is 20 in. to limit subgrade deformation to a tolerably small amount over the track life of 30 years.

10.2.3 Challenges encountered during design

10.2.3.1 Unstable slopes

Fills along existing track and fills required for new track design must be sufficiently wide to preclude loss of ballast shoulder material off the end of the tie, as shown in Figure 10.17. Note how the aerial view in Figure 10.17a indicates that ballast and roadbed material has been sliding into the water and produced a change in water color from the underwater "plume" of material local to the failure. Figure 10.17b shows the steepness of some slopes

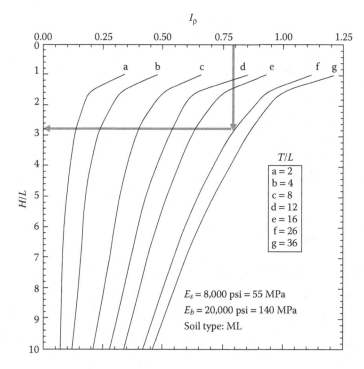

Figure 10.16 Chart A.3 from Appendix.

Figure 10.17 (a) Overhead photograph of unstable ballast shoulder; (b) oblique view at ground level at the same location.

and their inability to retain a ballast shoulder. These conditions are to be repaired by providing retaining walls at several locations. The existing substructure conditions at the location shown in Figure 10.17 were determined by GPR and are shown in Figure 10.18a. A strategy proposed for remediation is shown in Figure 10.18b.

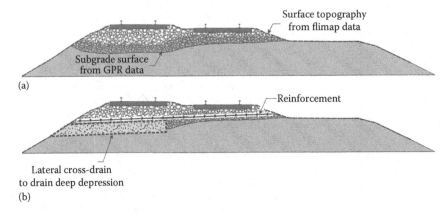

(a)

(b)

Lateral cross-drain
to drain deep depression

Figure 10.18 (a) Ballast pocket in soft subgrade location found by GPR; (b) proposed remediation.

10.2.3.2 Poor internal and external track drainage

A number of cut area locations had poor drainage, such as the location shown in Figure 10.19, with the corresponding substructure conditions indicated in Figure 10.20a. Note that in addition to poor external drainage conditions of inadequate ditches and insufficient longitudinal slope of the ditch, there are poor internal drainage conditions due to lack of a lateral escape path for water trapped in the roadbed. The proposed remedy is shown in Figure 10.20b.

Figure 10.19 Example of poor external track drainage.

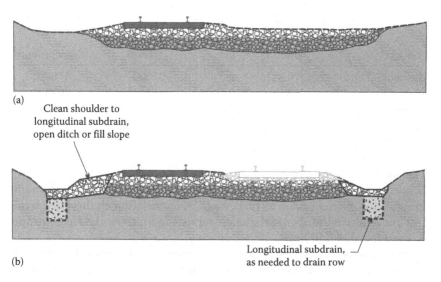

(a)

Clean shoulder to
longitudinal subdrain,
open ditch or fill slope

Longitudinal subdrain,
(b)   as needed to drain row

Figure 10.20 (a) Corresponding poor internal track drainage (upper); (b) remediation of drainage (lower).

10.3 TRACK DESIGN—NEW ROADBED

This case study examines the elements of new track design on an entirely new corridor alignment and presents the basic steps, concepts, and considerations that were undertaken to develop the project. The specific case is the 108 km high-speed Channel Tunnel Rail Link (CTRL) project, as discussed in O'Riordan and Phear (2001) and Phear and O'Riordan (2003).

This case study also illustrates an application of the observational method approach to design and construct, as described in Chapter 9. Site reconnaissance was used to identify the range of natural soil conditions and to allow the field staff and construction personnel to be prepared to install the appropriate track structure for the various subgrade conditions. This approach allows the subgrade variability to be accounted for in the design of the track substructure so that the subgrade settlement is minimized and the superstructure is uniformly supported.

10.3.1 Background

The CTRL is a double track operating at up to 300 km/h (188 mph) that had final design completed in the late 1990s with design and construction proceeding essentially until the line opened for service in late 2007.

Table 10.1 Loading information for track design

Car type	Axle load (kN)	Axle spacing (m)	Maximum speed (km/h)	Cycles per year
Eurostar	170	3.0	300	511,000
Commuter	130	2.5	160	240,900
Freight	225	1.8	150	273,750
UIC Design Load	250	1.6	–	–

The motivation for constructing this line was to connect the Channel Tunnel to the City of London at St. Pancras station. In addition to the high-speed service from London to Paris via the Channel Tunnel, this line carries 160 km/h maximum speed commuter service and intermodal freight trains at a top speed of 150 km/h.

Two sections of track design and construction comprise the CTRL project. Section 1 included the track from the channel tunnel terminal at Folkestone to Fawkham Junction in Kent and was 75 km in length and was constructed on ballasted track to a large extent, opening in 2003. Section 2 included the track from Fawkham Junction to St. Pancras station and was nearly 40 km in length and was built mainly in tunnels on bridges. The discussion of the CTRL project in this case study is focused solely on Section 1 and the design of the ballasted track along that section.

O'Riordan and Phear (2001) described the design of Section 1 of the CTRL project in detail including analysis of traffic loading. Table 10.1 shows the loading information used for track design.

The design number of equivalent axle loads required for the track design described in Chapters 4 and 5 is based on the largest percentage of traffic by applied tonnage. Completing this analysis indicates that the high-speed service applies approximately 30% more daily loading tonnage than the freight trains and should be used as the design axle load. Following the procedures described in Chapters 4 and 5, each of the other traffic types can be converted to this design axle load and an equivalent number of applied load cycles.

The determination of design dynamic loading is important to high-speed train loading analysis. As O'Riordan and Phear (2001) noted, the AREMA equation (Chapter 5) for dynamic loading prediction was developed based on mostly freight traffic data and likely overstates dynamic loading at speeds greater than 160 km/h. The data presented from several field studies indicated that the dynamic loading in high-speed train service is substantially lower than the AREMA equation might predict. This appears to make sense, but caution must be exercised in this regard because track geometry, track type, and train speed can all significantly affect dynamic loading. Field testing data from similar track, field conditions, and train design is a more prudent means to develop dynamic design load.

As an alternative, a validated vehicle/track interaction simulation model can be used for predicting design dynamic load (refer to Chapter 2).

For the CTRL project, based on available data (Woldringh and New 1999), the designers chose a design dynamic load of 350 kN, which is higher than the UIC design load of 250 kN, but is lower than what the AREMA equation would predict based on speed.

The determination of the number of applied load cycles is needed to accurately assess train loading conditions. Depending on the layer in question and the axle and truck spacing, the number of load cycles may change as described in Chapter 2. Widely spaced axles of a truck and a long spacing between the end trucks of two rail vehicles can create a condition under which each wheel will result in one load cycle for the track substructure, rather than the common group behavior where multiple axles act together to create one load cycle. On short freight cars, it was found that the two wheels of a truck acted together on the ballast and that the four axles around a coupler acted together when loading the subgrade. High-speed passenger equipment with long spacings between trucks at couplers almost universally results in independent stress pulses on the subgrade for each truck. Then the spacing of the axles on those trucks controls whether the axles act together or separately, but can often be taken as each axle being an individual load pulse on the ballast, while the two axles together create a single load pulse on the subgrade. This effectively doubles the number of applied load cycles for passenger traffic compared to freight traffic.

10.3.2 Site description and soil conditions

The site covers a wide variety of geologic settings, from the sand deposits near Folkestone, through the chalk deposits of southeast England, to the over-consolidated clays closer to London. This wide range of site conditions for natural soils is common for many railway corridors and illustrates the challenge of building a railway line across difficult soils. However, this challenge due to variable soil conditions is often assumed by designers to be less of a concern than choosing the proper alignment to minimize earthwork and bridge structures.

The most problematic soils along the CTRL corridor consisted of over-consolidated clay and chalk, which also had some high groundwater table-related challenges. Investigation of the spatial extent of these soils was conducted based on a combined desk and field study. The desk study identified problematic conditions and looked for alternative routes if possible. The field study consisted of performing reconnaissance over a long distance and in situ testing in more than 5,900 boreholes. Many of the boreholes were instrumented and monitored for up to 5 years to evaluate effects of seasonal variations (Phear and O'Riordan, 2003).

Although from an operational perspective a standard design cross section might be attractive, the variation in natural soil conditions will inevitably result in variations in the track settlement rate, track geometry deterioration, and track life. Instead of having a railway line that deteriorates more or less uniformly to maximize the life of the more expendable track superstructure components, the subgrade variability leads to premature failure of some sections well before the majority of track miles along the line.

10.3.3 Track foundation design

Available design procedures such as UIC Code 719R, as noted by O'Riordan and Phear (2001) and Phear and O'Riordan (2003), are based on experience rather than theory. To compensate for the limitations of this empirical approach and give more assurance to the extrapolated use of UIC Code for design, the project designers applied the track foundation design method developed by Li and Selig (Chapter 5).

The aim of track foundation design is to minimize the likelihood of subgrade failures over the design life, including failures due to progressive shear failure, excessive plastic deformation (ballast pocket), and subgrade erosion and mud pumping. A sufficiently thick granular layer and good drainage prevents the first two types of failures, and a properly designed subballast layer minimizes the likelihood of erosion and pumping failure.

An analysis performed during the CTRL project confirmed what was discussed in Chapter 4 regarding soil/subgrade stiffness and depth that has a dominant effect on track deformation. The design parameters described in O'Riordan and Phear (2001) include variations in resilient modulus for track substructure layers. For most ballast and subgrade materials, available data in the literature can be used to determine or infer resilient modulus for design of this project, as is the case for the project. However, a laboratory cyclic triaxial testing of recompacted chalk was undertaken for resilient modulus determination.

For a base case of clay subgrade with a compressive strength of 100 kPa (14 psi), with a design dynamic load of 350 kN, the design using the Li-Selig method produced the results shown in Figure 10.21. Note that 250 kN is the UIC design load.

The required layer thickness determined by design was found to vary substantially with applied wheel load as expected. Based on available data (Woldringh and New 1999), the designers chose to use dynamic wheel loading and an 80-year design life. The result of the design called for 0.9 m of granular layer thickness, which was found to be in general agreement with UIC 719R. The design deflection of the track became a construction control requirement.

The track was constructed with a maximum of 0.35 m of ballast, a 0.2 m of subballast, and a 0.35–0.7 m of prepared subgrade layer as shown in Figure 10.22. The prepared subgrade and subballast were all granular material placed on the natural subgrade. These three layers make up the

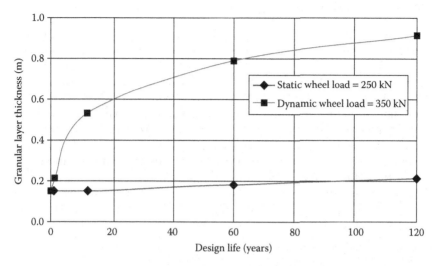

Figure 10.21 Required granular layer thickness versus design life for a clay subgrade. (Data from O'Riordan, N. and Phear, A., Design and construction control of ballasted track formation and subgrade for high speed lines. *Proceedings of Railway Engineering 2001*, London, 2001.)

granular layer thickness specified in the procedure described Chapter 5 for ballasted track. In the heavily over-consolidated clay sections, up to 0.65 m of clay was removed and replaced with compacted sand and geotextile layers to create a stable construction platform and to create a filter to provide separation of the clay from migration into the prepared subgrade. A drainage system to maintain drainage of water below the prepared subgrade was created and was 1.5–2 m deep in poorly drained cut locations along the right-of-way.

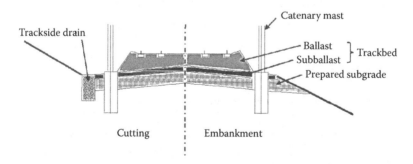

Figure 10.22 Schematic CTRL double-track cross section. (Data from O'Riordan, N. and Phear, A., Design and construction control of ballasted track formation and subgrade for high speed lines. *Proceedings of Railway Engineering 2001*, London, 2001.)

10.3.4 Construction control

Deformation and stiffness of track substructure layers can be specified by design, and monitoring of these substructure properties provides a construction control to assure that the design is being met. Density (unit weight) tests of compacted layers have traditionally been used for construction control; however, substructure layer stiffness provides a quality control measure that more directly relates to design. Ensuring uniform stiffness of each layer results in uniform track support that minimizes track stiffness variations and provides criteria for use during construction to identify appropriate methods to remediate locations that do not meet the specification.

For the CTRL project, the stiffness requirements for track substructure layers were monitored by using plate load test (600 mm diameter plate) on the top of the prepared subgrade and on the top of the subballast layer, using the following criteria:

$$E_{v2} = \frac{1.5r\Delta\sigma}{\Delta s} > 80\,\mathrm{MPa} \tag{10.1}$$

where:
   r is the plate radius (0.3 m)
   $\Delta\sigma$ is the increment of pressure under plate (200 kN/m²) between the first (250 kN/m²) and the second loading
   Δs is the increment of settlement of plate

The track design techniques described in the previous two cases were developed based on theory and field calibration. Results of these techniques should be monitored in service to verify performance of the track relative to design assumptions. Instrumentation to monitor layer performance and overall track settlement, while tracking applied load, would be beneficial to verify design performance and provide improved guidance on life cycle trade-offs to help designers judge the conservatism of their designs.

10.4 SLOPE INSTABILITY

Undertaking an investigation to study the condition of a failing embankment is critical to determining the root cause of the problem and the optimal solution. The geotechnical engineer's adage, "You pay for test borings whether you drill them or not," is never more true than in solving slope instability problems. An engineer cannot be certain that a solution will be effective and economical until the condition of the slope is determined and the mechanism of failure is well understood. This section presents a case study of an existing soil slope instability problem, describes the investigative approach used to understand the railway embankment condition, and discusses the remedial options available.

10.4.1 Investigation

The site is a three-track mainline right-of-way with a ±30 m high cut slope embankment to one side and an ±8 m high fill slope to the other (Figure 10.23). The high cut slope was a large fill embankment constructed for an interstate highway well after the tracks were in place. The crest of the fill slope is located immediately adjacent to one of the active mainline tracks, and the embankment fill slope terminates at a stream at the toe.

Routine track inspections at this location had long noted continual track settlement that required frequent geometry correction. Despite this repeated track maintenance, the track roughness deterioration was considered tolerable until a visual inspection of the embankment revealed indications of instability—the most prominent being a significant amount of tilting trees and bulging slope, as shown in Figure 10.24. Other indicators of instability present at the site were tension cracks and water seepage along the face of the slope, as well as surficial slumps and bulges on the slope surface. After the site reconnaissance, the investigation continued with a review of available background information, including railway information such as track geometry data and maintenance history, as well as geotechnical information such as topography, soil, ground water, rainfall, and geology. A digital terrain model (DTM) was developed for each site based on aerial LIDAR mapping.

Because this slope movement was active, a monitoring plan was quickly implemented. Soil borings and geotechnical instrumentation were used to document the extent and shape of the failure surface, its rate of movement, and the soil type and strength with depth.

The investigation phase consisted of three test borings and three slope inclinometers to measure the location and amount of lateral movement within the slope. After completion of the standard penetration test (SPT) and sampling, the test borings were converted to open-standpipe piezometers to monitor ground water level. The boring and instrumentation provided

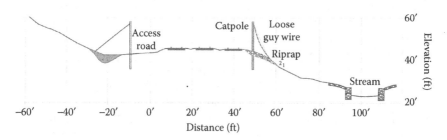

Figure 10.23 Cross section of slope case study.

Figure 10.24 Backward tilting of trees and soil bulge at the toe of slope.

information on the existing condition and behavior of the embankment including the nature and location of the subsurface movement. To obtain information at critical locations along the embankment, efforts were made to conduct the test boring and install the instrumentation at the crest of the embankment slope, within the active track envelope and at mid-slope locations (Figure 10.25).

Although slope movement was suspected at the crest and mid-slope borings, a boring, an inclinometer, and a piezometer were located at the toe of the adjacent highway embankment to determine if the track problem was associated with the highway embankment and if the failure was so deep-seated that remediation would need to extend into the highway embankment. Resisting the movement of such a large soil mass could substantially increase the remediation cost, time, and effort compared to a shallower failure. As the borings were advanced, the SPT blow counts were recorded, which indicated the relative density and consistency. The borings extended to a sufficient depth into the lower stiffer soil layers to characterize the resistance these layers can offer. Laboratory soil index property tests were conducted to supplement the field data. GPR surveys and dynamic cone penetration tests were also conducted at the site.

To understand the mechanism contributing to the slope instability, computer models were developed corresponding to the slope geometry and subsurface conditions developed for the specific site. Slope stability analyses were also conducted on models reflecting remedial methods. The analysis showed the extent of the remedial work needed and the sensitivity of the slope to the remedial work and ground water lowering.

(a)

(b)

Figure 10.25 Drilling on-track at (a) the crest and (b) Off-track at the mid-slope of railway embankment.

10.4.2 Site conditions

Based on the information collected during the investigation phase of the project, typical cross sections of the substructure conditions were developed. Figure 10.26 is a typical cross section for the project site.

Measurement of the slope inclinometer over 3 months indicated an abrupt distortion in the crest and mid-slope boring locations (B and C in Figure 10.26) but no change in the boring at the toe of the highway embankment (location A). Monitoring of the inclinometers and piezometers showed no evidence of movement near the highway embankment and that the slip surface was confined to the fill slope from the outermost track at the crest of the fill slope to the toe. The failure surface and zone of affected soil is shown in Figure 10.27. The "zone of instability" is the location within the slope where the strength of the soil is overcome by the forces trying to make the soil mass move. These driving forces are primarily from gravity and water pressure. The zone of instability is also the location where the unstable soil mass on the outer part of the slope moves away from the stable soil on the inner part of the slope.

It was also determined that a 15-ft deep ballast pocket had formed under the track due to repeated placement of ballast intended to restore track geometry as movement occurred. As this ballast material accumulated, the ballast pocket grew and was apparently moving along the failure surface and with the failing soil mass. The SPT "blow counts" indicated that the upper fill material of clayey sand to sandy clay was relatively soft and overlaid a stiff to hard clay. The stabilization method can make use of this stiff layer to resist embankment movement by anchoring or embedding into it to a sufficient depth to mobilize the needed resistance.

Water often plays a key role in driving slope movement due to its weight and ability to permeate cracks, weaken, and lubricate sliding surfaces. Based on the measured depth to water at these locations, the phreatic surface was determined, as shown in Figure 10.28. An artesian condition exists at the toe of the highway embankment due to elevated recharge from the highway embankment. The water surface elevation suddenly decreases as it approaches, and drains into, the very permeable ballast pocket at Boring B. The water level then slowly decreases in elevation as it approaches the level of the stream to the right.

Consideration was given to lowering the water level in the embankment by installing drains to intercept the ground water at the foot of the highway embankment. By doing so, the water surface elevation would drop, as shown in Figure 10.29. However, this was not expected to be sufficient to provide stability because the water level in the failure zone would not be substantially diminished after the water table was lowered, and the phreatic surface in the failing soil mass would remain essentially unchanged, as shown in Figure 10.29. Instead, stabilization is aimed at directly resisting the driving forces.

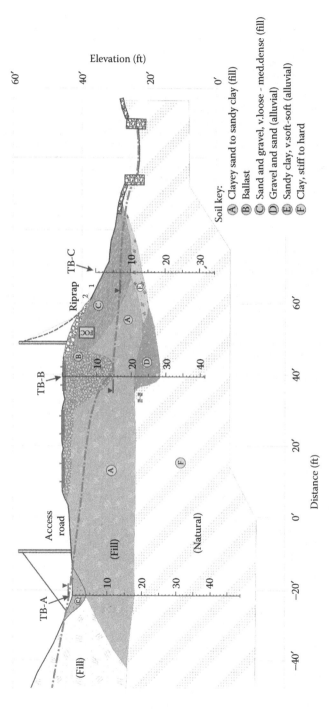

Figure 10.26 Cross section of site conditions for slope case study.

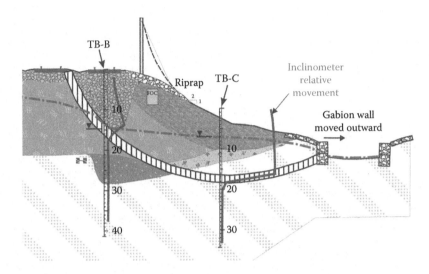

Figure 10.27 Zone of instability.

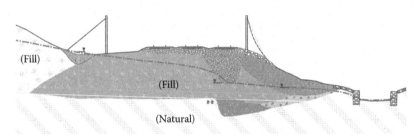

Figure 10.28 Phreatic surface.

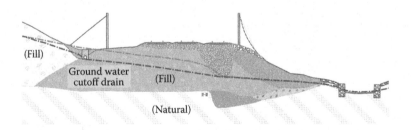

Figure 10.29 Lowering of phreatic surface from drain at toe of highway embankment.

10.4.3 Potential solutions

Once the mechanism driving the embankment instability was understood, remedial alternatives were considered. Potential solutions for the site conditions included

- Reconstruction of the embankment
- Use of retaining structure
- Buttressing or flattening of the slope
- In-place reinforcement of the embankment soils

Determination of the most appropriate and economical remedial alternative was aided by a systematic approach using a decision matrix, as shown in Figure 10.30. The decision matrix is built by first listing the required solution characteristics (top horizontal axis) and then assigning a weighting factor to these solution characteristics based on relative importance. The relative importance of each characteristic is based on how important the characteristic is to the success of the project, or in other words, how necessary or desirable it is. The weighting factor varied from ten (critical/necessary) to zero (not necessary). The next step in using the decision matrix is listing the potential solutions (left-side vertical axis) and then deciding how likely the potential solution will be to achieving the desired outcome. The result of the decision matrix is the total composite score of the potential solution, which is calculated by multiplying each solution characteristic by the weighting factor, with the product then summed horizontally. The most desirable solution has the highest total score. The decision matrix is intended to be iterated among various people within the decision-making hierarchy.

As shown in Figure 10.30, several remedial options were considered, and each of these are discussed next.

10.4.3.1 Flattening/buttressing

This option would require extending beyond property boundaries and also, more critically, rerouting a stream or putting the stream in a closed culvert.

10.4.3.2 Retaining wall with anchored tie backs

Consideration was given to a retaining wall installed into the stiffer soil layer with lateral tie-backs, as shown in Figure 10.31. Driving of the piling vertically would likely provide large vibrations that could produce settlement of the tracks. The piles would need to be installed by first pre-auguring holes.

10.4.3.3 Soil reinforcement

It was considered to combine permeation grouting of the ballast pocket together with placing reinforced compaction grout columns near the toe of the slope, as shown in Figure 10.32. The soil improvement provided by permeation grouting is expected to stabilize the ballast pocket on the upper

Scale to rate characteristics -->	Disruption to traffic	Stability improvement	Cost of construction	Conventionality of approach	Access/Setup	Follow-up	Ancillary benefits	Total score
Weighting factor for solution characteristic; relative importance (rank 10 = high; 0 = low)	8	10	6	3	5	2	5	
Potential solutions								
Flattening/Buttressing	6	1	8	10	7	9		189
Piling retaining wall	8	10	3	9	4	6		241
Gravity retaining wall	8	7	6	9	6	8		243
Anchored blocks	10	10	7	3	4	5		261
Soil dowels	10	10	9	3	7	8		294
Micropile with capbeam	9	8	6	4	3	6		227
								0
								0

Solution characteristics

Column descriptions:

- Disruption to traffic: 10 = Can utilize short work windows, track remains in-place, no post construction delay (minutes) // 0 = Requires extended track outage, track completely removed, 'cure' time needed (days)
- Stability improvement: 10 = Provides good, long-term stability improvement // 0 = Provides minimal, shortterm stability improvement
- Cost of construction: 10 = Low cost of actual construction work // 0 = High cost of actual construction work
- Conventionality of approach: 10 = There is common experience with solution, contractor/equipment are widely available, can be performed by RR forces, high probability of success // 0 = There is limited experience with solution, requires specialty contractor, unknown success rate
- Access/Setup: 10 = Easy mobilization to inaccessible sites, utilizes small and agile equipment // 0 = Requires large equipment, much material, limited equipment agility
- Follow-up: 10 = No follow-up treatment or special inspections needed // 0 = Requires followup treatments and/or frequent monitoring
- Ancillary benefits: 10 = Potential widespread applicability, provides other secondary benefits // 0 = Unique solution to specific problem, provides no secondary benefits

Figure 10.30 Decision matrix for selecting solution to the embankment instability.

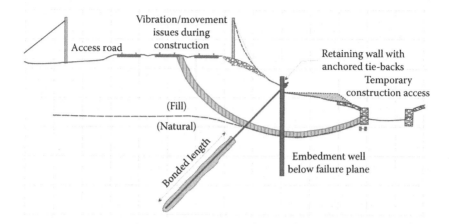

Figure 10.31 Retailing wall solution.

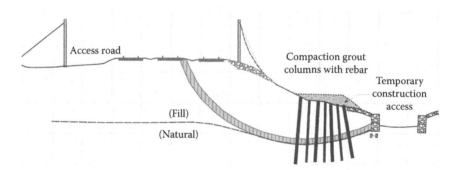

Figure 10.32 Reinforced compaction grout column solution.

portion of the slide by making it a solid mass that would also seal off the water to the failure surface. On the lower portion of the slide, the series of grout columns would intersect the failure surface and be embedded into the stiffer soil below to provide resistance to further movement. Although this solution is certainly tenable, the expense of the permeation grouting was considered excessive, and it was thought that a better method of stabilizing the lower portion of the slope was possible.

10.4.3.4 Micropiles with cap beam

Placement of micropiles that are angled, as shown in Figure 10.33, was considered. Micropiles are most effective for relatively shallow instabilities,

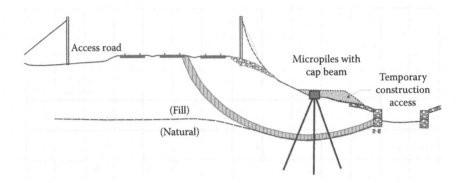

Figure 10.33 Micropile solution.

and due to the size of the unstable mass of soil, micropiles scored relatively low on the decision matrix owing to effectiveness and cost.

10.4.3.5 Anchored blocks

Anchored blocks were another alternative that were considered for slope stability improvement. They have tendons that extend a sufficient depth to develop the needed anchoring and tension to compress and hold the sliding volume of soil together as one mass (Figure 10.34). On the surface of the slope, concrete blocks provide the needed reaction to tighten the tendon rods to the required design post-tension force. This method scored relatively high in the decision matrix and was considered to be one of the two preferred alternatives.

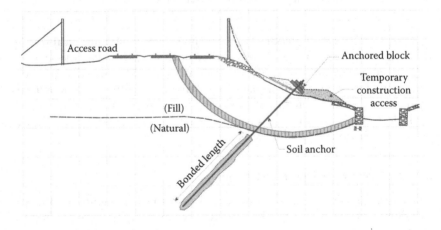

Figure 10.34 Anchored block solution.

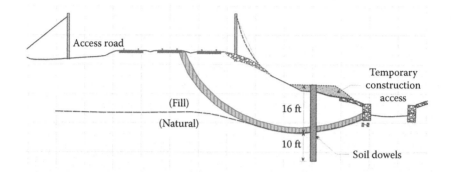

Figure 10.35 Soil dowel solution.

10.4.3.6 Soil dowels

The final remediation considered was placement of soil dowels to a design depth of embedment into the stiff soil layer, as shown in Figure 10.35. The dowels are placed after excavating the soil by auger drilling and placement of a cage of reinforcing bars then filling with concrete. The longitudinal spacing of the dowels was determined based on the amount of arching of the soil pressure distribution between dowels. The soil strength properties affect how wide the spacing can be with softer soils offering less arching of stress around them and hence a smaller required spacing. The soil dowel alternative scored highest in the decision matrix and is expected to offer a practical method of stabilizing the embankment slope against further movement. Therefore, soil dowels and anchor blocks appear to offer the best two methods for slope stabilization.

10.4.4 Extent of remediation

The length of track affected by the sliding mass of soil was determined based on track geometry measurements and by GPR data, as shown in Figure 10.36.

The upper three images in Figure 10.36 are the GPR data from the left, center, and right antennas positioned laterally across the track closest to the crest of the slope. The deep ballast pocket is shown by the distortion of the subgrade surface centered about track location MP 105 + 1,800' and extending approximately 125 feet in either direction. However, the lower image in Figure 10.36 shows that a significant amount of geometry error from track settlement is also centered about the ballast pocket location and extends over a length of approximately 400 feet in either direction of the deepest portion of the ballast pocket. But the final decision on remediation length came from a further review of the GPR data in Figure 10.37. It was determined that the best indication of a deforming substructure is found in the length over which the subgrade layer is

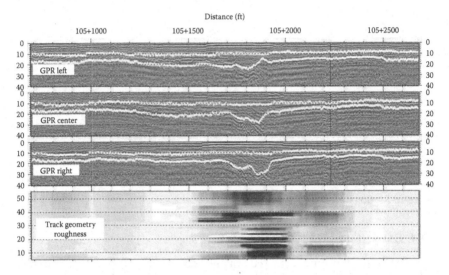

Figure 10.36 GPR and track geometry results for the site.

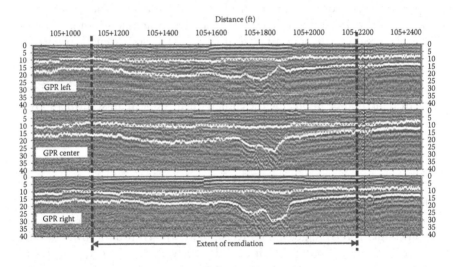

Figure 10.37 Remediation length determined from GPR results.

distorted compared to the subgrade profile shape some distance away where it appears to be stable. From the GPR reflection pattern shown in Figure 10.37, the distortion of the subgrade layer appears to extend from MP 105 + 1,100' to MP 105 + 2,200', resulting in a remediation length of 1,100 ft along the track.

Appendix

A.1 ALLOWABLE CUMULATIVE STRAIN VERSUS ALLOWABLE DEVIATOR STRESS

a: $N = 10^8$ b: $N = 10^7$ c: $N = 10^6$ d: $N = 10^5$

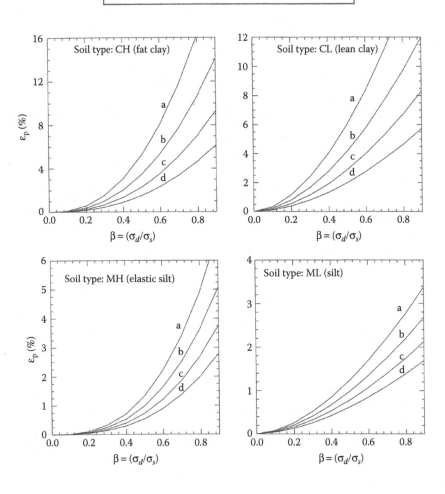

A.2 DESIGN CHARTS BASED ON STRAIN INFLUENCE FACTOR

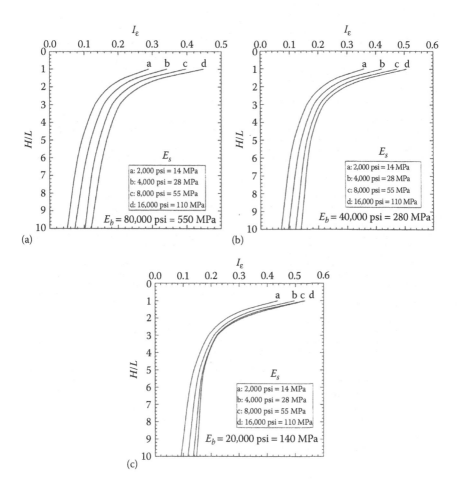

A.3 DESIGN CHARTS BASED ON DEFORMATION
INFLUENCE FACTOR

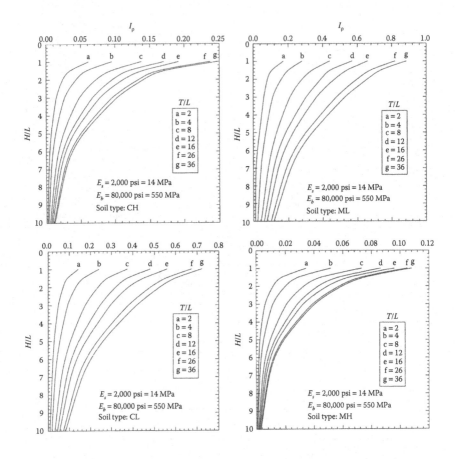

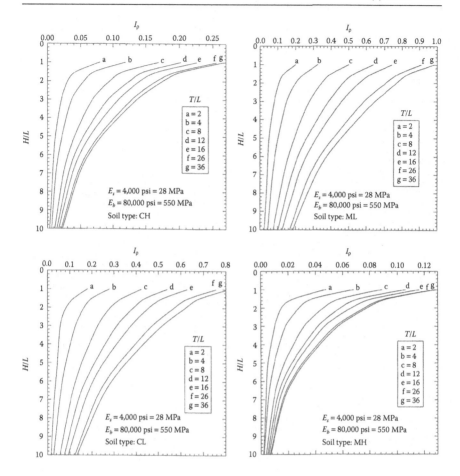

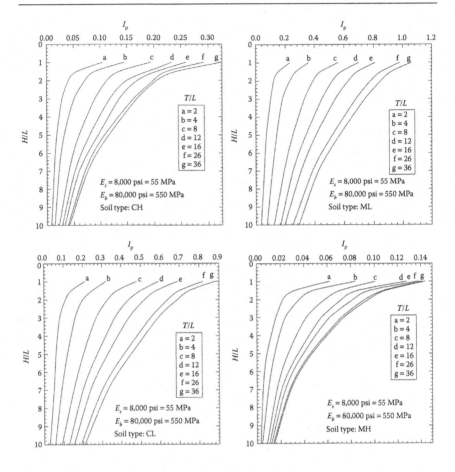

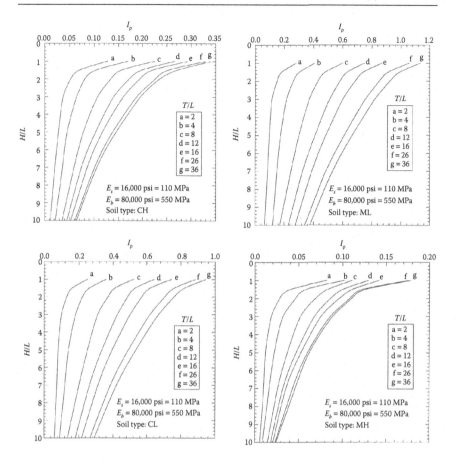

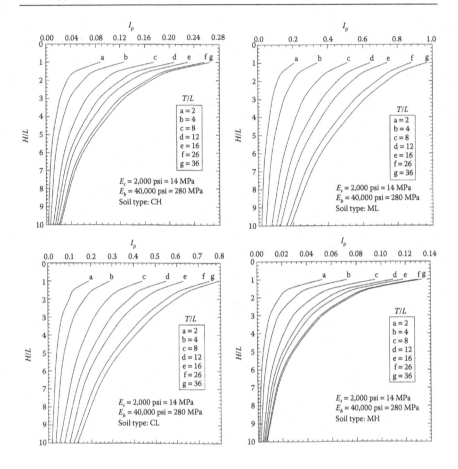

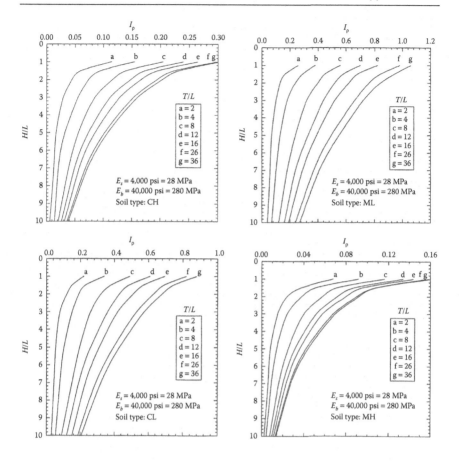

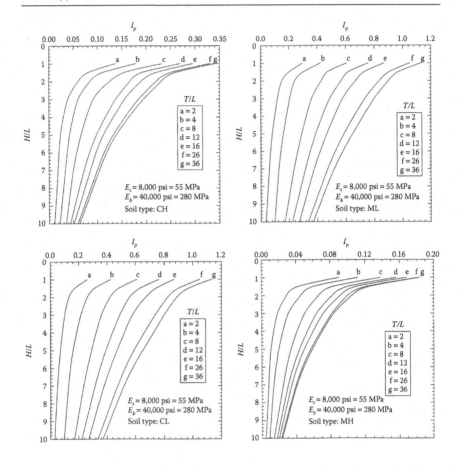

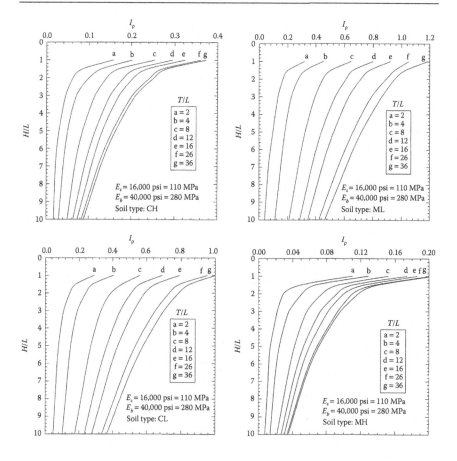

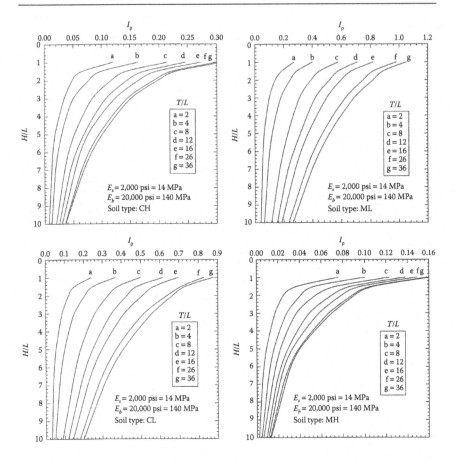

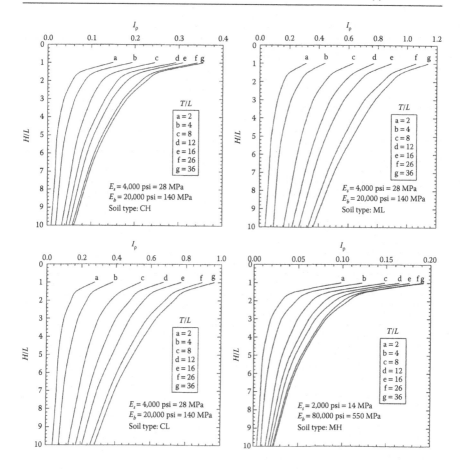

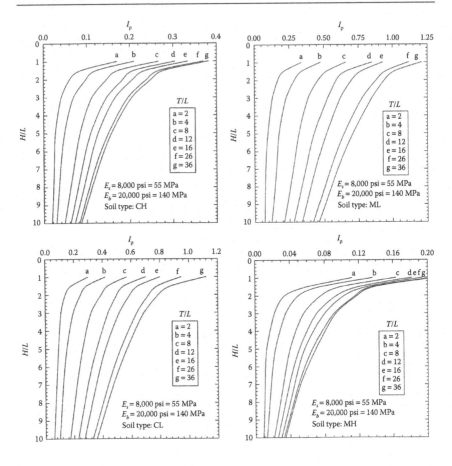

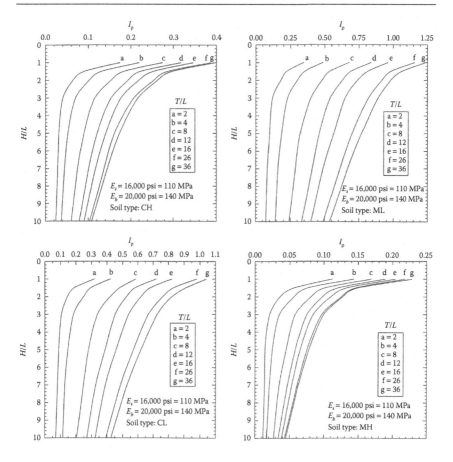

References

Aashto guide for design of pavement structures. (1986). American Association of State Highway and Transportation Officials, Washington, DC.

Ahlf, R. E. (March 1975). M/W costs: How they are affected by car weight and the track structure, *Railway track and structures*, pp. 34–37; 90–92.

Ahmed, S. B. and Larew, H. G. (1962). A study of the repeated load strength moduli of soils. In: *Proceedings of the 1st International Conference on the Structural Design of Asphalt Pavements*. Michigan University, pp. 637–648.

Al-Nuaimy, W., Eriksen, A., and Gasgoyne, J. (2004). Train-mounted GPR for high-speed rail trackbed inspection. *Proceedings of the 10th International Conference on Ground Penetrating Radar*, Delft, the Netherlands.

Al-Qadi, I., Wie, X., and Roberts, R. (2007). Scattering analysis of railroad ballast using ground penetrating radar. *Transportation Research Board Annual Meeting*, Washington, DC.

Ambrose, S. E. (2000). *Nothing Like it in the World*. Simon and Schuster, New York.

American Society for Testing and Materials (ASTM) D2434. (2002). Standard test method for permeability of granular soils (constant head). In: *Annual Book of Standards*, Vol. 04.10. ASTM, West Conshohocken, PA.

Andersen, K. H., Brown, S. F., Foss, I., Pool, J. H., and Rosenbrand, W. F. (1976). Effect of cyclic loading on clay behaviour. In: *Proceedings of the Conference on Design and Construction of Offshore Structures*. Institute of Civil Engineers, London, pp. 1–5.

AREMA. (2012). *Manual for Railway Engineering*. American Railway Engineering and Maintenance of Way Association, Lanham, MD.

Asphalt Institute (1998). HMA trackbeds-hot mix asphalt for quality railroad and transit trackbeds. Information series, IS-137.

Baladi, G. Y., Vallejo, L. E., and Goitom, T. (1983). *Normalized Characterization Model of Pavement Materials*, ASTM Special Technical Publication 807, pp. 55–64.

Barksdale, R. D. (1975). *Test Procedures for Characterizing Dynamic Stress-Strain Properties of Pavement Materials*, Special Report 162. Transportation Research Board, Washington, DC.

Barksdale, R. D. and Alba, J. L. (1993). *Laboratory Determination of Resilient Modulus for Flexible Pavement Design*, Interim Report No. 2. NCHRP, National Research Council, Washington, DC.

Barry, C., Schwartz, C., and Boudreau, R. (May 2006). *Geotechnical Aspects of Pavements*, Report No. FHWA NHI-05-037. Federal Highway Administration.

Bennett, K. C., Ho, C. L., and H. Q. Nguyen (2011). Verification of box test model and calibration of finite element model, *Transportation Research Record 2261*, Transportation Research Board, Washington, DC, pp. 171–177.

Berggren, E. G., Smekal, A., and Silvast, M. (June 4–8, 2006). Monitoring and substructure condition assessment of existing railway lines for upgrading to higher axle loads and speeds. *Proceedings of 7th World Congress on Railway Research*, Montreal, Canada.

Bian, X., Chen, Y., and Hu, T. (2008). Numerical simulation of high-speed train induced ground vibrations using 2.5D finite element approach. *Science in China Series G: Physics, Mechanics & Astronomy*, Vol. 51, No. 6, 632–650.

Bieniawski, Z. (1989). *Engineering Rock Mass Classifications: A Complete Manual for Engineers and Geologists in Mining, Civil and Petroleum Engineering*. John Wiley & Sons, New York.

Bilow, D. and Randich, G. (2000). Slab track for the next 100 years. *Proceedings of the 2000 AREMA Conference*, Chicago, IL.

Bishop, A. (1955). The use of the slip circle in the stability analysis of slopes. *Geotechnique*, Vol. 5, No. 1. London.

Boucher, D. L. and Selig, E. T. (1987). Application of petrographic analysis to ballast performance evaluation. *Transportation Research Record 1131*. Performance of aggregates and other track performance issues, Transportation Research Board, Washington, DC, pp. 5–25.

Briaud, J.-L., James, R. W., and Hoffman, S. B. (1997). *Settlement of Bridge Approaches (The Bump at the End of the Bridge)*, National Cooperative Highway Research Program, Synthesis 234. Transportation Research Board, Washington, DC.

British Railways Board, Group Standard GM/TT0088 Issue 1 Revision A. (October 1993). *Permissible Track Forces for Railway Vehicles*.

Brough, M., Stirling, A., Ghataora, G., and Madelin, K. (2003). *Evaluation of Railway Trackbed and Formation: A Case Study*. NDT & E International, London.

Brown, S. F. (1974). Repeated load testing of a granular material. *Journal of the Geotechnical Engineering Division*, ASCE, Vol. 100, No. GT7, 825–841.

Brown, S. F. (1979). The characterization of cohesive soils for flexible pavement design. In: *Proceedings of the Design Parameters in Geotechnical Engineering*. British Geotechnical Society, London, Vol. 2, pp. 15–22.

Brown, S. F. (2007). The effects of shear stress reversal on the accumulation of plastic strain in granular materials under cyclic loading. In: *Design and Construction of Pavements and Rail Tracks: Geotechnical Aspects and Processed Materials*. Taylor & Francis, London.

Brown, S. F., Bell, C. A., and Brodrick, B. V. (1977). *Permanent Deformation of Flexible Pavements*, Final Technical Report, Grant Number DA ERO 75 G 023. University of Nottingham, Nottingham.

Brown, S. F., Lashine, A. K. F., and Hyde, A. F. L. (1975). Repeated load triaxial testing of a silty clay. *Geotechnique*, Vol. 25, No. 1, 95–114.

Brown, S. F. and Selig, E. T. (1991). The design of pavement and rail track foundations. In: *Cyclic Loading of Soil: From Theory to Design*, M.P. O'Reilly and S.F. Brown, eds. Van Nostrand Reinhold, New York, pp. 249–305.

Bureau of Reclamation. (2004). *Guidelines for Performing Foundation Investigations for Miscellaneous Structures*. US Department of the Interior.

Castro, G. and Poulos, S. J. (1977). Factors affecting liquefaction and cyclic mobility. *Journal of the Geotechnical Engineering Division*, ASCE, Vol. 103, No. GT6, 501–516.

Cedergren, H. R. (1989). *Seepage, Drainage and Flow Nets*, 3rd Edition. John Wiley & Sons, New York.

Chang, C. S., Adegoke, C. W., and Selig, E. T. (1980). The GEOTRACK model for railroad track performance. *Journal of Geotechnical Engineering Division*, ASCE, Vol. 106, No. GT11, 1201–1218.

Chang, C. S. and Z. Yin (2011). Micromechanical modeling for behavior of silty sand with influence of fine content, *International Journal of Solids and Structures*, Vol. 48, No. 19, pp. 2655–2667.

Chiang, C. C. (August 1988). *Effects of Water and Fines on Ballast Performance in Ballast Box*, Geotechnical Report No. AAR89-366p. Department of Civil Engineering, University of Massachusetts, Amherst, MA.

Chipman, P. (1930). A quest for ideal track. In: *Engineers and Engineering*. The Engineers Club, Philadelphia, PA. Reprint by the Portland Cement Association, Skokie, IL.

Chrismer, S. (1985). *In-Track Performance Test of Geotextiles at Caldwell, Texas*, Report No. R-611. Association of American Railroads, AAR Technical Center, Chicago, IL.

Chrismer, S. (August 4, 2011). Personal Communication with Alexander Smekal of the Swedish Transport Administration, Trafikverket, Sweden.

Chrismer, S., Johnson, D. M., Trevizo, C., and LoPresti, J. (1994). *Design Over Lift Tamping*, Technology Digest, TD-800. Association of American Railroads, Chicago, IL.

Chrismer, S. and Li, D. (April 1997). Cone penetrometer testing for track substructure design and assessment. *Proceedings of the 6th International Heavy Haul Conference*, Vol. 1, Cape Town, South Africa, pp. 113–122.

Chrismer, S. and Li, D. (November 8–9, 2000). Considerations in slab track foundation design. *Proceedings of the Roadbed Stabilization and Ballast Symposium*. American Railway Engineering and Maintenance of Way Association, St. Louis, MO.

Chrismer, S.M. and Selig, E.T., (July 1994). *Mechanics-Based Model to Predict Ballast-Related Maintenance Timing and Costs*. Report No. R-863, Association of American Railroads, Chicago, IL.

Chrismer, S., LoPresti, J., Read, D., and Li, D. (October 1995). Cone penetrometer testing on North American railroads. *Proceedings of the International Symposium on Cone Penetration Testing*, Vol. 2, Sweden, pp. 1–7.

Chrismer, S. M. (March 1990). *Track Surfacing with Conventional Tamping and Stone Injection*, Report No. R-719. Association of American Railroads, Chicago, IL.

Chrismer, S. M. (February 1994). *Mechanics-Based Model to Predict Ballast-Related Maintenance Timing and Costs*. PhD dissertation, Department of Civil Engineering, University of Massachusetts, Amherst, MA.

Chrismer, S. M., and LoPresti, J. (1996). *Track Substructure Assessment with the Cone Penetrometer*, TD-96-002. Transportation Technology Center, Association of American Railroads, Pueblo, Colorado.

Chrismer, S. M., Selig, E. T., and Laine, K. J. (December 1992). *Investigation of Ballast Conditions Before and After Undercutting*, Report No R-818. Association of American Railroads, Chicago, IL.

Chrismer, S. M., Terrill, V., and Read, D. (July 1996). *Hot Mix Asphalt Underlayment Test on Burlington Northern Railroad*, R-892. Association of American Railroads, Transportation Technology Center, Pueblo, CO.

Chrismer, S. M. and Wnek, J. M. (2011). A method to characterize cyclic error in the track geometry waveform. *ASME Joint Rail Conference*, Pueblo, CO.

Christopher, B. R., Schwartz, C. S., and Boudreau. (2006). *Geotechnical Aspects of Pavements*, Contract No. DTFH68-02-P-00083, Central Federal Lands Highway Division, Federal Highway Administration, Lakewood, CO. http://www.fhwa .dot.gov/engineering/geotech/pubs/05037.

Clark, M., Gordon, M., Glannopoulos, A., and Forde, M. (2004). Advanced analysis of ground penetrating radar signals on railway. *Proceedings of the 7th International Conference on Railway Engineering*, London.

Cruden, D. and Varnes, D. (1996). Landslide types and processes, Chapter 3. In: *Landslides: Investigation and Mitigation*, A. Turner and R. Schuster, eds. Special Report 247, National Academy Press, Washington, DC.

Culley, R. W. (1971). Effect of freeze-thaw cycling on stress-strain characteristics and volume change of a till subjected to repetitive loading. *Canadian Geotechnical Journal*, Vol. 8, No. 3, 359–371.

Department of Transportation, FRA. (April 2000). Track Safety Standards Part 213, Subpart G, Class of Track 6 and Higher.

Drumm, E. C., Boateng-Poku, Y., and Pierce, T. J. (1991). Estimation of subgrade resilient modulus from standard tests. *Journal of Geotechnical Engineering*, ASCE, Vol. 116, No. 5, 774–789.

Dumas, C., Mansukhani, S., Porbaha, A., Short, R., Cannon, R., McLain, K., Putcha, S. et al. (September 2003). *Innovative Technology for Accelerated Construction of Bridge and Embankment Foundations in Europe*, Federal Highway Administration Report Number FHWA-PL-03-014.

Duncan, J. (July 1996). State of the art: Limit equilibrium and finite-element analysis of slopes. *Journal of Geotechnical Engineering*, ASCE, Vol. 122, No. 7, 577–596.

Dunnicliff, J. (1988). *Geotechnical Instrumentation for Monitoring Field Performance*. John Wiley & Sons, New York.

Ebersöhn, W. (1995). *Substructure Influence on Track Maintenance Requirements*. PhD dissertation, Department of Civil Engineering, University of Massachusetts, Amherst, MA.

Ebersöhn, W. and Selig, E. T. (1994a). Use of track geometry measurements for maintenance planning. *Transportation Research Board, Journal of the Transportation Research Board*, No. 1470, 84–92.

Ebersöhn, W. and Selig, E. T. (1994b). Track modulus measurements on a heavy haul line, *Transportation Research Board, Journal of the Transportation Research Board*, No. 1470, 73–83.

Ebersöhn, E. (1997). Track maintenance management philosophy. *Proceedings of the 6th International Heavy Haul Railway Conference*, International Heavy Haul Association, Cape Town, South Africa, April.

Ebrahimi, A. (2011). *Deformational Behaviour of Fouled Railway Ballast*. PhD dissertation, Department of Civil Engineering, University of Wisconsin-Madison, Madison, WI.

Edil, T. B. and Motan, S. E. (1978). Soil water potential and resilient behavior of subgrade soils, pp. 54–63.

Edil, T. B., and Motan, S. E. (1979). Soil-water potential and resilient behaviour of subgrade soils. *Transportation Research Record*, Vol. 705, 54–63.

Edris, E. Y., Jr. and Lytton, R. L. (May 1976). *Dynamic Properties of Subgrade Soils, Including Environmental Effects*, TTI-2-18-74-164-3. Texas Transportation Institute, Texas A&M University, College Station, TX.

Eid, H. T. and Stark, T. D. (1998). Undrained shear strength from cone penetration test. In: *Geotechnical Site Characterization*, Robertson and Mayne, eds. Balkema, Rotterdam, the Netherlands.

Elfino, M. K. and Davidson, J. L. (1989). Modeling field moisture in resilient moduli testing. In: *Resilient Moduli of Soils: Laboratory Conditions*, ASCE Geotechnical Special Publication No. 24. ASCE, New York, pp. 31–51.

Eriksen, A., Venables, B., Gascoyne, J., and Bandyopadhyay, S. (June 2006). Benefits of high speed GPR to manage trackbed assets and renewal strategies. *Permanent Way Institute Conference Proceedings*, Brisbane, Australia.

ERRI. (1999). *Bridge Ends Embankment Structure Transition State of the Art Report*, Report D230.1/RP3. European Rail Research Institute.

Esveld, C. (2001). *Modern Railway Track*. MRT-Productions, Zaltbommel, the Netherlands.

Farritor, S. and Fateh, M. (2013). *Measurement of Vertical Track Deflection from a Moving Rail Car*, Report No. DOT/FRA/ORD-13/08. Federal Railroad Administration.

FHWA. (1980). *Highway Subdrainage Design*, Report No. FHWA-TS-80-244. Federal Highway Administration, Offices of Research and Development, Washington, DC.

Ford, R. (1995). Differential ballast settlement, and consequent undulations in track, caused by vehicle-track interaction. *Interaction of Railway Vehicles with the Track and Its Substructure*, Knothe, Grassie, and Elkins, eds. Swets and Zeitlinger, pp. 222–233.

Forrester, K. (2001). *Subsurface Drainage for Slope Stabilization*. ASCE Press, Reston, VA.

Fredlund, D. G., Bergan, A. T., and Sauer, E. K. (1975). Deformation characterisation of subgrade soils for highways and runways in northern environments. *Canadian Geotechnical Journal*, Vol. 12, No. 2, 213–223.

Fredlund, D. G., Bergan, A. T., and Wong, P. K. (1977). Relation between resilient modulus and stress conditions for cohesive subgrade soils. *Transportation Research Record*, Vol. 642, 73–81.

Frohling, R. D., Scheffel, H., and Ebersohn, W. E. (August 1996). The vertical dynamic response of a rail vehicle caused by track stiffness variations along the track. In: *The Dynamics of Vehicles on Roads and On Tracks: Proceedings of the 14th IAVSD Symposium*, L. Segel, ed. Swets and Zeitlinger, Ann Arbor, MI.

Garber, N. J. and L. A. Hoel (2015). *Traffic and Highway Engineering*, 5th Edition, Cengage Publishing, Independence, KY.

Gilchrist, A. O., ed. (2000). *A History of Engineering Research on British Railways*. Institute of Railway Studies, York University, York.

Grabe, P. J. (2002). *Resilient and Permanent Deformation of Railway Foundations under Principal Stress Rotation*. PhD thesis, University of Southampton, Southampton.

Grabe, P. J. and Shaw, F. J. (2009). Design life prediction of a heavy haul track formation. *Proceedings of the 9th Internation Heavy Haul Conference*, Shanghai, China, pp. 22–29.

Greisen, C., Lu, S., Duan, H., Farritor, S., Arnold, R., Sussmann, T., GeMeiner, W. et al. (March 2009). Estimation of rail stress from real-time vertical track deflection measurement. *Proceedings of the ASME Joint Rail Conference*, Pueblo, CO.

Griffiths, D., Fenton, G., and Martin, T., eds. (2000). *Slope Stability 2000—Proceedings of Sessions of Geo-Denver 2000*, Geotechnical Special Publications No. 101. ASCE Press, Reston, VA.

Haestad Methods and Durrans, S. (2003). *Stormwater Conveyance Modeling and Design*. Haestad Press, Waterbury, CT.

Han, X. and Selig, E. (1997). Effects of fouling on ballast settlement. *Proceedings of the 6th International Heavy Haul Railway Conference*, Cape Town, South Africa, pp. 257–268.

Hay, W. W. (1982). *Railroad Engineering*, 2nd Edition. John Wiley & Sons, New York.

Haynes, J. H. and Yoder, E. J. (1963). *Effects of Repeated Loading on Gravel and Crushed Stone Base Course Materials Used at the AASHO Road Test*, Highway Research Record 39, National Academy of Sciences, Washington, DC.

Heath, D. L., Shenton, M. J., Sparrow, R. W., and Waters, J. M. (1972). Design of conventional rail track foundations. *Proceedings of the Institution of Civil Engineering*, Vol. 51, 251–267.

Hering, H. (1989). *Application and Testing of the Dynamic Track Stabilizer on New Lines*. Railway Engineering International, No. 2.

Heyns, F. J. (May 2000). *Railway Track Drainage Design Techniques*. Doctoral dissertation, University of Massachusetts, Amherst, MA.

Heyns, F. J. (2001). *Railway Track Drainage Design Techniques, Doctoral dissertation*, University of Massachusetts, Amherst, MA.

Hoek, E. and Bray, J. (1981). *Rock Slope Engineering*, 3rd Edition. Elsevier Science Publishers, Essex.

Holm, G., Andreasson, B., Bengtsson, P., Bodare, A., and Eriksson, H. (2002). *Mitigation of Track and Ground Vibration by High Speed Trains at Ledsgard, Sweden*, Swedish Deep Stabilization Research Centre, Linkoping, Sweden.

Holtz, R., Kovacs, W., and Sheahan, T. (2011). *An Introduction to Geotechnical Engineering*, 2nd Edition, Pearson, Upper Saddle River, NJ.

Huang, Y. H. (1993). *Pavement Analysis and Design*. Prentice Hall, Englewood Cliffs, NJ.

Huang, Y. H., Lin, C., and Deng, X. (April 1984). *KENTRACK: A Computer Program for Hot Mix Asphalt and Conventional Ballast Railway Trackbeds*, Research Report RR-84-1. The Asphalt Institute, College Park, MD.

Huang, Y. H., Rose, J. G., and Khoury, C. J. (February 23–25, 1987). Thickness design for hot-mix asphalt railroad trackbeds. *Asphalt Paving Technology, 1987 Annual Technical Sessions*, Vol. 56, Reno, NV, pp. 427–453.

Hungerford, E. (1911). *The Modern Railroad*. A. C. McClurg, Chicago, IL. http://www.gutenberg.org/files/40242/40242-h/40242-h.htm.

Hunt, G. A. (December 1999). Track damage models as tools for track design optimization. In: *Proceedings of the Conference on Innovations in the Design and Assessment of Railway Track*. Technical University of Delft, Delft, the Netherlands.

Hyde, A. F. L. and Brown, S. F. (1976). The plastic deformation of a silty clay under creep and repeated loading. *Geotechnique*, Vol. 26, No. 1, 173–184.

Hyslip, J. P. (2002). *Fractal Analysis of Geometry Data for Railway Track Condition Assessment*. PhD dissertation, Department of Civil Engineering, University of Massachusetts, Amherst, MA.

Hyslip, J. P. (September 2007). Substructure maintenance management: Its time has come. *Proceedings of the Annual Conference of the American Railway Engineering and Maintenance-of-Way Association*, Chicago, IL.

Hyslip, J., Li, D., and McDaniel, R. (June 29–July 2, 2009). Railway bridge transition case study. In: *Proceedings of the 8th International Conference on the Bearing Capacity of Roads, Railways and Airfields*, E. Tutumluer and I. Al-Qadi, eds. University of Illinois at Urbana-Champaign, Champaign, IL, Taylor & Francis Group, London.

Hyslip, J. P. and McCarthy, B. M. (September 2000). Substructure investigation and remediation for high tonnage freight line. *Proceedings of the 2000 Annual Conference American Railway Engineering and Maintenance-of-Way Association*, Dallas, TX.

Hyslip, J. P., Olhoeft, G. R., Smith, S. S., and Selig, E. T. (October 2005). *Ground Penetrating Radar for Railroad Track Substructure Evaluation*, NTIS Report No. FRA/ORD-05/04. Federal Railroad Administration Office of Research and Development, US Department of Transportation, Washington, DC.

Hyslip, J. P., Smith, S., Olhoeft, G. R., and Selig, E. T. (June 2004). Ground penetrating radar for railway substructure condition assessment. *Proceedings Railway Engineering 2004 Conference*, London.

Hyslip, J., Trosino, M., and Faircloth, J. (June 2011). Railway embankment assessment and case studies. *Proceedings of International Heavy Haul Association Specialty Conference*, Calgary, Canada.

Hyslip, J. P. and Vallej, L. E. (1997). Fractal analysis of the roughness and size distribution of granular materials. *Engineering Geology*, Vol. 48, No. 3–4, 231–244.

Indraratna, B. and Salim, W. (2005). *Mechanics of Ballasted Rail Tracks, A Geotechnical Perspective*. Taylor & Francis, London.

Indraratna, B., Thakur, P. K., and Vinod, J. S. (2010). Experimental and numerical study of railway ballast behavior under cyclic loading. *International Journal of Geomechanics*, ASCE, Vol. 10, No. 4, 136–144.

International Union of Railways. (January 1, 1994). *Earthworks and Track-Bed Layers for Railway Lines*, UIC Code 719R, 2nd Edition.

Janbu, N. (1968). *Slope Stability Computations*, Soil Mechanics Foundation Engineering Report. The Technical University of Norway, Trondheim, Norway.

Jenkins, H. H., Stephenson, J. E., Clayton, G. A., Morland, G. W., and Lyon, D. (1974). The effect of track and vehicle parameters on wheel/rail vertical dynamic forces. *Railway Engineering Journal*, 2–16.

Jovanovic, S. (2006). Rail track quality assessment and decision making. *AREMA Proceedings*, Louisville, KY.

Judge, A. (2013). *Measurement of the Hydraulic Conductivity of Gravels Using a Laboratory Permeameter and Silky Sands Using Field Testing with Observation Wells*. Doctoral dissertation, University of Massachusetts, Amherst, MA.

Judge, A., Ostendorf, D., DeGroot, D., and Zlotnik, V. (2014). Pneumatic permeameter for transient tests on coarse gravel. *Journal of Hydrologic Engineering*, Vol. 19, No. 2, 319–327.

Kallas, B. F. and Riley, J. (1967). Mechanical properties of asphalt pavement materials. In: *Proceedings of the 2nd International Conference on Structural Design of Asphalt Pavements*. University of Michigan, pp. 931–952.

Kallianen, A. and Kolisoja, P. (June 2013). A new method to determine the desired minimum railway embankment dimensions. *Proceedings of the International Conference on the Bearing Capacity of Roads, Railroads and Airfields 2013*, Trondheim, Norway.

Karol, R. H. (1983). *Chemical Grouting*, Marcel Dekker, New York.

Kaye, B. (1994). *A Random Walk Through Fractal Dimensions*, 2nd Edition. VCH Publishers, New York.

Kerr, A. D. (2003). *Fundamentals of Railway Track Engineering*. Simmons Boardman Publishing, Omaha, NE.

Kerr, A. D. and Moroney, B. E. (October 1993). Track transitions problems and remedies. *AREA Bulletin* 742, American Railway Engineering Association, Washigton, DC, Vol. 94, pp. 267–298.

Kirwan, R. W., Farrell, E. R., and Maher, M. L. J. (1979). Repeated load parameters of a glacial till related to moisture content and density. *Proceedings of the 7th European Conference on Soil Mechanics and Foundation Engineering, England*, Vol. 2, 69–74.

Klassen, M. J., Clifton, A. W., and Watters, B. R. (1987). *Track Evaluation and Ballast Specification*, Transportation Research Board Record No. 1131, National Academies of Sciences, Washington, DC.

Klauder, L. T., Chrismer, S. M., and Elkins, J. (2002). Improved spiral geometry for high speed rail and predicted vehicle response. *Journal of Transportation Research Board*, Vol. 1785, Washington, DC.

Knutson, R. M. and M. R. Thompson (1978). Resilient response of railway ballast, *Transportation Research Record* No. 651, Transportation Research Board, Washington, DC, pp. 31–39.

Knutson, R. M., Thompson, M. R., Mullin, T., and Tayabji, S. D. (1977). *Materials Evaluation Study Ballast and Foundation Materials Research Program*, FRA OR&D 77 02. University of Illinois at Urbana-Champaign, Champaign, IL.

Koerner, R. (1990). *Designing with Geosynthetics*. 2nd Edition. Prentice Hall, Englewood Cliffs, NJ.

Kondapalli, S. and Bilow, D. (2008). *Life Cycle Benefit of Concrete Slab Track*. Portland Cement Association, Skokie, IL.

Krumbein, W. C. (1941). Measurement and geological significance of shape and roundness of sedimentary particles. *Journal of Sedimentary Petrology*, Vol. 11, No. 2, 64–72.

Ladd, C. (1991). Stability evaluation during staged construction. 22nd Terzaghi Lecture. *Journal of Geotechnical Engineering*, ASCE, Vol. 117, No. 4.

Lambe, T. W. and Whitman, R. V. (1969). *Soil Mechanics*. John Wiley & Sons, New York.

Lane, E. W. (1953). Progress report on studies on the design of stable channels of the bureau of reclamation. *Proceedings of the ASCE*, Vol. 79 Reston, VA.

Larew, H. G. and Leonards, G. A. (1962). A strength criterion for repeated loads. *Proceedings of the Highway Research Board*, No. 41, pp. 529–556.

Leroueil, S. (2001). Natural slopes and cuts: Movement and failure mechanisms. *Geotechnique*, Vol. 51, No. 3, 197–243.

Leroueil, S. (2004). Geotechnics of slopes before failure. In: *Landslides: Evaluation and Stabilization*, Lacerda, Ehrlich, Fortoura, and Sayao, eds. Taylor & Francis, London.

Li, D. (1985). Research on wheel/track dynamic responses in the rail joint region, (in Chinese, English abstract). *Journal of Changsha Railway Institute*, No. 4, 46–59.

Li, D. (1987). Wheel/track vertical dynamic interaction and responses, (in Chinese, English abstract). *Journal of the China Railway Society*, Vol. 9, No. 1, 1–8.

Li, D. (1989). The study of lumped-parameter model for railroad track dynamics, (in Chinese, English abstract). *Journal of the China Railway Society*, Special Issue on Civil Engineering, 82–90.

Li, D. (May 1991). *Effect of Strain Rate on Strength and Stiffness Properties of Soils*, Geotechnical Report No. AAR 91-384R. Department of Civil Engineering, University of Massachusetts, Amherst, MA.

Li, D. (1994). *Railway Track Granular Layer Thickness Design Based on Subgrade Performance under Repeated Loading*. PhD dissertation, Department of Civil Engineering, University of Massachusetts, Amherst, MA.

Li, D. (June 28–29, 1998). How to design ballast depths for new track construction. In: *Proceedings of the Roadbed Stabilization and Ballast Symposium*. The American Railway Engineering and Maintenance of Way Association (AREMA), Ypsilanti, MI.

Li, D. (August 5–8, 2000). *Deformations and Remedies for Soft Railroad Subgrades Subjected to Heavy Axle Loads*, Geotechnical Special Publication No. 103, Advances in Transportation and Geoenvironmental Systems Using Geosynthetics, ASCE, Denver, CO, pp. 307–321.

Li, D. (August 2010). *Slab Track Field Test and Demonstration Program for Shared Freight and High-Speed Passenger Service*, DOT/FRA/ORD-10/10, Final Report. U.S. Department of Transportation, Federal Railroad Administration, Office of Research and Development, Washington, DC.

Li, D. and Bilow, D. N. (2008). Testing of slab track under heavy axle loads, Transportation Research Record. *Journal of the Transportation Research Board*, No. 2043, 55–64, US National Academies.

Li, D., Bilow, D., and Sussmann, T. R. (April 2010). Slab track for shared freight and high speed passenger service. *ASME, Proceedings of the Joint Rail Conference on High Speed and Intercity Rail*, Urbana, IL.

Li, D. and Chrismer, S. (October 1999). *Soft Subgrade Remedies Under Heavy Axle Loads. Railway Track and Structures*, 15–18.

Li, D. and Davis, D. (2005). Transition of railroad bridge approaches. *Journal of Geotechnical and Geoenvironmental Engineering*, ASCE, Vol. 131, No. 11, 1392–1398.

Li, D., Duran, C., and McDaniel, R. (August 2006). *Investigation of Open Deck Bridge Transition Issues at the Eastern Mega Site*, Technology Digest TD-06-022. Transportation Technology Center, Association of American Railroads, Pueblo, CO.

Li, D., Elkins, J. A., Otter, D. E., and Wilson, N. G. (June 1996). Vehicle/track dynamic models for wheel/rail forces and track response. *Proceedings of International Conference on Vehicle-Infrastructure Interaction IV*, San Diego, CA.

Li, D., Elkins, J., Wu, H., and Singh, S. (May 1999). *Characterization of Track Stiffness and Damping Parameters*, Report No. R-930. Association of American Railroads.

Li, D., Guins, T., and Kalay, S. (May 5–9, 2003). Performance-based track geometry assessment for network efficiency. *Proceedings of the International Heavy Haul Association 2003 Specialist Technical Session*, Dallas, TX, pp. 5.25–5.30.

Li, D., Harbuck, R., Morgart, D., and Mischke, M. (September 2005). BNSF experience: Track geometry inspection taking into account vehicle performance. *Railway Track and Structures*, Vol. 101, No. 9, 19–23.

Li, D. and Hyslip, J. (2008). *Geotechnical Considerations for Slab Track*, P-07-035. Prepared for the Portland Cement Association, Transportation Technology Center, Pueblo, CO.

Li, D., LoPresti, J., and Davis, D. (January 2002). *Application and Performance of Hot-Mix-Asphalt Trackbed Over Soft Subgrade. Railway Track and Structures*, pp. 13–15.

Li, D., McDaniel, R., Clark, D., and Maal, L. (May 2010). Assessing the effects of heavy-axle-load traffic on track components: Update of experiments at revenue service mega sites. *Railway Track and Structures*, Vol. 106, No. 5, 17–20.

Li, D., McDaniel, R., and GeMeiner, B. (July 2007). Study of bridge approaches at revenue service mega sites. *Railway Track and Structures*, Vol. 103, No. 7, 18–21.

Li, D., Meddah, A., GeMeiner, B., and Mattson, S. (December 2007). *Dynamic Load and Track Response: Bridge Approach Test at the Western Mega Site*, Technology Digest TD-07-042. Transportation Technology Center, Association of American Railroads, Pueblo, CO.

Li, D., Meddah, A., GeMeiner, B., and McDaniel, R. (July 2009). *Bridge Approach Remedies in Revenue Service*, Technology Digest TD-09-018. Transportation Technology Center, Association of American Railroads, Pueblo, CO.

Li, D., Meddah, A., Hass, K., and Kalay, S. (2006). Relating track geometry to vehicle performance using neural network approach. *Journal of Rail and Rapid Transit*, Vol. 220, No. 3, 273–282.

Li, D., Meddah, A., Lundberg, W., and Kalay, S. (June 22–25, 2009). Latest development in performance-based track geometry inspection technologies. *Proceedings of 9th International Heavy Haul Conference*, Shanghai, China, 113–120.

Li, D., Otter, D., and Carr, G. (2010). Railway bridge approaches under heavy axle load traffic: Problems, causes, and remedies. *Journal of Rail and Rapid Transit*, Vol. 224, No. F5, 383–390. doi:10.1243/09544097JRRT345.

Li, D. and Read, D. (June 25–27, 2013). Heavy axle load track substructure research by TTCI. *Proceedings of the 9th International Conference on the Bearing Capacity of Roads, Railways and Airfields*, Trondheim, Norway, Vol. 2, pp. 879–888.

Li, D., Read, D., and Chrismer, S. (July 1997). *Effects of Heavy Axle Loads on Soft-subgrade Performance*, Technology Digest 97-020. Association of American Railroads, Pueblo, CO.

Li, D., Rose, J., Lees, H., and Davis, D. (July 2001). *Hot-Mix Asphalt Trackbed Performance Evaluation at Alps, New Mexico*, Technology Digest 01-015. Transportation Technology Center, Association of American Railroads, Pueblo, CO.

Li, D., Rose, J., and LoPresti, J. (April 2001). *Test of Hot-Mix Asphalt Trackbed over Soft Subgrade under Heavy Axle Load*, Technology Digest 01-009. Transportation Technology Center, Association of American Railroads, Pueblo, CO.

Li, D. and Selig, E. T. (November 1991). *Evaluation of Resilient Modulus for Fine-Grained Subgrade Soils*, Geotechnical Report No. AAR91-392R. Department of Civil Engineering, University of Massachusetts, Amherst, MA.

Li, D. and Selig, E. T. (1994a). Resilient modulus for fine-grained subgrade soils. *Journal of Geotechnical Engineering*, ASCE, Vol. 120, No. 6, 939–957.

Li, D. and Selig, E. T. (July 1994b). *Investigation of BN Track Subgrade Problems Near Alps, NM.* Internal report to the Association of American Railroads and the Burlington Northern Railroad, Amherst, MA.

Li, D. and Selig, E. T. (December 1994c). *Investigation of AMTRAK Track Foundation Problems in Edgewood and Aberdeen, MD.* Internal report to the Association of American Railroads and AMTRAK, Amherst, MA.

Li, D. and Selig, E. T. (1995a). Evaluation of railway subgrade problems. *Transportation Research Record* 1489. Transportation Research Board, National Research Council, Washington, DC, pp. 17–25.

Li, D. and Selig, E. T. (July 1995b). *Evaluation and Remediation of Potential Railway Subgrade Problems under Repeated Heavy Axle Loads*, R-884. Association of American Railroads, Transportation Technology Center, Pueblo, CO.

Li, D. and Selig, E. T. (1995c). Wheel/track dynamic interaction: Track substructure perspective. *Vehicle System Dynamics*, Vol. 24, 183–196.

Li, D. and Selig, E. T. (1996). Cumulative plastic deformation for fine-grained subgrade soils. *Journal of Geotechnical Engineering*, ASCE, Vol. 122, No. 12, 1006–1013.

Li, D. and Selig, E. T. (1998). Method for railroad track foundation design I: Development. *Journal of Geotechnical Engineering*, ASCE, Vol. 124, No. 4, 316–322.

Li, D. and Selig, E. T. (April 1998). Method for railroad track foundation design II: Applications. *Journal of Geotechnical Engineering*, ASCE, Vol. 124, No. 4, 323–329.

Li, D. and Shust, W. (December 1997). *Investigation of Lateral Track Strength and Track Panel Shift Using AAR's Track Loading Vehicle*, R-917. Association of American Railroads, Pueblo, CO.

Li, D. and Shust, W. (June 1999). *Automated Measurements of Lateral Track Panel Strength and Examinations of Track Maintenance Effects Using AAR's Track Loading Vehicle*, Report No. R-918. Association of American Railroads.

Li, D., Shust, W., and Bowman, R. (March 1998). The effects of maintenance operations: determining how maintenance affects track lateral strength. *Railway Track and Structures*, pp. 14–17.

Li, D., Stabler, L., Harbuck, R., Mischke, M., and Kalay, S. (2005). Implementation of performance-based track geometry inspection on North American railroads. *Proceedings of the 8th International Heavy Haul Conference*, Rio de Janeiro, Brazil, pp. 553–560.

Li, D., Sussmann, T. R., and Selig, E. T. (October 1996). *Procedure for Railway Track Granular Layer Thickness Determination*, R-898. Association of American Railroads, Transportation Technology Center, Pueblo, CO.

Li, D., Thompson, R., and Kalay, S. (July 3–4, 2002b). Development of continuous lateral and vertical track stiffness measurement techniques. *Proceedings of the Railway Engineering*, London.

Li, D., Thompson, R., Marquez, P., and Kalay, S. (2004). Development and implementation of a continuous vertical track support testing technique, Transportation Research Record. *Journal of the Transportation Research Board*, No. 1863, 68–73, US National Academies.

Loo, P. J. (1976). A practical approach to the prediction of rutting in asphalt pavements, The shell method. *Annual Meeting of TRB*. Washington, DC.

LoPresti, J. and Li, D. (2005). *Long-Term Performance of Hot-Mix Asphalt over Soft Subgrade at FAST*, Technology Digest TD-05-035. Transportation Technology Center, Association of American Railroads, Pueblo, CO.

Lotfi, H. R. and Oesterle, R. G. (2005). *Slab track for 39-ton axle load, structure design, PCA R & D Serial no. 2832*, Portland Cement Association, Skokie, IL.

Lu, S. (2008). *Real Time Vertical Track Deflection Measurement System*. Doctoral dissertation, University of Nebraska, Lincoln, NE.

Luo, W. (1973). The characteristics of soils subjected to repeated loads and their applications to engineering practice. *Soils and Foundations*, Vol. 13, No. 1, 12–26.

Majidzadeh, K., Bayomy, F., and Khedr, S. (1978). Rutting evaluation of subgrade soils in Ohio. *Transportation Research Record*, No. 671, 75–91.

Majidzadeh, K., Khedr, S., and Guirguis, H. (1976). Laboratory verification of a mechanistic subgrade rutting model. *Transportation Research Record*, No. 616, 34–37.

Manacorda, G., Morandi, D., Sarri, A., and Staccone, G. (April 2002). A customized GPR system for railroad tracks verification. In: *Proceedings of GPR2002, the 9th International Conference on Ground Penetrating Radar*, Santa Barbara, CA, S.K. Koppenjan, and H. Lee, eds., Proceedings of SPIE, Vol. 4758.

Marriott, D. C. (1996). Improving the performance of track lining and leveling machines. *Proceedings of the Institution of Civil Engineers: Transport*, Vol. 117, Paper 11031.

Martland C. (Summer 2013). Introduction of heavy axle loads by the North American rail industry. *Journal of the Transportation Research Forum*, Vol. 52, No. 2.

Matsui, T., Ohara, H., and Ito, T. (1980). Cyclic stress strain history and shear characteristics of clay. *Journal of the Geotechnical Engineering Division*, ASCE, Vol. 106, No. GT10, 1101–1119.

Miller, G. A., Teh, S. Y., Li, D., and Zaman, M. M. (2000). Cyclic shear strength of a soft railroad subgrade. *Journal of Geotechnical Engineering*, ASCE, Vol. 126, No. 2, pp. 139–147.

Mishra, D., Tutumluer, E., Boler, H., Hyslip, J., and T. Sussmann (2014). Railroad track transitions with multidepth deflectometers and strain gages, *Transportation Research Record: Journal of the Transportation Research Board*, Vol. 2448, 105–114.

Mitchell, R. J. and King, R. D. (1977). Cyclic loading of an Ottawa area champlain sea clay. *Canadian Geotechnical Journal*, Vol. 14, 52–63.

Moaveni, M., Wang, S., Hart, J. M., Tutumluer, E., and Ahuja, N. (2013). *Aggregate Size and Shape Evaluation Using Segmentation Techniques and Aggregate Image Processing Algorithms*, Paper No. 13-4167, Transportation Research Record 2335, Washington, DC, pp. 50–59, Journal of the Transportation Research Board.

Momoya, Y., Watanabe, K., Sekine, E., Tateyama, M., Shinoda, M., and Tatsuoka, F. (2007). Effects of continuous principal stress axis rotation on the deformation characteristics of sand under traffic loads. In: *Design and Construction of Pavements and Rail Tracks: Geotechnical Aspects and Processed Materials*. Taylor & Francis, London.

Monismith, C. L. (1976). Rutting prediction in asphalt concrete pavements: A state of the art. *1976 Annual Meeting of the TRB*, Washington, DC.

Monismith, C. L., Ogawa, N., and Freeme, C. R. (1975). Permanent deformation characteristics of subgrade soils due to repeated loading. *Transportation Research Record*, No. 537, 1–17.

Moossazadeh, J. and Witczak, M. W. (1981). Prediction of subgrade moduli for soil that exhibits nonlinear behavior. *Transportation Research Record*, No. 810, 9–17.

Morgenstern, N. and Price, V. (1967). The analysis of the stability of general slip surfaces. *Geotechnique*, Vol. 15, No. 1, London.

Moroney, B. E. (December 1991). *A Study of Railroad Track Transition Points and Problems*. Master of Civil Engineering Thesis, University of Delaware, Newark, DE.

McCullough, D. (1977). *The Path Between the Seas*. Simon and Schuster, New York.

Norrish, N. and Wyllie, D. C. (1996). Rock slope stability analysis, Chapter 15. In: *Landslides: Investigation and Mitigation*, A. Turner and R. Schuster, eds., Special Report 247, National Academy Press, Washington, DC.

Olhoeft, G. R. (2000). Maximizing the information return from ground penetrating radar. *Journal of Applied Geophysics*, Vol. 43/2–4, 175–187.

Olhoeft, G. R. and Selig, E. T. (April 2002). Ground penetrating radar evaluation of railroad track substructure conditions. In: *Proceedings of the 9th International Conference on Ground Penetrating Radar*, Santa Barbara, CA, S.K. Koppenjan and H. Lee, eds., Proceedings of SPIE, Bellingham, WA, Vol. 4758, pp. 48–53.

Olhoeft, G. R. and Smith III, S. S. (May 2000). Automated processing and modeling of GPR data for pavement thickness and properties. In: *Proceedings of the 8th International Conference on Ground Penetrating Radar*, Gold Coast, Australia, D.A. Noon, G.F. Stickley, and D. Longstaff, eds., SPIE, Bellingham, WA, Vol. 4084, pp. 188–193.

Olhoeft, G. R., Smith III, S., Hyslip, J. P., and Selig, E. T. (June 2004). GPR in railroad investigations. *Proceedings of the 10th International Conference On Ground Penetrating Radar, 21–24*, Delft, the Netherlands, pp. 635–638.

ORE. (October 1970) Repeated loading of clay and track foundation design, D 71/ RP 12. Office for Research and Experiments of the International Union of Railways, Utrecht, the Netherlands.

Oregon Department of Transportation. (March 1990). *Proceedings of the Workshop on Resilient Modulus Testing*, FHWA-TS-90-031. Oregon State University, Corvallis, OR.

O'Reilly, M. P. (1991). Cyclic loading testing of soils. In: *Cyclic Loading of Soils: from Theory to Design*, O'Reilly and Brown, eds., Blackie, London, pp. 70–121.

O'Reilly, M. P. and Brown, S. F., eds. (1991). *Cyclic Loading of Soils: From Theory to Design*. Van Nostrand Reinhold, New York.

O'Reilly, M. P., Brown, S. F., and Overy, R. F. (1991). Cyclic loading of silty clay with drainage periods. *Journal of Geotechnical Engineering*, ASCE, Vol. 117, No. 2, 354–362.

O'Riordan, N. and Phear, A. (2001). Design and construction control of ballasted track formation and subgrade for high speed lines. *Proceedings of Railway Engineering 2001*, London.

Palynchuk, B., Sather, S., and Bunce, C. (October 2007). Approaches to mitigating decayed buried timber within railway embankments. *Proceedings of the 60th Annual Canadian Geotechnical Conference*, Ottawa, Canada.

Panuccio, C. M., Wayne, R. C., and E. T. Selig (December 1978). Investigation of a plate index test for railroad ballast, *Geotechnical Testing Journal, American Society for Testing and Materials (ASTM)*, Vol. 1, No. 4, 213–222.

Pariseau, W. (2011). *Design Analysis in Rock Mechanics*, 2nd Edition. Taylor & Francis, London.

Parsons, B. K. (February 1990). Hydraulic conductivity of railroad ballast and track structure drainage, Geotechnical Report No AAR90-372P. Department of Civil Engineering, University of Massachusetts, Amherst, MA.

Parsons, B. K. (1990). *Hydraulic Conductivity of Railroad Ballast and Track Substructure Drainage*. Master's thesis, University of Massachusetts, Amherst, MA.

Peck, R. B. (1969). Advantages and limitations of the observational method in applied soil mechanics. *Geotechnique*, Vol. 19, No. 2, 171–187.

Penny, C. and Ingham, D. (2003). The search for the ideal trackform. *Proceedings of the Annual AREMA Conference*, Washington, DC.

Pezo, R. F., Kim, D.-S., Stokoe, K. H., and Hudson, W. R. (January 1991). Aspects of a reliable resilient modulus testing system. *Transportation Research Board Preprint, 70th Annual Meeting*, Washington, DC.

Phear, A. and O'Riordan, N. (July 27–31, 2003). Earthworks for the channel tunnel rail link high speed railway, UK. In: *Proceedings of the Geotechnical Engineering for Transportation Projects*, Los Angeles, CA, pp. 1621–1628.

Pierson, L. A. (2012). Rock fall hazard rating system. In: *Rockfall: Characterization and Control*, A.K. Turner and R.L. Schuster eds. Transportation Research Board, National Academy of Sciences, Washignton, DC.

Plotkin, D. and Davis, D. (February 2008). *Bridge Approaches and Track Stiffness*, DOT/FRA/ORD-08-01, Washington, DC.

Poulsen, J. and Stubstad, R. N. (November 1977). *Laboratory Testing of Cohesive Subgrade: Results and Implications Relative to Structural Pavement Design and Distress Models*. The National Danish Road Laboratory.

PRC. (2010). *Code for Design of High-Speed Railway (Trial)*. China Railway Publishing House, Beijing, China.

Presle, G. (2000). The EM 250 high-speed track recording coach and the EM-SAT 120 track survey car, as networked track. *Geometry Diagnosis and Therapy Systems*, Rail Engineering International, Edition 2000, No. 3, De Rooi Publications, Veenendaal, the Netherlands.

PRR. (1911). *Report of the General Managers Committee Appointed to Detemine by Experiment the Necessary Depth of Ballast Stone*. Pennsylvania Railroad, Philadelphia, PA.

Raymond, G. (1997). *Stresses and Deformation in Railway Track*. Canadian Institute of Guided Ground Transport, Report No. 77-15. Queens' University at Kingston, Ontario, Canada.

Raymond, G. P. (1985). Design for railroad ballast and subgrade support. *Journal of the Geotechnical Engineering Division*, ASCE, Vol. 104, No. GT1, 45–59.

Raymond, G. P., Gaskin, P. N., and Addo-abedi, F. Y. (1979). Repeated compressive loading of Leda clay. *Canadian Geotechnical Journal*, Vol. 16, No. 1, pp. 1–10.

Read, D. and Li, D. (1995). *Subballast Considerations for Heavy Axle Load Trackage*. Technical Digest 95-028, Association of American Railroads, Transportation Technology Center, Pueblo, CO.

Read, D. and Li, D. (October 2006). *Design of Track Transitions*, Research Results Digest 79. Transit Cooperative Research Program, Transportation Research Board, National Academies, 37p.

Read, D., Li, D., Gehringer, E., and Tutumluer, E. (2012). Evaluation of ballast under heavy-axle-loads. *Railway Track and Structures*, Vol. 108, No. 8, 13–17.

Read, D., Li, D., Thompson, H., Sussmann, T., and Nui, W. (2011). Ground penetrating radar technologies evaluation and implementation. *Railway Track and Structures*, Vol. 107, No. 9, 21–24.

Read, D., Meddah, A., Li, D., and Mui, W. (May 2013). *Ground Penetrating Radar Technology Evaluation on the High Tonnage Loop at the Transportation Technology Center*, DTFR53-00-C-00012, Task Order 248. Federal Railroad Administration.

Redden, J. W., Selig, E. T., and Zaremsbki, A. M. (February 2002). Stiff track modulus considerations. *Railway Track and Structures*, 25–30.

Ridgeway, H. H. (November 1982). *Pavement Subsurface Drainage Systems*, NCHRP Synthesis of Highway Practice, Report 96, National Academies of Sciences.

Roberts, R., Al-Qadi, I., Tutumluer, E., and Boyle, J. (September 2008). *Subsurface Evaluation of Railway Track Using Ground Penetrating Radar*, Report # DOT/FRA/ORD-09/08. U.S. DOT, Federal Railroad Administration.

Roberts, R., Al-Qadi, I., Tutumluer, E., Boyle, J., and Sussmann, T. (June 2006). Advances in railroad ballast evaluation using 2-GHz horn antennas. *Proceedings of the 11th International Conferences on Ground Penetrating Radar*, Columbus, OH.

Robertson, P. K. and Campanella, R. G. (1983). Interpretation of cone penetration tests Parts 1 and 2. *Canadian Geotechnical Journal*, Vol. 20, 718–745.

Robnett, Q. L. and Thompson, M. R. (1976). Effect of lime treatment on the resilient behavior of fine-grained soils. Transportation Research Record, Vol. 560, No. 1, 1–20.

Robnett, Q. L., Thompson, M. R., Knutson, R. M., and Tayabji, S. D. (May 1975). Development of a structural model and materials evaluation procedures. *Ballast and Foundation Materials Research Program*, report to FRA, Report No. DOT-FR-30038. University of Illinois, Champaign, IL.

Rose, J. G. (April 1987). *Simplified Design Guide for Hot Mix Asphalt (HMA) Railroad Trackbeds*. National Asphalt Pavement Association (NAPA) and The Asphalt Institute (TAI), pp. 1–8.

Rose, J. G. (April 15–18, 2013). Selected in-track applications and performances of hot-mix asphalt trackbeds. *Proceedings of the 2013 Joint Rail Conference*, Knoxville, TN.

Rose, J. G. and Huang, Y. H. (April 17–19, 1990). Hot-mix asphalt railroad trackbed systems. *Presented at the ASME/IEEE Joint Railroad Conference*, Technical Papers, Palmer House, Chicago, IL, pp. 85–90.

Rose, J. G., Huang, Y. H., Drnevich, V. P., and Lin, C. (July 1983). *Hot Mix Asphalt Railroad Trackbed: Construction and Performance*. National Asphalt Pavement Association QIP 104-7/83, and The Asphalt Institute TAI-RR 83-2.

Rose, J. G., Li, D., and Walker, L. A. (September 2002). Tests and evaluations of in-service asphalt trackbeds. *Presented at the American Railway Engineering and Maintenance-of-Way Association 2002 Annual Conference and Exposition*, Washington, DC.

Rose, J. G., Liu, S., and Souleyrette, R. (April 2–4, 2014). KENTRACK 4.0: A railway trackbed structural design program. *Proceedings of the 2014 Joint Rail Conference*, Colorado Springs, CO.

Rose, J. G. and Tucker, P. M. (September 2002). Quick-fix, fast-track road crossing renewals using panelized asphalt underlayment system. *Presented at the American Railway Engineering and Maintenance-of-Way Association 2002 Annual Conference and Exposition*, Washington, DC.

Roner, C. J. (1985). *Some Effects of Shape, Gradation, and Shape on the Performance of Railroad Ballast*, Master's Thesis, University of Massachusetts, Amherst, MA.

RTR. (2006). *Slab Track*, Special State of the Art Report. Eurailpress, Hamburg, Germany.

Saarenketo, T., Silvast, M., and Noukka, J. (April 30–May 1, 2003). Using GPR on railways to identify frost susceptible areas. *Proceedings of 6th International Conference on Railway Engineering*, London, 11p.

SAE Fatigue Design Handbook, AE-22, 3rd Edition (1997).

Sangrey, D. A., Castro, G., Poulos, S. J., and France, J. W. (June 1978). Cyclic loading of sand, silts and clays. *Proceedings of ASCE Special on Earthquake Engineering and Soil Dynamics*, 836–851.

Sangrey, D. A. and France, J. W. (January 7–11, 1980). Peak strength of clay soils after a repeated loading history. *Proceedings of the International Symposium on Soils under Cyclic and Transient Loading*, Swansea, pp. 421–430.

Sauer, E. K. and Monismith, C. L. (1968). Influence of soil suction on behaviour of a glacial till subjected to repeated loading. *Highway Research Record*, Vol. 215, 8–23.

Saussine, G., Allain, E., Paradot, N., and Gaillot, V. (May 2011). *Ballast Flying Risk Assessment Method for High Speed Line*, World Congress on Railway Research, Lille, France.

Seed, H. B., Chan, C. K., and Lee, C. E. (1962). Resilience characteristics of subgrade soils and their relation to fatigue failures in asphalt pavement. In: *Proceedings of First International Conference on the Structural Design of Asphalt Pavements*, University of Michigan, Ann Arbor, MI.

Seed, H. B., Chan, C. K., and Monismith, C. L. (1955). Effects of repeated loading on the strength and deformation of compacted clay. *Highway Research Record*, Vol. 34, 541–558.

Seed, H. B. and McNeill, R. L. (1956). Soil deformation in normal compression and repeated loading tests. *Highway Research Board Bulletin*, No. 141, 44–53.

Selig, E. T. (April 1980). *FAST Substructure Static Strain and Deflection from 175 to 422 MGT*, Prepared for U. S. Department of Transportation, Transportation Systems Center, Cambridge, MA.

Selig, E. T. (April 1997). Substructure maintenance management. *Proceedings of the 6th International Heavy Haul Association*, Cape Town, South Africa.

Selig, E. T. (April 1998). *Oral Communication*. Hadley, MA.

Selig, E. T. (1987). Tensile zone effects on performance of layered systems. *Geotechnique*, Vol. 37, No. 3, 247–254.

Selig, E. T. (March 1989). *Compilation of Field Measurements on Ballast State of Compactness*, Bulletin No 728, Vol. 89. American Railway Engineering Association, Washington, DC.

Selig, E. T. and Chang, C. S. (1981). Soil failure modes in undrained cyclic loading. *Journal of the Geotechnical Engineering Division*, ASCE, Vol. 107, No. GT5, 539–551.

Selig, E. T. and Hyslip, J. P. (May 2003). Effects of heavy axle loads on track substructure. Proceedings of the International Heavy Haul Association Specialty Conference, Dallas, TX.

Selig, E. T., Hyslip, J. P., Olhoeft, G. R., and Smith, S. (May 2003). Ground penetrating radar for track substructure condition assessment. *Proceedings of Implementation of Heavy Haul Technology for Network Efficiency*, Dallas, TX, pp. 6.27–6.33.

Selig, E. T. and Li, D. (1994). Track modulus—Its meaning and factors influencing it, Transportation Research Record. *Journal of the Transportation Research Board*, No. 1470, 47–54, US National Academies.

Selig, E., Parsons, B., and Cole, B. (1993). Drainage of railway ballast. *Proceedings of the 5th International Heavy Haul Railway Conference*, International Heavy Haul Association, Beijing, China, pp. 200–206.

Selig, E. T. and Waters, J. M. (1994). *Track Geotechnology and Substructure Management*. Thomas Telford, London.

Selig, E. T., Lin, H., Dwyer, L. J., Duann, S. W., and Tzeng, H. (April 1986). *Layered System Performance Evaluation, Phase II Final Report,* Report No. OUR85-329F, Office of University Research, U.S. DOT, Washington, DC.

Shackel, B. (1973). The derivation of complex stress-strain relations. *Proceedings of the 8th International Conference on Soil Mechanics and Foundation Engineering*, Moscow, Russia, pp. 353–359.

Shackel, B. (1973). Repeated loading of soils—A review. *Australian Road Research*, Vol. 5, No. 3, 22–49.

Sharp, M. K., Olsen, R. S., and Kala, R. (1992). *Field Evaluation of the Site Characterization and Analysis Penetrometer System at Philadelphia Naval Shipyard*. Department of the Army, Waterways Experiment Station, Corps of Engineers, Vicksburg, MS.

Sharpe, P. (2011). Roadbed assessment with deflection-based measurement technology. Presentation at the Annual Meeting of the Transportation Research Board, Washington, DC.

Shifley, L. H., Jr. and Monismith, C. L. (August 1968). *Test Road to Determine the Influence of Subgrade Characteristics on the Transient Deflections of Asphalt Concrete Pavements*. Report No. TE 68-5. Office of Residential Services, University of California, Berkeley, CA.

Silvast, M., Levomäki, M., Nurmikolu, A., and Noukka, J. (June 4–8, 2006). NDT techniques in railway structure analysis. *Proceedings of the 7th World Congress on Railway Research*, Montreal, Canada.

Silvast, M., Nurmikolu, A., Kolisoja, P., and Levomäki, M., (September 24–27, 2007). GPR technique in the analysis of railway track structure. *Proceedings of the 14th European Conference on Soil Mechanics and Geotechnical Engineering*, Madrid, Spain.

Silvast, M., Nurmikolu, A., Wiljanen, B., and Levomaki, M. (2010). An inspection of railway ballast quality using ground penetrating radar in Finland. *Proceedings of the IMechE, Part F: Journal of Rail and Rapid Transit*, Vol. 224, No. F5, 345–351. doi:10.1243/09544097JRRT367.

Simmons, D. and Şentürk, F. (1992). *Sediment Transport Technology.* Water Resources Publications, Littleton, CO.

Singh, S. P., Davis, D. D., Guillen, D., Williams, D. (April 2004). *Reducing the Adverse Effects of Wheel Impacts on Special Trackwork Foundations,* Report No. DOT/FRA/ORD 04/08. US Department of Transportation, Federal Railroad Administration, Washington, DC.

Sluz, A., Sussmann, T. R., and Samavedam, G. (2003). Railroad embankment stabilization demonstration for high speed rail corridors, ASCE, Geotechnical Special Publication 120. *Grouting and Ground Treatment,* Vol. 2, 803–812.

Soil Conservation Service (SCS). (1986). *Urban Hydrology for Small Watersheds.* Technical Release 55 (TR-55). U.S. Department of Agriculture, Washington, DC.

Spencer, E. (1967). A method of analysis of the stability of embankments assuming parallel interslice forces. *Geotechnique,* Vol. 17, No. 1, London.

Staplin, D., Harding, C., Pagano, W., and Chrismer, S. (2013). Designing durable track support for higher-speed trains using track geotechnology. *AREMA 2013 Conference Proceedings,* Indianapolis, IN.

Stevens, F. W. (1926). *Beginnings of the New York Central Railroad.* Knickerbocker Press, New York.

Suiker, A. S. J. (2002). *The Mechanical Behaviour of Ballasted Railway Tracks,* Doctoral dissertation, Technical University of Delft, the Netherlands.

Sussmann, T. R. (1999). *Application of Ground Penetrating Radar to Railway Track Substructure Maintenance Management.* Doctoral dissertation, University of Massachusetts, Amherst, MA.

Sussmann, T. R., Ebersohn, W. E, and Selig, E. T. (2001a). Fundamental non-linear track load-deflection behavior for condition evaluation. *Journal of the Transportation Research Board,* 61–67, Transportation Research Record 1742, Intercity Passenger Rail; Freight Rail; and Track Design and Maintenance, Washington, DC.

Sussmann, T. R., Ebersohn, W., Tomas, M., and Selig, E. T. (2004). Trackbed deflection under combined freight and high speed passenger service, ASCE, Geotechnical Special Publication No. 126, *Geotechnical Engineering for Transportation Projects,* pp. 1602–1609.

Sussmann, T. R., Heyns, F. J., and Selig, E. T. (1999). Characterization of track substructure performance. In: *Recent Advances in the Characterization of Pavement Geomaterials,* Geotechnical Special Publication No. 89, ASCE Amherst, WA.

Sussmann, T. R. and Hyslip, J. (April 2010). Track substructure design methodology and data, *ASME, Proceedings of the 2010 Joint Rail Conference on High Speed and Intercity Rail,* Urbana, IL.

Sussmann, T. R., Hyslip, J. P., and Selig, E. T. (2003a). Railway track condition indicators from ground penetrating radar. *NDT&E International,* Vol. 36, 157–167.

Sussmann, T. R., Kish, A., and M. J. Trosino (2003). Investigation of the influence of track maintenance on the lateral resistance of concrete tie track, *Journal of the Transportation Research Board, Transportation Research Record 1825,* Railroad Research: Intercity Passenger Transportation, Track Design and Maintenance, and Hazardous Materials Transport, Washington, DC, pp. 56–63.

Sussmann, T. R., O'Hara, K. R., Kutrubes, D., Heyns, F. J., and Selig, E. T. (March 2001b). Development of ground penetrating radar for railway infrastructure condition detection. *Proceedings of the Symposium on the Application of Geophysics to Engineering and Environmental Problems,* Denver, CO.

Sussmann, T. R., O'Hara, K. R., and Selig, E. T. (2002). Development of material properties for railway application of ground penetrating radar. *Proceedings of the 9th International Conference on Ground Penetrating Radar*, S. Koppenjan and H. Lee, eds., Proceedings of the Society of Photo-Optical Instrumentation Engineers (SPIE), Vol. 4758, Santa Barbara, CA.

Sussmann, T. R., Ruel, M., and Chrismer S. (2012). Sources of ballast fouling and influence considerations for condition assessment criteria. *Journal of the Transportation Research Board*, Vol. 2289, 87–94. Transportation Research Record, National Academies of Sciences, Washington, DC.

Sussmann, T. R. and Selig, E. T. (April 1997). Lime stabilization of railway track subgrade. *Proceedings of the 6th International Heavy Haul Railway Conference*, Cape Town, South Africa.

Sussmann, T. R. and Selig, E. T. (2000). Resilient modulus backcalculation techniques for track, ASCE, Geotechnical Special Publication No. 94. *Proceedings of the Performance Confirmation of Constructed Geotechnical Facilities*, pp. 401–410, Amherst, MA.

Sussmann, T. R. (2007a). *Application of Track Load-Deflection Data to Safety Evaluation, Maintenance, and Rehabilitation*, Railroad Engineering 2007, London, UK.

Sussmann, T. R. (2007b). *Track Geometry and Deflection from Unsprung Mass Acceleration Data*, Railroad Engineering 2007, London, UK.

Takahashi, M., Hight, D. W., and Vaughan, P. R. (January 1980). Effective stress changes observed during undrained cyclic triaxial tests on clay. *Proceedings of the International Symposium on Soils under Cyclic and Transient Loading*, Swansea, pp. 201–209.

Talbot, A.N. 1918. Stresses in Railroad Track. Reports of the Special Committee on Stresses in Railroad Track, *Proceedings of the AREA*, First Progress Report, Vol. 19, pp. 873–1062.

Tandon, V., Picornell, M., and S. Nazarian (1996). *Evaluation and Guidelines for Drainable Bases*, Report No. TX-96 1456-1F, Texas Department of Transportation in Cooperation with the Federal Highway Administration, Washington, DC.

Tanimoto, K. and Nishi, M. (1970). On resilience characteristics of some soils under repeated loading. *Soils Foundation* Vol. 10, No. 1, 75–92.

Taylor, D. W. (1948). *Fundamentals of Soil Mechanics*. John Wiley & Sons. New York.

Terzaghi, K., Peck, R., and Mesri, G. (1996). *Soil Mechanics in Engineering Practice*, 3rd Edition. John Wiley & Sons, New York.

Thom, N. H. and S. F. Brown (1987). The effect of moisture on the structural performance of a crushed limestone road base, *Transportation Research Record 1121*, Transportation Research Board, Washington, DC, pp. 50–56.

Thompson, M. R. (1984). Important properties of base and subbase materials. *Proceedings of the Conference on Crushed Stone for Road and Street Construction and Reconstruction*, Arlington, VA.

Thompson, M. R. (December 1990). *Results of Resilient Modulus of Vicksburg Clay for FAST Test Center*. Internal Report for American Association of Railroads, University of Illinois, Urbana, IL.

Thompson, M. R. and LaGrow, T. G. (December 1988). *A proposed Conventional Flexible Pavement Thickness Design Procedure*. FHWA-IL-UI-223, Illinois Cooperative Highway and Transportation at Urbana-Champaign, Champaign, IL.

Thompson, M. R. and Robnett, Q. L. (June 1976). *Final Report, Resilient Properties of Subgrade Soils.* FHWA-IL-UI-160, University of Illinois, Urbana, IL.

Thompson, M. R. and Robnett, Q. L. (1979). Resilient properties of subgrade soils. *Journal of Transportation Engineering*, ASCE, Vol. 105, No. 1, 71–89.

Townsend, F. C. and Chisolm, Ed. E. (November 1976). *Plastic and Resilient Properties of Heavy Clay Under Repetitive Loadings.* Technical Report S 76 16, Soils and Pavement Laboratory, U.S. Army Engineer Waterways Experiment Station.

TREDA Version. (December 2001). *Program and User's Guide*, Volpe National Transportation, Systems Center, Cambridge, MA.

Turner, A. K. and Schuster, R. (1996). *Landslides Investigation and Mitigation*, Transportation Research Board Special Report 247. National Academy of Sciences, Washington, DC.

Turner, A. K. and Schuster, R. (2012). *Rockfall Characterization and Control.* Transportation Research Board Miscellaneous Publication, National Academy of Sciences, Washington, DC.

Tutumluer, E., Dombrow, W., and Huang, H. (September 21–24, 2008). Laboratory characterization of coal dust fouled ballast behavior. *Proceedings of the AREMA Annual Conference*, Salt Lake City, UT.

Tutumluer, E., Huang, H., Hashash, Y., and Ghaboussi, J. (September 17–20, 2006). Aggregate shape effects on 28 ballast tamping and railroad track lateral stability. *Proceedings of AREMA Annual Conference*, Louisville, KY.

Tutumluer, E., Stark, T., Mishra. D., and Hyslip, J. (2012). *Investigation of Differential Movement at Railroad Bridge Approaches through Geotechnical Instrumentation.* University of Illinois, Champaign, IL.

UIC Code 719. (2008). *Earth and Track Bed for Railway Lines*, 3rd Edition. International Union of Railways.

Unified Facilities Criteria. (2001). *Pavement Design for Airfields*, UFC 3-260-02. US Department of Defense.

Unified Facilities Criteria. (January 2004). *Soil Stabilization for Pavements*, UFC 3-250-11. US Department of Defense.

Unified Facilities Criteria. (June 2005). *Soil Mechanics*, UFC 3-220-10N. US Department of Defense.

U.S. Department of Transportation, Volpe National Transportation Systems Center. (1996). CWR-Buckle Version 1.00 User's Guide.

U.S. Department of Transportation, Volpe National Transportation Systems Center. (2001). Track Residual Deflection Analysis (TREDA), Program and User's Guide.

Uzan, J. (1985). Characterization of granular materials. *Transportation Research Record*, Vol. 1022, 52–59.

Vaughan, V. E. (1960). The behaviour of subgrade under repetitive loadings. *Proceedings of Canadian Good Roads Association*, Vol. 41.

Volpe National Transportation. (March 1996). *CWR-Buckle Version 1.0 User's Guide*, U.S Department of Transportation, Systems Center, Cambridge, MA.

Vorster, D. J. (2012). *The Use of Ground Penetrating Radar for Track Substructure Characterization*, Masters of Engineering dissertation, University of Pretoria, Pretoria, South Africa.

Watters, B. R., Klassen, M. J., and Clifton, A.W. (1987). Evaluation of ballast materials using petrographic criteria. *Transportation Research Board 1131*, Performance of aggregates and other track performance issues, Washington, DC, pp. 45–63.

Wellington, A. M. (1887). *The Economic Theory of the Location of Railways*. Wiley, New York.

Wightman, W., Jalinoos, F., Sirles, P., and Hanna, K. (2003). *Application of Geophysical Methods to Highway Related Problems*, Report No. NHI-05-037, Federal Highway Administration, Washington, DC. http://www.cflhd.gov/resources/geotechnical/documents/geotechpdf.pdf.

Wilson, N. E. and Greenwood, J. R. (1974). Pore pressure and strains after repeated loading of saturated clay. *Canadian Geotechnical Journal*, Vol. 11, 269–277.

Wnek, J. M. and Chrismer, S. M. (2012). Comparing geometry correction capability of AGGS and TGCS tamper control systems on Amtrak's northeast corridor. *Proceedings of the ASME/ASCE/IEEE 2012 Joint Rail Conference*, Philadelphia, PA.

Wnek, M. (2013). *Investigation of Aggregate Properties Influencing Railroad Ballast Performance*. Master's thesis, University of Illinois at Urbana-Champaign, Urbana, IL.

Woldringh, R. F. and New, B. M. (1999). Embankment design for high speed trains on soft soils, Geotechnical Engineering for Transport Infrastructure. *Proceedings of the XII European Conference on Soil Mechanics and Geotechnical Engineering*, Amsterdam, the Netherlands, pp. 1703–1712.

Wood, D. M. (1982). Laboratory investigations of the behavior of soils under cyclic loading: A review, Chapter 20. In: *Soil Mechanics Transient and Cyclic Loads*, G.N. Pande and O.C. Zienkiewicz, eds. Wiley, New York, pp. 513–582.

Wu, T. (1996). Soil strength properties and their measurement, Chapter 12. In: *Landslides Investigation and Mitigation*, A. Turner and R. Schuster, eds. Transportation Research Board Special Report 247, National Academy of Sciences, Washington, DC.

Wyllie, D. and Norrish, N. (1996). Stabilization of rock slopes, Chapter 18. In: *Landslides Investigation and Mitigation*, A. Turner and R. Schuster, eds., Transportation Research Board Special Report 247. National Academy of Sciences, Washington, DC.

Yang, C. W. (1992). *The Subgrade Deformation for a Heavy Haul Line Must be Within an Acceptable Level*, (in Chinese), Research Report. The Chinese Academy of Railway Science.

Yasuhara, K., Hirao, K. and Aoto, H. (January 7–11, 1980). A simplified strain time relation for soils subjected to repeated loads. *Proceedings of the International Symposium on Soils under Cyclic and Transient Loading*, Swansea, pp. 791–800.

Yoder, E. J. and Witczak, M. W. (1975). *Principles of Pavement Design*, John Wiley and Sons, New York.

Yoo, T. S. (March 1978). Railroad ballast density measurements, *Geotechnical Testing Journal, American Society for Testing and Materials (ASTM)*, Vol. 1, No. 1, 41–54.

Yuan, D., Nazarian, S., and Chen, D. (2000). Use of seismic methods in monitoring pavement deterioration during accelerated pavement testing with TxMLS. *Proceedings International Conference on Accelerated Pavement Testing*, Reno, NV.

Zoeteman, A. and Esveld, C. (1999). *Evaluating Track Structures: Life Cycle Costs Analysis as a Structured Approach*. World Congress on Railway Research, Tokyo, Japan.

Index

Printed in the United States
by Baker & Taylor Publisher Services